MW00761214

Number Theory:
A Programmer's Guide

Related Titles:

0-07-026205-5	HAREL/POLITI, *Modeling Reactive Systems with Statecharts*
0-07-913094-1	JONES C., *Estimating Software Costs*
0-07-032826-9	JONES C., *Applied Software Measurement, Second Edition*
0-07-032880-3	JONES P., *Handbook of Team Design*
0-07-043196-5	MONTGOMERY, *Building Object-Oriented Software*
0-07-913271-5	MUSA, *Software Reliability Engineering*
0-07-057923-7	SODHI/SODHI: *Software Reuse*
0-07-061719-8	STONE, *Developing IT Applications in a Changing IT Environment*

Number Theory:
A Programmer's Guide

Mark A. Herkommer

McGraw-Hill
New York   San Francisco   Washington, D.C.   Auckland   Bogotá
Caracas   Lisbon   London   Madrid   Mexico City   Milan
Montreal   New Delhi   San Juan   Singapore
Sydney   Tokyo   Toronto

Library of Congress Cataloging-in-Publication Data

McGraw-Hill

A Division of The McGraw-Hill Companies

Copyright © 1999, by The McGraw-Hill Companies, Inc. All rights reserved. Printed
in the United States of America. Except as permitted under the United States
Copyright Act of 1976, no part of this publication may be reproduced or distributed in
any form or by any means, or stored in a data base or retrieval system, without the
prior written permission of the publisher.

1 2 3 4 5 6 7 8 9 0 AGM/AGM 9 0 3 2 1 0 9 8

ISBN 0-07-913074-7

*The sponsoring editor for this book was Simon Yates, and the production
supervisor was Tina Cameron. It was set in Century Schoolbook by Editorial
Compuvision, Mexico.*

Printed and bound by Quebecor / Martinsburg.

*McGraw-Hill books are available at special quantity discounts to use as premiums and
sales promotions, or for use in corporate training programs. For more information,
please write to the Director of Special Sales, McGraw-Hill, 11 West 19th Street, New
York, NY 10011. Or contact your local bookstore.*

for
my
loving
parents
Rosemarie
and
Andrew

Contents

1

Introduction to Number Theory

"All is number"

MOTTO OF THE PYTHAGOREAN SCHOOL

1.0 Introduction

Computer technology is a new tool, unique in character, selfless at best, useless at worst. But the nature of numbers, their theory, is much much older than computers. In fact, the study of numbers is older than recorded history — first steps toward modern mathematics being taken 300,000 years ago. Indeed, the concept of a number and magnitude are as ancient as the human race, in some ways perhaps even older. It is generally believed that once genuine number awareness developed, humans counted on their fingers and toes until the number of objects exceeded the number of digits. Using the principle of one-to-one correspondence, man was able to enumerate the elements of larger sets using collections of stones or shells. Recording of numbers was achieved by notching sticks and bones. From this point it's easy to see how the record keeper could group individuals according to a base. Historically (before computers) the most common bases observed are 3, 5, 10, and 20.

Geometry represents one of the best preserved early applications of mathematical thought. The pottery, weaving, and basketry crafts of Neolithic people show concern for congruence and symmetry (Boyer and Merzbach, 1991): the building blocks of geometry. There is little doubt that ancient Egyptians used geometry to survey fields whose markers were moved or destroyed by the annual floods along the Nile.

It was from Egypt that the best early written documents demonstrating abstract mathematical thought were obtained.

The theory of numbers is that branch of mathematics concerned with the behavior of the natural numbers: that is, 0, 1, 2, ... Historically zero was not included within the definition of natural numbers. The term "whole numbers" was generally reserved to mean natural numbers and zero. More recent math books, however, tend to include zero with the natural numbers, and following this trend, I include it in our definition. Natural numbers and their additive inverses are called the integers.

But what of the integers? And why is their study important? Without trying to convince you that knowledge is a benefit unto itself, let me assert that numbers are as much a part of our lives as our wallets and purses. If you consider virtually any human endeavor and strip away that part which is numeric, all that would remain is a flimsy scaffold that could not even bear up under its own weight. Numbers give our lives meaning — they sustain us in all we do.

1.1 About the Use of Proofs in the Book

Number theory has two faces, or at least two sides on a single face. One is that of a pure science, profound and often inscrutable. The other is empirical science, accessible but not particularly revealing of the underlying structure of numbers. In this book I try to take both perspectives whenever possible, particularly when the deeper theory sheds light on the ultimate application of the method. Where theory does not afford any insight to a problem, the type of brutal force that only a computer can deliver will be used without remorse.

Throughout the book I have tended to view number theory from the perspective of applications; as a tool useful for solving certain programming problems one is likely to encounter. Even so, the classical mathematical tools are used frequently to simplify problems and prove more general statements. Thus for any problem we have two logical approaches: induction and exhaustion.

One of the more potent weapons in the arsenal of number theory is proof by induction. Induction includes the logical steps we take that lead from the specific to the more general. Andrews (1971) stated it more formally and with respect to number theory.

The Principle of Proof by Mathematical Induction. A statement about integers is true for all integers greater than or equal to 1 if (1) it is true for integer 1, and (2) whenever it is true for all integers, 1, 2, ..., k, then it is true for the integer $k + 1$.

The "statement about integers" is an assertion that is believed to be true for $n = 1, 2, \ldots, k$ and is more generally referred to as the *induction hypothesis*. However, not all problems in number theory contain an induction hypothesis. In fact, many defy hypothetical statements altogether, and here's the rub.

The roots of the pure science of number theory are quite deep. Investigations along these lines are encumbered by two potential problems: The first is the theoretical background that each investigator must have (fairly extensive and rigorous); the other is the absence of direct applicability of the results (logical conclusions bring their own reward). In fact, before 1950, mathematicians working problems in number theory would routinely brag that their results were *more* interesting because they were untainted by the stigma of merchantability.

Times have changed. Doing things that may actually be useful (or indeed profitable) is no longer necessarily less worthy than more high-minded and unselfish mathematical endeavors. Why there is a new philosophy has to do with the introduction of the computer. It has reawakened us to integer arithmetic. Now just about anything you do with integers can be construed to be a real-world application because you can write a program to do it. That is not to say that we no longer need theory — quite the opposite. Rather, now theory and practice augment and intensify each other.

In general I have steered clear of proofs in this book for several reasons. First, most of the results presented have been known for quite some time, and proofs for theorems given here can be found in just about every book on elementary number theory. Second, in many cases the proof does not enhance one's understanding of how to use the result. For our purposes and in the context of this book it is more important to use the result than to merely have it. Lastly, computer technology has illuminated some short cuts. We can now test theories with many numbers will little effort beyond compiling and linking a simple program. And when we find a way to limit our cases we arrive at "proof by exhaustion."

The Principle of Proof by Exhaustion. A statement about integers is true if the number of integers in the domain is finite and the statement is shown to be true for each integer in the domain.

Proof by exhaustion has had some important successes, including the proof of the Four-Color Map Theorem (see Chapter 11). It, together with induction and deductive reasoning, serves as the framework in which we can solve a wide variety of problems in number theory.

1.2 Compiling and Linking Programs and Notation Used

In this chapter and throughout this book we will be creating a library of C functions that perform various number theoretic computations. After testing a function and finding it to be bug free, you should save it for future use with an object librarian. Microsoft's LIB, Absoft's LB, and Unix's AR programs are examples of object librarians. The disk that comes with this book includes source code and executable code for each of the programs presented. The individual functions have also been separated, compiled independently, and put into an object library.

If you are using the Microsoft C/C++ compiler (any version 6.0 or higher), you can use the following command line to compile and link Program 1_2_1:

```
C> CL 1_2_1.C
```

The CL command will produce an executable called 1_2_1.EXE that runs from the DOS command line. Some compilations may require additional stack space. A stack error will occur when the program is executed. To reserve additional stack space use the following command when compiling:

```
C> CL 1_2_1.C -F4000
```

The above command will produce an executable called 1_2_1.EXE with a stack size of 4Kbytes. Lastly, if you want to compile without linking, you should use the command line option "-c". For example:

```
C> CL -c GCD.C
```

will make GCD.OBJ but will not attempt to link it. This is useful for compiling functions that are not main programs.

The programs are numbered by chapter, section, and order. Program listings are shown in the Courier font. The following program is called Program 1_2_1, for Chapter 1, Section 2, and first program in this section. It can be compiled and linked by the command

```
C> CL 1_2_1.C
```

using Microsoft C/C++.

Program 1_2_1.c. Prints a decimal number in octal and hexadecimal notation

```
/*
** Program 1_2_1 - Print a decimal number in octal and hexadecimal
*/
#include "numtype.h"
void main(void)
{
  INT a;

  printf("decimal     = ");
  scanf("%ld", &a);
  printf("octal       = %lo\n", a);
  printf("hexadecimal = %lx\n", a);
}
```

Example execution of the programs presented in the book is shown in gray and is also printed in the Courier font. What you, gentle reader, must type as input to the program is printed in **bold italic**. The rest is program output. For example, the execution of the program shown above is shown as:

```
C> 1_2_1
decimal     = 123456789
octal       = 726746425
hexadecimal = 75bcd15
```

1.3 Numerical Precision

Its an unfortunate fact but computers are general-purpose tools and as a result seek the lowest common denominator in their functionality. It is always more desirable to perform computations in hardware; the numerical precision available to us for integers is limited to 1, 2, and 4 bytes. Table 1.1 shows the unsigned data ranges available using standard integer precision:

Table 1.1. Standard Integer Precision for Unsigned Data

Data type	Number of bytes	Unsigned data range	Signed data range
Char	1	0–255	– 128–127
Short int	2	0–65,535	– 32,768–32,767
Long int	4	0–4,294,967,295	– 2,147,483,648 2,147,483,647

There is another class of integers referred to as multiple precision integers. Their length is limited by software and computer memory rather than the standard precision of the computer. In the programs in this book I have used a typedef to name the integer types. The def-

inition can be found in the file numtype.h (a C header file). The type
INT is a long, signed integer and NAT is a long-unsigned integer; integer and natural number respectively. I have included a chapter on
multiple-precision integers (Chapter 12) and related arithmetic functions. Although these functions are designed for 1000-digit integers,
this limit can be increased easily to accommodate larger numbers,
should you so choose.

You can use the algorithms presented in Chapters 1–11 of this book
to create programs that work with multiple-precision integers
described in Chapter 12. This will require some programming effort,
however. Of course you can use the programs as they are presented
and get the standard precision of your computer hardware.

1.4 Algorithmic Complexity

Throughout this book, references are made to the *complexity* of a particular algorithm and strategies for reduction. Generally, complexity
is classified according to space or time complexity. We will not concern
ourselves with space complexity. For our purposes, complexity is
meant to mean the number of computational steps it takes to trans-
form the input data to the output data. The measure of complexity is
a function of the quantity of input data, called the problem size.
Knowledge of both the complexity of a particular algorithm and the
problem size will permit us to estimate the running time for a particular application.

Counting the number of instructions is among the most important
consideration in designing efficient algorithms. Since the number of
instructions a computer can process per second is relatively fixed, the
only way we can introduce more efficient algorithms is by reducing
the number of instructions executed.

When considering the nature of complexity we can consider the
worst-case or the average-case complexity. Although both are useful,
average-case complexity will be more important to us than the worst
case. Comparisons of this nature are useful for evaluating the
strength of a particular algorithm.

The notation most often used for expressing algorithmic complexity
is called "big O" —the order of magnitude of the computational complexity. The big O notation was originally introduced by P. Bachman
(1894). For further discussion of this and other notations for quanti-
fying complexity, see Knuth (1976).

Order recognizes only the term that grows fastest in the measure of
an algorithm's complexity. For example, if the complexity of a computation is determined by the polynomial $(2n^3 + 5n^2 + n + 4)$, then the
complexity is on the order of n^3; $O(n^3)$.

When comparing the algorithmic complexity of two functions, say f and g, it is sometimes expedient to use the following definition:

$$\lim_{n \to \infty} O(f) / O(g) = L$$

These cases arise:

1. if L = a constant, then the order of f and g are equal,

2. if $L = 0$, then the order of f is less than the order of g,

3. if $L = \infty$, then the order of g is less than the order of f.

Although this definition is useful for making general comparisons, in applications the problem size does not necessarily tend to infinity.

A trial division algorithm for testing primality can give us a sense of how complexity is measured. Consider the following code fragment for testing the primality of n:

```
for (i = 2; i*i <= n; i++)
 if ((i % n) == 0) break;

if (i*i > n)
 printf("%d is prime\n", i);
else
 printf("%d is composite\n", i);
```

The trial division is not a polynomial time algorithm, even though it may give that appearance. Although primality can be determined in $\sqrt{n}$ steps, its running time is exponential in the length of n. This leads us to a definition of what polynomial time computability is.

Definition. Given a number n, we say a function f is *computable in polynomial time* if there exists an algorithm that takes at most $p(\log n)$ bit operations to compute f, for some polynomial p.

Using various time complexities and problem sizes, we can create Table 1.2. For the estimation of the run time, a bit operation rate of 10^6 bit/s was used.

Table 1.2. Complexity and Problem Size Run Times

Complexity	$n = 1000$	Runtime	$n = 1,000,000$	Run Time
$O(1)$	1	1 ?s	1	1 μs
$O(\log_2 n)$	10	6 ?s	20	20 μs
$O(n)$	10^3	0.001 s	10^6	1 s
$O(n \sum \log_2 n)$	10^4	0.01 s	$10^6 \cdot 20$	20 s
$O(n^2)$	10^6	1 s	10^{12}	12 days
$O(n^3)$	10^9	17 min	10^{18}	32,000 years
$O(2^n)$	10^{301}	10^{12} AU	$10^{301,030}$	$10^{301,006}$ AU
$O(n!)$	10^{2568}	10^{134} AU	$10^{5,565,709}$	$10^{5,565,685}$ AU

The time units AU stands for age of the universe, roughly 15 billion years. It's a safe bet that we will need to see some significant improvement in processor speeds before we attack those problems with $O(n!)$ complexity.

Generally, any $O(P)$ algorithm, where P is a polynomial, is an efficient algorithm. These problems fall into the P class. Intractable problems are those for which a polynomial-time algorithm does not (yet) exist. The NP class (nondeterministic polynomial time) includes problems that can be solved in polynomial time if educated guesses are permitted and necessary. NP class problems abound in number theory and its applications, as we shall see.

Challenge

1. Show that for very small n an $O(2^n)$ algorithm is actually more efficient than an $O(n^3)$ algorithm. Write a program to find the point where the algorithms cross each other.

2

Divisibility

What would life be without arithmetic, but a scene of horrors.

SIDNEY SMITH, *LETTERS* 1835

2.0 Introduction

The property of divisibility seems almost magical: an undeniable yet inexplicable phenomenon. Many mathematicians prefer to characterize divisibility as a random event, but I do not. Divisibility is as regular as clockwork and as infinite as the ages. Certainly one of oldest branches of number theory concerns itself with the identification of those numbers that are divisible by other numbers and those that are not. It seems natural that we should begin our investigation here.

Arithmetic was learned by each of us at a very early age. So much so that we may be inclined to take some of its propositions for granted as being merely common sense. However, in order to systematize our study I will define several terms and restate age-old concepts so that we will share a common vocabulary. Below are the basic arithmetic operations and the names of the operands:

Addition:	augend	plus	addend	equals	sum
	3	+	4	=	7
Multiplication:	multiplicand	times	multiplier	equals	product
	3	·	4	=	12
Subtraction:	minuend	minus	subtrahend	equals	difference
	7	−	4	=	3

Division:	dividend	divided by	divisor	equals	quotient	plus	remainder
	14	/	4	=	3	+	2

Governing the arithmetic operations are laws. These laws provide the framework upon which we build our analysis of numbers. If this book were a legal contract, you might think of the laws or arithmetic as the what is commonly referred to as "boilerplate": standard provisions that must always be included.

Basic laws of arithmetic:

$a + b = b + a$	commutative law of addition
$a + (b + c) = (a + b) + c$	associative law of addition
$a \cdot b = b \cdot a$	commutative law of multiplication
$a \cdot (b \cdot c) = (a \cdot b) \cdot c$	associative law of multiplication
$a \cdot (b + c) = a \cdot b + a \cdot c$	distributive law
if $a < b$ and $b < c$, then $a < c$	ordering law
if $a < b$, then $a + c < b + c$	monotonic law of addition
if $a < b$ and $c > 0$, then $a \cdot c < b \cdot c$	monotonic law of multiplication
if $a + x = a + y$, then $x = y$	addition cancellation
if $a \cdot x = a \cdot y$ and $a \neq 0$, then $x = y$	multiplication cancellation

The Concept of a Group and a Ring

While this book should be considered a practical guide for programmers, we cannot, nor do we want to, avoid the underlying concepts of algebraic number theory.

Definition. A *group* is a set G and a binary operation $\oplus$ such that

1. there exists an element $e \in G$, such that for each $a \in G$, $e \oplus a = a$;

    e is a left neutral element for G.

2. for each $a \in G$ there is an element $b \in G$, such that $b \oplus a = e$;

    b is the left inverse of a with respect to e

3. if $a, b, c \in G$, then $a \oplus (b \oplus c) = (a \oplus b) \oplus c$.

If we consider the integers, usually designated by the letter Z, together with the operation of addition, we would say that the structure constitutes a *group* because

1. there exists an integer 0, such that for each $a \in Z$, $0 + a = a$,

2. for each integer a there exists an integer b, such that $b + a = 0$; b is $-a$,

3. the associative law holds.

On the other hand, the integers Z together with the operation of multiplication are not a group. Although multiplication is associative and 1 is a left neutral element, integers do not have a left inverse in Z (with the exceptions of 1 and -1).

An *abelian group* or *commutative group* is a group in which the additional axiom:

4. $a \oplus b = b \oplus a$, for all $a, b \in G$

also holds.

Definition. A *ring* $(R; + \cdot)$ is a set R is a nonempty set R closed under two binary operations ($+$ and $\cdot$), such that $(R; +)$ is abelian and $\cdot$ is associative and has an identity. Also, $\cdot$ is distributive over $+$, thus:

5. $a \cdot (b + c) = (a \cdot b) + (a \cdot c)$.

A *commutative ring* is a ring whose elements have the property of being commutative with respect to its binary operations. An *integral domain* is a commutative ring that satisfies:

6. $a \cdot b = b \cdot c$, and $a \neq 0$ implies $b = c$, for all $a, b, c \in R$.

Although with respect to numbers we associate $+$ with addition and $\cdot$ with multiplication, you should think of these as simply binary operations possessing certain properties. With respect to $+$, the identity is usually represented by 0, with respect to $\cdot$ by 1.

Definition. A *field* $(F; + \cdot)$ is a commutative ring in which every nonzero element has a multiplicative inverse.

By way of example, the set of integers form an integral domain denoted by the letter Z. Throughout this book the numbers that we will be considering will be the integers, sometimes called the *rational integers*. It is often useful to think of Z as a set of numbers { ... , -3, $-2, -1, 0, 1, 2, 3, ...$ } over which certain arithmetic operations are performed.

The rational numbers, numbers that can be represented and the quotient of two integers, constitute a field. It is indicated by the letter Q. Other familiar examples of fields are the set of real numbers R and complex numbers C.

2.1 Euclid's Division Lemma

Many of the underpinnings of number theory are provided by Euclid of Alexandria, the great Greek mathematician and author of the book *Elements*. Euclid's Division Lemma is the first of his contributions we will look at. It asserts what we intuitively know to be true. The division of one integer by another positive integer gives an integral quotient and a remainder smaller than the divisor. Expressed in terms of multiplication and addition, we can restate Euclid's Division Lemma.

Euclid's Division Lemma 2.1. Given that a is arbitrary and b is positive, there exist unique integers q and r, such that:

$$a = b \cdot q + r \qquad\qquad 0 \leq r < b$$

The idea of basis representation, or radix, is also an important part of number theory. There are essentially two ways to represent numbers: addition and positional. I am sure that both types are familiar to you.

The Roman method, that is, Roman numerals, of writing numbers is classified as an addition system for representing numbers. Each letter represents a number: I = 1, V = 5, X = 10, L = 50, C = 100, D = 500, and M = 1000. The number is determined by adding the individual letter values. To reduce the number of letters used, the exception to this rule is that if a smaller number stands to the left of a larger number, then the value is subtracted. There is obvious difficulty in representing large numbers. Calculations are worse. What is L minus VI? Does XV divide either LX or CM? We could use Roman numerals for studying number theoretic problems, but do we really want to?

Far more useful is the position system for numeric representation, which employs the concept of digits. It comes to us from India by way of Arabia. Of the positional systems, the decimal system offers the significant advantage of familiarity. However, many problems in number theory are better solved with another system, or even a more general, base independent system.

Using Euclid's Division Lemma, we can develop the concept of base (or radix) and uniqueness of representation. If g is greater than 1, then any integer a greater than 0 can be represented uniquely in the form

$$a = c_0 + c_1 \cdot g + c_2 \cdot g^2 + \ldots + c_n \cdot g^n,$$

where c_n is positive and $0 \le c_m < g$. g is called the *base*. If you let $g = 10$, then $c_0, c_1, c_2, \ldots$ are the digits that we commonly associate with numbers. For example:

$$1638_{10} = 1 \cdot 10^3 + 6 \cdot 10^2 + 3 \cdot 10^1 + 8 \cdot 10^0$$

$$= 1000 \quad + 600 \quad + 30 \quad + 8.$$

Everybody who uses computers knows (or should know) that the computer does its work in base 2 (binary), rather than base 10 (decimal). We often refer to base 8 (octal) or base 16 (hexadecimal) because it is generally more convenient to use groups of bits rather than address each bit individually. Base 16 has come to be preferred over base 8 because it divides a byte (i.e., 8 bits) evenly into two 4-bit groups (which are sometimes called "nibbles").

In the binary system, $g = 2$: that is, base 2:

$$100111_2 = 1 \cdot 2^5 + 0 \cdot 2^4 + 0 \cdot 2^3 + 1 \cdot 2^2 + 1 \cdot 2^1 + 1 \cdot 2^0$$

$$= 32 \qquad\qquad\qquad\qquad + 4 \quad + 2 \quad + 1.$$

Throughout most of this book, we will be manipulating the data type called *long integers*. They are 32 bits in length. On some occasions we will use *short integers* having a length of 16 bits. When we use them as *unsigned long integers* (which we shall prefer to do) the largest number that can be represented is 4,294,967,295 ($2^{32} - 1$); the smallest is 0. Similarly the range of unsigned short integers is from 0 to 65,535 ($2^{16} - 1$).

Signed long integers have the 32nd bit reserved for the sign of the number, positive or negative. If we choose to use signed integers, our data range shifts from strictly positive values to half positive and half negative. Thus, the largest possible value for a signed integer is 2,147,483,647 ($2^{31} - 1$) and the smallest is −2,147,483,648 (-2^{31}). *Signed short integers* have a range from −32,768 (-2^{16}) to 32,767 ($2^{16} - 1$).

Unquestionably the most common basis representations are 2, 8, 10, and 16. To compare representations, consider the following table.

Because we do not have any numerical digits to represent numbers greater than 9, we commonly use as digits a–f, either upper- or lowercase. Older books on number theory and numerical analysis, particularly those that use terminology such as "electronic computers" and "computing machines," tend to use Greek letters to represent hexadecimal digits.

Table 2.1. Representations of Numbers 0 Through 20 in Base 2, Base 8, and Base 16

Base 10	Base 2	Base 8	Base 16
0	0	0	0
1	1	1	1
2	10	2	2
3	11	3	3
4	100	4	4
5	101	5	5
6	110	6	6
7	111	7	7
8	1000	10	8
9	1001	11	9
10	1010	12	a
11	1011	13	b
12	1100	14	c
13	1101	15	d
14	1110	16	e
15	1111	17	f
16	10001	20	10
17	10001	21	11
18	10010	22	12
19	10011	23	13
20	10100	24	14

It is likewise possible to represent fractions using a base other than 10: for example, in decimal notation where we have each digit to the right of the decimal point representing tenths, hundredths, thousandths, and so on. In base 2 we get halves, quarters, eighths, sixteenths, etc. Base 16 has 16ths, 256ths, 4096ths, etc. For example,

$$0.1_{10} \quad = \quad 0.0001\ 1001\ 1001\ 1001\ ..._2 \quad = \quad 0.1999\ ..._{16}$$

$$0.5_{10} \quad = \quad 0.1_2 \qquad\qquad\qquad\qquad\quad = \quad 0.8_{16}$$

$$\pi \quad = \quad 11.0010\ 0100\ 0011\ 1111\ ..._2 \quad = \quad 3.243f\ ..._{16}$$

The algorithm for converting between bases actually has two parts: conversion of the part to the left of the decimal and conversion of the part to the right of the decimal. Given u and the desired radix b, we get the digits one at a time, first to the left of the decimal:

$$v_0 = [\ u\ ] \bmod b$$

$$v_1 = [\ u\ /\ b\ ] \bmod b$$

$$v_1 = [[\ u\ /\ b\ ]\ /\ b\ ] \bmod b$$

...

to yield the integer number with the digits $v_n \ldots v_2 v_1 v_0$. To the right of the decimal, the computations proceed as follows:

$$v_{-1} = [\, u \cdot b \,]$$

$$v_{-2} = [\, (u \cdot b - [\, u \cdot b \,]) \cdot b \,]$$

$$v_{-3} = [\, (\, [\, (u \cdot b - [\, u \cdot b \,]) \cdot b - [\, (u \cdot b - [\, u \cdot b \,]) \cdot b \,]) \cdot b \,]$$

...

The next applications illustrate how these algorithms can be implemented.

Application: Converting Integers Between Base 2^n Radices

Base, or radix, converts easily from decimal to octal and hexadecimal integers with the help of a short program and a few less used `printf` format specification flags; 'o' and 'x'. Alternatively, I could use the Microsoft C/C++ runtime function `_ltoa()` for converting to any radix. However, to illustrate the algorithm and the general process of radix conversion, I will not use either of these and instead use a simple binary shift. Because I am using long integers, I'll use the '1' prefix letter. Also, when printing unsigned long integers in decimal notation I need to use the 'u' format specification instead of 'd' to avoid printing the sign.

To convert from decimal to binary representation, we mask all of the bits except one and check its value. For octal conversion, we mask 3 bits, and for hexadecimal, we mask 4. The following discussion is with respect to binary but the process is the same for a 2^n base. We will start with the highest bit and work our way to the lowest. With respect to a single bit, its value is called parity. If the parity of the bit extracted is odd, we print a 1, if even then we print a 0. The actual digits printed come from the array called `digit[]`. It contains the hexadecimal digits and, as a subset, octal and binary.

The binary mask for a long integer that we will use to exclude all 32 of the bits except the lowest is 00000000000000000000000000000001, or simply 1. Using this mask, we employ the bitwise-AND operator, &, to extract a bit.

To position the bit to be extracted, we use the right-shift operator. We begin by shifting the bit in the first position to the right 31 places, mask all of the other bits and print that single bit's value. Next we shift the second bit 30 places to the right and, like before, mask all of the other bits and print that bit's value. This process continues until all 32 bits have been printed.

Program 2_1_1.c. Convert a decimal integer to binary, octal, and hexadecimal representations

```
/*
/*
** Program 2_1_1 - Print a decimal number in binary, octal,
**                 and hexadecimal
*/
#include "numtype.h"

void main(int argc, char *argv[])
{
  NAT a, n;
  int i;
  char digit[16] = {  '0', '1', '2', '3', '4', '5', '6', '7',
                      '8', '9', 'a', 'b', 'c', 'd', 'e', 'f'};

  if (argc == 2) {
    n = atol(argv[1]);
  } else {
    printf("Convert a number to binary, octal, and hexadecimal\n\n");
    printf("decimal     = ");
    scanf("%ld", &n);
  }

  printf("hexadecimal = ");
  for ( i = 28, a = n; i >= 0; i -= 4)
    printf( "%c", digit[(a >> i) & 15]);
  printf("\n");

  printf("octal       = ");
  for ( i = 30, a = n; i >= 0; i -= 3)
    printf( "%c", digit[(a >> i) & 7] );
  printf("\n");

  printf("binary      = ");
  for ( i = 31, a = n; i >= 0; i--)
    printf( "%c", digit[(a >> i) & 1] );
  printf("\n");
}
```

Example execution of the programs presented in the book is shown in gray. What you, gentle reader, type as input to the program is printed in **_bold italic_**. The rest is program output. Upon executing Program 2_1_1, we get

```
C> 2_1_1
Convert a number to binary, octal, and hexadecimal

decimal     = 123456789
hexadecimal = 075bcd15
octal       = 00726746425
binary      = 00000111010110111100110100010101
```

Printing numbers that have decimal places is a little more difficult. Using the algorithm previously described, we can write a program that will print floating-point numbers in the corresponding hexadecimal notation. I prefer to use bit shifting instead of the modulus because, in the latter case, the digits come in reverse order. Again, the array digit[] is used to hold the digits of base 16.

Program 2_1_2.c. Convert an floating-point number into hexadecimal

```
/*
** Program 2_1_2 - Print a floating point number in hexadecimal
*/
#include "numtype.h"

void main(int argc, char *argv[])
{
  NAT a;
  int i;
  double n;
  char digit[16] = { '0', '1', '2', '3', '4', '5', '6', '7',
                     '8', '9', 'a', 'b', 'c', 'd', 'e', 'f'};

  if (argc == 2) {
    n = atof(argv[1]);
  } else {
    printf("Print a floating point number with hexadecimal
            digits\n\n");
    printf("decimal   = ");
    scanf("%lf", &n);
  }

/* integer part */

  printf("hexadecimal = ");
  for ( i = 28, a = n; i >= 0; i -= 4)
    printf( "%c", digit[(a >> i) & 15] );
  printf(".");

/* fractional part */

  for ( i = 0; i < 12; i++)
  {
    n = (n - floor(n)) * 16;
    printf( "%c", digit[(int) n] );
  }
  printf("\n");
}
```

The output from Program 2_1_2 is:

```
C> 2_1_2
Print a floating point number with hexadecimal digits

decimal     = 3.141592653589793
hexadecimal = 00000003.243f6a8885a3
```

Application: Converting IEEE Floats to IBM Mainframe Floats

When writing applications for more than one computer, it is often necessary to convert data from one floating-point format to another. This is especially true manipulating data in binary files, where we do not often have the luxury of an American Standard Code for Information Interchange (ASCII) representation. In this application we are going to convert Institute of Electrical and Electronics Engineers (IEEE) floating-point numbers into IBM mainframe floating-point numbers and back. What we will be doing in essence is converting from base 2 to base 16 — what could be easier?

Unfortunately, we must understand the something about the nuances of floating-point representation on both systems. In the IEEE standard, binary floating-point numbers, owing to fixed precision, approximate real numbers by values:

$$(-1)^s \cdot b_0.b_1b_2 \ldots b_{p-1} \cdot 2^e,$$

where s is the sign bit (0 or 1), $b_0.b_1b_2 \ldots b_{p-1}$ is the mantissa (each b_i is a binary digit), and e is the exponent. The mantissa is always normalized so b_0 is always 1. Because b_0 is known implicitly to be 1, it is left off, and 23-bit precision can then be used to represent 24 bits. This gives us 1 more bit for the exponent, e. The exponent 8 bits is biased by +127 so that it is always positive.

In converting from IEEE to IBM floats, the sign bit is simply extracted—no conversion is necessary. Once the mantissa has been extracted, we put b_0 back in to create a 24-bit mantissa. However, the important parts for converting the base 2 floats to base 16 is (1) aligning the mantissa based on the value of the exponent, and (2) converting the exponent properly.

The only unfinished business is to reverse the bytes so that they are in the correct order for the IBM floating-point format. It is worth noting that it is not be necessary to reverse the byte order if you are working on a computer that employs big-end order. The ordering nature is based on the type of processor used by the computer. Table 2.2 shows the byte order nature for several processors:

Table 2.2. Comparison of Byte and Bit Order for Various Processors

Processor	Byte order	Bit order
Intel 80xxx	Little	Little
Motorola 680xx	Big	Little
DEC VAX	Little	Little
IBM mainframes	Big	Big

To facilitate the bit manipulations, we work with unsigned long integers. This allows us easy access to the bits without any worries about sign bit propagation during rotation operations on negative numbers. Bit order conversion and mantissa bias are taken care of in the arithmetic operations.

Converting from IBM floats to our computer is significantly easier because we do not need to know the floating-point format of our processor. We can cheat a little because we know that:

$$\text{value} = \pm\,\text{mantissa} \cdot 16^{(\text{exponent}-70.0)},$$

where the mantissa and exponent are extracted from the IBM float. This way we can employ the C/C++ pow() function and avoid having to build up the floating-point format. The following code performs conversion between IBM floats and your host computer system.

Program 2_1_3.c. Convert between IEEE floating-point format and IBM mainframe

```
/*
** Program 2_1_3 — Convert IEEE floating-point format (MS C) to
**                 IBM mainframe format and back
*/
#include "numtype.h"

void main(int argc, char *argv[])
{
  unsigned long i, fp2ibm();
  float     x, ibm2fp();

  if (argc == 2) {
    x = atof(argv[1]);
  } else {
    printf("Convert IEEE float to IBM mainframe float\n\n");
    printf("x = ", x);
    scanf("%f", &x);
  }

  printf("fp = %f\n",        x);
  i = fp2ibm(&x);
  printf("ibm = %8.8lx\n",    i);
  x = ibm2fp(&i);
  printf("fp = %f\n",        x);
}

/*
** fp2ibm — convert a IEEE (Microsoft) floating point number to
**          IBM floating point format
*/
unsigned long fp2ibm(value)
float *value;
{
  unsigned long e, m, s, val, reverse_long();
/* extract the sign bit, exponent, mantissa */
```

```
    memcpy(&val, value, 4);
    s = (val & 0x80000000);                    /* sign */
    e = (val & 0x7f800000) >> 23;              /* exponent */
    m = (val & 0x007fffff) | 0x00800000;       /* mantissa */

/* assemble the representation and reverse bytes */

    val = s | (((e + 133) >> 2) << 24) | (m >> ((258 - e) & 3));
    reverse_long(&val);

    return(val);
}

/*
** ibm2fp — convert a IBM floating-point number to
**          the host computer's floating-point format
*/
float ibm2fp(value)
unsigned long *value;
{
  unsigned long iexpon, mantis, negval, reverse_long();
  float val;

/* reverse bytes and extract the sign bit, exponent, mantissa */

  reverse_long(value);
  negval = (*value & 0x80000000);
  iexpon = (*value & 0x7f000000) >> 24;
  mantis = (*value & 0x00ffffff);

/* compute the floating-point value */

  val = (double) mantis * pow((double)16.0, (double) (iexpon - 70.0));
  if (negval) val = - val;

  return(val);
}


unsigned long reverse_long(a)
unsigned long *a;
{
  unsigned char s[4], c;

#ifdef MSDOS
  memcpy( s, a, 4);
  c = s[0]; s[0] = s[3]; s[3] = c;
  c = s[1]; s[1] = s[2]; s[2] = c;
  memcpy( a, s, 4);
#endif

  return(*a);
}
```

The input is simply a floating-point number. The IBM mainframe format is presented in hexadecimal notation. The program output demonstrates the conversion process in both directions.

```
C> 2_1_3
Convert IEEE float to IBM mainframe float

x   = 1023.5
fp  = 1023.500000
ibm = 00f83f43
fp  = 1023.500000

C> 2_1_3
Convert IEEE float to IBM mainframe float

x   = -1023.5
fp  = -1023.500000
ibm = 00f83fc3
fp  = -1023.500000
```

I grant you that although the output is not much to look at, its extremely useful when exchanging floating-point data between an IBM mainframe and another computer.

Application: Signed-Digit Number Systems

In general, the allowable digits in a given base g take the following values: $\{0, 1, 2, \ldots , (g - 1)\}$. As odd as it may sound, it is sometimes useful to represent numbers using a number system that allows signed digits. In this case the allowable digits are $\{ -(g-1), -(g-2), \ldots , -2, -1, 0, 1, 2, \ldots , (g-1)\}$. Using this digit set, we can create what is called a signed-digit (SD) number system.

Consider the following example: let $g = 10$. Our digit set consists of 19 digits: $\{-9, -8, \ldots , -1, 0, 1, \ldots , 8, 9\}$. For notational convenience we will place a sign in front of every SD number; $\{-9, -8, \ldots , -1, +0, +1, \ldots , +8, +9\}$. This notation may be slightly different from the one in other books, which may use an underline or an overline to indicate a negative digit.

For a two-digit SD number, x, we have $-9 - 9 \leq x \leq + 9\ +9$. This yields 199 numbers. However, since each digit has 19 possibilities, the two digits of x, that is, $x_1 x_0$, give 192 or 361 different representations of our 199 numbers. Obviously, some of the numbers can be represented in more than one way; others cannot. For example, $1 = + 0 + 1 = + 1 - 9$ and $-1 = + 0 - 1 = -1 + 9$. There is only one representation for a number that is evenly divisible by 10 or whose absolute value is greater than 90.

A simple program can write out a table of the values and their redundant values.

Program 2_1_4.c. Prints a table of signed-digit numbers for –99 to 99

```
/*
** Program 2_1_4 — Print a table of signed-digit numbers
**                      from -99 to 99
*/
#include "numtype.h"

void main(void)
{
  int a, b, i;

  for (i = -99; i <= 99; i++)
  {
    a = i / 10;
    b = i % 10;

    printf("%5d\t\t", i);
    printf("%+2d%+2d\t\t", a, b);

    if ((b == 0) || (a == -9) || (a == 9))
      printf("\n");
    else if (i < 0)
      printf("%+2d%+2d\n", a-1, b+10);
    else if (i > 0)
      printf("%+2d%+2d\n", a+1, b-10);
  }
}
```

Part of the output has been deleted to conserve space. I will do this from time to time in this book for programs that have relatively long streams of output. When a part has been deleted, it will be indicated by ellipses dots (...). You can run the program from the disk provided to view all of the output.

```
C> 2_1_4
  -99        -9-9
  -98        -9-8
  -97        -9-7
. . .
   -5        +0-5        -1+5
   -4        +0-4        -1+6
   -3        +0-3        -1+7
   -2        +0-2        -1+8
   -1        +0-1        -1+9
    0        +0+0
    1        +0+1        +1-9
    2        +0+2        +1-8
    3        +0+3        +1-7
    4        +0+4        +1-6
    5        +0+5        +1-5
. . .
   97        +9+7
   98        +9+8
   99        +9+9
```

You may rightly wonder about the utility of SD numbers. They are not purely mathematical contrivances but actually are useful entities with a purpose in this universe. The original motivation for their development was to eliminate carry propagation chains in addition and subtraction operations (Koren, 1993). More recently SD numbers have direct application in certain algorithms that perform multiplication and division.

Challenges

1. Modify Program 2_1_1 so that a is defined as a NAT rather than INT. Explain what is happening to negative numbers when they are converted.

2. Modify Program 2_1_1 so that it will not print leading zeroes.

3. Create a new program that will print (without redundant code) a single table of numeric values for every base from 2 to 16. (Hint: Use % and / with the base instead of masking bits.)

4. It is possible to construct a number system using a negative base. Write a program that prints a table of base −10 values from 1 to 1000.

5. With help from Roman numerals, answer the age-old riddle "When is five minus four equal to two?"

6. Investigate the nature of four-digit SD numbers. What is the greatest number of representations for any four-digit SD number? The fewest? Write a program to generate the numbers with all of the possible representations for each number.

2.2 Factors and Divisors, Primes and Composites

When considering factors and divisors, it is common to consider only positive integers, otherwise known as the natural numbers. Although zero is sometimes not included within the definition of natural numbers, as a matter of convenience I include it in our definition. The integers are the set of natural numbers and their additive inverses.

It is well known that it is not always possible to divide one number by another and have the quotient be a natural number. What does this mean in real life? Opa can't divide $10 between his three grandchildren without a pocket full of change.

Definition. Natural number a is said to be a *factor* of b if $a \cdot c = b$ for some natural number c. If this is true, we write $a \mid b$ (read a divides

b). Given that $a \cdot c = b$ for some c is true, we can say that a is a *divisor or factor* of b.

Definition. Two elements a, $b \in D$ are said to be *associates* if $a \mid b$ and $b \mid a$.

Definition. An element $u \in D$ is called a *unit* if u has a multiplicative inverse in D.

We know that all numbers are divisible by 1, a unit in the natural numbers. A factor other than 1 is called a *proper factor*. There are some numbers that do not have any proper factors; these are called *prime numbers* or, more simply, *primes*. The first few primes are well known: 2, 3, 5, 7, 11, 13, ...

A negative integer can be thought to have one additional factor: –1. In all other ways then, its factoring behaves the same as its positive counterpart. The units of the integers are 1 and –1.

Any natural number that is not a prime or 1 is called a *composite*. All composites are the product of two or more primes.

Although Chapter 4 treats the subject of primes more fully, we can understand divisibility better if we look at one of the direct methods for finding prime numbers — trial division. Simply stated, we divide a number by all the numbers less than or equal to the square root of that number, and if none divides evenly, then the number must be prime.

Below is a program that finds the primes and composites less than 1000. It prints out two columns. In the first column are primes and in the second are composites.

Program 2_2_1.c. Makes a table prime and composite numbers using trial division

```
/*
** Program 2_2_1 - Find primes and composites by trial division
*/
#include "numtype.h"

void main(void)
{
  int i, j;

  printf("prime    composite\n");

  for (i = 2; i <= 1000; i++)
  {
    for (j = 2; j*j <= i; j++)
      if ((i % j) == 0) break;

    if (j*j > i)
      printf("%lu\n", i);          /* prime */
    else
      printf("      %lu\n", i);    /* composite */
  }
}
```

If you know your prime numbers, there won't be too many surprises in the output:

```
C> 2_2_1
prime    composite
2
3
         4
5
         6
7
         8
         9
         10
11
         12
13
         14
         15
         16
17
         18
19
         20
...
```

In the example shown above, only the first 20 numbers are output. You can view the output one page at a time using the `more` command:

```
C> 2_2_1 | more
```

Also, you can redirect the output to a file so that it can be examined using a text editor or browser.

```
C> 2_2_1 > output.txt
```

From the program, it can be seen that to determine if a number is a composite, we only need to find one divisor. This is where the modulus operator, %, is of great service to number theory and to us. The modulus operator returns the remainder after performing division. Any divisor that leaves 0 remainder *must* be a factor of the dividend. Any dividend that has a factor must be a composite.

2.3 Greatest Common Divisor

The greatest common divisor or GCD is the largest natural number that will divide (without remainder) two (or more) numbers. The GCD of 30 and 42 is 6, since 6 is the largest number that evenly divides 30 and 42. Of the other factors we can also say this; 5 is a factor of 30 but not of 42 and 7 is a factor of 42 but not 30.

Another way to look at the GCD is from the point of view of simple operations between sets. Define two sets, A and B, and the set of prime divisors of 30 and 42, respectively. Next determine the intersection of A and B:

$$A = \{2, 3, 5\}, B = \{2, 3, 7\}, A \cap B = \{2, 3\}.$$

The GCD is the product of all the factors in A ∩ B, the intersection of A and B. For 30 and 42, the numbers are 2 and 3, the product of which is 6.

In the case of 30 and 42, we determined the GCD by inspection. However, you probably partially factored one number mentally and test divided it against the other number. We could implement such an approach in a program, but it is not necessary. Over two thousand years ago, this problem was attacked and defeated by Euclid. Today, the celebrated procedure bears his name.

Euclid's Algorithm

To determine the greatest common divisor for any two numbers, we can use Euclid's algorithm. Euclid gave a description of this process in Proposition 2 of Book VII of the *Elements*, around 300 B.C. What is most remarkable about this algorithm is that it yields the greatest common divisor of two numbers without actually factoring either one!

Euclid's algorithm is quite simple and at the same time quite remarkable. Perhaps it is its simplicity that makes it remarkable. To find the greatest common divisor of a and b, we begin a process of division with remainders. We note first that $a > b$, and then divide out as many bs as possible from a with a positive remainder being less than b. b becomes the new dividend and the remainder becomes the new divisor. The process is then repeated. When the remainder is finally reduced to zero, the GCD has been found.

Shown as a series of indefinite but finite number of steps,

$$a = b \cdot q_1 + r_1 \qquad\qquad 0 < r_1 < b$$

$$b = r_1 \cdot q_2 + r_2 \qquad\qquad 0 < r_2 < r_1$$

$$r_1 = r_2 \cdot q_3 + r_3 \qquad\qquad 0 < r_3 < r_2$$

$$r_2 = r_3 \cdot q_4 + r_4 \qquad\qquad 0 < r_4 < r_3$$

$$\cdots$$

$$r_{n\text{-}2} = r_{n-1} \cdot {}_1 q_n + r_n \qquad\qquad 0 < r_n < r_{n\text{-}1}$$

$$r_{n\text{-}1} = r_n \cdot q_{n+1} \qquad\qquad r_{n+1} = 0$$

The GCD of a and b corresponds to r_n, which is usually written as

$gcd(a,b) = r_n$

Why does this algorithm work? Because any common divisor of both b and r_1 must also be a divisor of a. Furthermore, any common divisor of both a and b must also be a divisor of r_1. The problem of finding the gcd is now reduced to finding the GCD for b and r_1. Proceeding in this way the process is repeated until no remainder remains. We know that we will eventually get there because $b > r_1 > r_2 > r_3 > ... > r_n > 0$. The GCD is then the number that will evenly divide $r_n \cdot 1$, and that number is r_n.

Consider the following example for finding the GCD of 13,020 and 5797:

$13020 = 5797 \cdot 2 + 1426$

$5797 = 1426 \cdot 4 + 93$

$1426 = 93 \cdot 15 + 31$

$93 = 31 \cdot 3 + 0$

Thus, according to Euclid's algorithm, the greatest common divisor of 13,020 and 5 797 is 31. We can verify this the hard way by factoring each number and seeing which factors are common:

$13,020 = 2 \cdot 2 \cdot 3 \cdot 5 \cdot 7 \cdot 31$

$5,797 = 11 \cdot 17 \cdot 31$

The only factor in common is 31, just as we determined previously.

Number pairs that do not have any common factors have a GCD equal to 1. They are said to be *relatively prime* or *coprime*. Here is another example, 512 and 135:

$512 = 135 \cdot 3 + 107$

$135 = 107 \cdot 1 + 28$

$107 = 28 \cdot 3 + 23$

$28 = 23 \cdot 1 + 5$

$23 = 5 \cdot 4 + 3$

$5 = 3 \cdot 1 + 2$

$3 = 2 \cdot 1 + 1$

$2 = 1 \cdot 2 + 0$

From this example we see that 512 and 135 are relatively prime since the last nonzero remainder is equal to 1. You may wish to factor these numbers to confirm this fact. When you have finished, examine Program 2_3_1 below that implements Euclid's algorithm for two variables. The order of the operations has been shuffled a little so we can use a while loop for program control.

Program 2_3_1.c. Finds the greatest common divisor for two numbers using Euclid's algorithm

```
/*
** Program 2_3_1 - Find the greatest common divisor using
**                 Euclid's algorithm (prompted input)
*/
#include "numtype.h"

void main(void)
{
  NAT a, b, r;

  printf("Greatest common divisor (Euclid's algorithm)\n\n");
  printf("a = ");
  scanf("%lu", &a);
  printf("b = ");
  scanf("%lu", &b);

/* work until remainder is 0 */

  while (b > 0)
  {
    r = a % b;
    a = b;
    b = r;
  }

  printf("gcd = %lu", a);
}
```

Here is the output from two runs of Program 2_3_1 taken from the examples above:

```
C> 2_3_1
Greatest common divisor (Euclid's algorithm)

a = 13020
b = 5797
gcd = 31

C> 2_3_1
Greatest common divisor (Euclid's algorithm)

a = 512
b = 135
gcd = 1
```

It is noteworthy that $gcd(a,b,c) = gcd(gcd(a,b),c)$. This means that you can nest the operations when you have to compute the GCD for more than two numbers. Procedurally, the operations are to

1. compute the GCD for the first two numbers

2. for each additional number compute a new GCD using the old GCD and another number

3. the final GCD is the (greatest common division) for all the numbers

You may have noticed that Euclid's algorithm, as implemented in Program 2_2_1, has been modified so that it does not actually use a division operation. Since the algorithm terminates when a remainder of 0 is obtained, we really only need to retain the last remainder and not keep any divisors at all. That is why only the modulus operation is used.

To build up our library of number theoretic routines, the GCD computation of Program 2_3_1 has been converted into a function. Also, as you can see in the output, instead of prompting for the integers, all the numbers are arranged on the command line. This allows our new program to operate on any number of input numbers.

Program 2_3_2.c. Find the greatest common divisor using Euclid's algorithm. Numbers are accepted as command-line arguments

```
/*
** Program 2_3_2 - Find the greatest common divisor using
**                  Euclid's algorithm
*/
#include "numtype.h"

void main(int argc, char *argv[])
{
  NAT a, b;
  int i;

  printf("Greatest common divisor (Euclid's algorithm)\n\n");
  if (argc < 3) {
    printf("Usage: 2_3_2 number number [number ...]\n");
    exit(1);
  }

  a = atol(argv[1]);
  for (i = 2; i < argc; i++)
  {
    b = atol(argv[i]);
    a = GCD(a, b);
  }
  printf("gcd = %lu\n", a);
}
```

```
/*
** GCD - find the greatest common divisor using Euclid's algorithm
*/
#include "numtype.h"

NAT GCD(NAT a, NAT b)
{
  NAT r;

  while (b > 0)
  {
    r = a % b;
    a = b;
    b = r;
  }

  return(a);
}
```

To find the GCD of 420, 442, 462, and 504, type:

```
C> 2_3_2 420 441 462 504
Greatest common divisor (Euclid's algorithm)

gcd = 21
```

Next I will look at GCD algorithms designed with binary representations in mind.

The Least Remainder Algorithm

The least remainder algorithm (LRA) is a somewhat obscure variation to Euclid's algorithm developed, Leopold Kronecker (1823–1891) to save computational steps (Uspensky and Haeslet 1939). From Euclid's division lemma we know that given an arbitrary a and positive b, there exist unique integers q and r such that

$$a = b \cdot q + r,$$

where $0 \leq r < b$. If we permit negative remainders, then the division can be rearranged to

$$a = b \cdot (q + 1) - (b - r),$$

with the negative remainder $\mid b - r \mid < b$ when $r > 0$. In fact, we can devise the division in such a way that the remainder is always less than $b / 2$ by choosing $\varepsilon = \pm 1$

$$a = a \cdot q + \varepsilon \cdot r.$$

The modified Euclid's algorithm is now

$$a = b \cdot q_1 + \varepsilon_1 \cdot r_1 \qquad\qquad 0 < r_1 \le b/2$$

$$b = r_1 \cdot q_2 + \varepsilon_2 \cdot r_2 \qquad\qquad 0 < r_2 \le r_1/2$$

$$r_1 = r_2 \cdot q_3 + \varepsilon_3 \cdot r_3 \qquad\qquad 0 < r_3 \le r_2/2$$

$$r_2 = r_3 \cdot q_4 + \varepsilon_4 \cdot r_4 \qquad\qquad 0 < r_4 \le r_3/2$$

$$\cdots$$

$$r_{n \cdot 2} = r_{n-1 \cdot 1} q_n + \varepsilon_n \cdot r_n \qquad\qquad 0 < r_n \le r_{n \cdot 1}/2$$

$$r_{n \cdot 1} = r_n \cdot q_{n+1} \qquad\qquad r_{n+1} = 0,$$

where ε_1 , ε_2, ... ,ε_n, are positive or negative units.
Consider the example from before, comparing the two algorithms:

Euclid's algorithm	Least remainder algorithm
$512 = 135 \cdot 3 + 107$	$512 = 135 \cdot 4 - 28$
$135 = 107 \cdot 1 + 28$	$135 = 28 \cdot 5 - 5$
$107 = 28 \cdot 3 + 23$	$28 = 5 \cdot 6 - 2$
$28 = 23 \cdot 1 + 5$	$5 = 2 \cdot 1 + 1$
$23 = 5 \cdot 4 + 3$	$2 = 1 \cdot 2 + 0$
$5 = 3 \cdot 1 + 2$	
$3 = 2 \cdot 1 + 1$	
$2 = 1 \cdot 2 + 0$	

Again we see that 512 and 135 are relatively prime but in three
fewer steps. This is the advantage of the LRA. The LRA will never be
longer than Euclid's algorithm and will generally be shorter. From a
programming point of view, we have a trade off. To take fewer steps
we need a test to see if a negative remainder should be used. The test
is implemented two steps, first using a bit shift to get $b/2$. Next, the
test is performed.

Program 2_3_3.c. Finds the greatest common divisor using the least
remainder algorithm

```
/*
** Program 2_3_3 -  Find the greatest common divisor using the
**                  Least Remainder Algorithm
*/
#include "numtype.h"
```

```
void main (int argc, char *argv[])
{
  NAT a, b;
  int i;

  printf("Greatest common divisor (least remainder algorithm)\n\n");
  if (argc < 3) {
    printf("Usage: 2_3_3 number number [number ...]\n");
    exit(1);
  }

  a = atol(argv[1]);
  for (i = 2; i < argc; i++)
  {
    b = atol(argv[i]);
    a = LRA(a, b);
  }

  printf("gcd = %lu\n", a);
}

/*
** LRA - find the greatest common divisor using the
**       Least Remainder Algorithm
*/
#include "numtype.h"

NAT LRA(a, b)
NAT a, b;
{
  NAT r;

  while (b > 0)
  {
    r = a % b;
    if (r > (b >> 1)) r = b - r;
    a = b;
    b = r;
  }

  return(a);
}
```

As before, to find the greatest common divisor of 420, 442, 462, and 504, type

```
C> 2_3_3 420 441 462 504
Greatest common divisor (least remainder algorithm)

gcd = 21
```

The Right-Shift Binary GCD Algorithm

Binary shifting greatest common divisor (BS GCD) algorithms were developed to take computational advantage of the computer's numerical environment (Stein, 1967). This class of algorithms exploits the fact that a bit shift or multiple bit shifts execute much faster than division. In fact, the Intel DIV (integer division) instruction can use as much as 5 *times* as many central processing unit (CPU) clock cycles than the SHR (right bit shift) instruction, meaning that it can be 5 *times* as slow.

As we observed in the preceding sections, the modified Euclid's algorithm and LRA use the modulus operator and division has been avoided. However, in number-theoretic applications, it is often necessary to work with numbers well beyond the numeric range of the built-in data types. If integers larger than the representative capacity of unsigned long integers are to be used, then we must resort to special representations called multiple precision integers (see Chapter 9). With respect to multiple precision integers, even though binary shift algorithms have more main loop iterations, because they use only bit-shift and subtraction they are generally more efficient (Sorenson, 1994).

The right-shift binary greatest common divisor (RSB GCD) algorithm is based on the fact that if $a > b$, then $gcd(a, b) = gcd(b, (a - b) / 2)$. Recognizing that division by two is the same as a right bit shift, the following four-step algorithm can be developed:

1. while a and b are both even

   then shift a and b right g places until one is not

2. while a is even

   shift a right until a is not

   while b is even

   shift b right until b is not

3. while $a \neq b$

   if $a < b$ then swap(a, b)

   $a = a - b$

   while a is even

   shift a right until a is not

4. return$(a \cdot 2^g)$

This algorithm has been implemented in Program 2_3_4.

Program 2_3_4.c. The right-shift binary greatest common divisor algorithm

```
/*
** Program 2_3_4 — Find the greatest common divisor using the
**                 Right-Shift Binary algorithm
```

```
*/
#include "numtype.h"
void main (int argc, char *argv[])
{
  NAT a, b;
  int i;

  printf("Greatest common divisor (right-shift binary
  algorithm)\n\n");
  if (argc < 3)
  {
    printf("Usage: 2_3_4 number number [number ...]\n");
    exit(1);
  }

  a = atol(argv[1]);
  for (i = 2; i < argc; i++)
  {
    b = atol(argv[i]);
    a = RSBGCD(a, b);
  }

  printf("gcd = %lu\n", a);
}

/*
** RSBGCD — find the greatest common divisor using the
**          Right-Shift Binary algorithm
*/
#include "numtype.h"

NAT RSBGCD(NAT a, NAT b)
{
  NAT c, g, t;

/* find the even part of the gcd */

  g = 0;
  while (!((a | b) & 1))
  {
    a >>= 1;
    b >>= 1;
    g++;
  }
/* remove extra factors of 2 */

  while (!(a & 1)) a >>= 1;
  while (!(b & 1)) b >>= 1;

/* find the odd part of the gcd */

  while (a != b)
  {
    if (a < b) { t = a; a = b; b = t; }
    a -= b;
    while (!(a & 1)) a >>= 1;
  }

  return(a << g);
}
```

It should be comforting to see that the output from Program 2_3_4 is the same as Program 2_3_2:

```
C> 2_3_4 420 441 462 504
Greatest common divisor (right-shift binary algorithm)

gcd = 21
```

Benchmark tests between the modified Euclid's and BS GCD algorithms using random 4-byte integers indicate that the modified Euclid's GCD algorithm is about 18% faster. This may be the result of the fact that the modulus operator is implemented in hardware.

However, the situation reverses for large multiple-precision integers. Below are two tables comparing several GCD algorithms applied to multiple-precision integers (Shallit and Sorenson, 1994). The first table shows the average running times in CPU seconds; the second shows the average number of main loop iterations.

Table 2.3. A Comparison of Greatest Common Divisor Algorithms for Multiple Precision Integers (Shallit and Sorenson, 1994)

Algorithm	Input size in decimal digits				
	100	250	500	1000	
Euclid's algorithm	0.073	0.357	1.28	4.85	Seconds
Right-shift binary	0.049	0.201	0.660	2.38	Seconds
Left-shift binary	0.049	0.211	0.714	2.64	Seconds
Algorithm	Input size in decimal digits				
	100	250	500	1000	
Euclid's algorithm	191	494	980	1938	Iterations
Right-shift binary	469	1173	2341	4682	Iterations
Left-shift binary	219	550	1102	2212	Iterations

The right-shift binary algorithm is the clear winner in spite of having the highest iteration count. According to Shallit and Sorenson, it easily beats both Euclid's algorithm and the left-shift binary algorithm.

My personal experience has not been as good as that of others with the left-shift binary algorithm (see Challenge 3, below). When applied to 4-integers, I found the LSB GCD algorithm to be significantly worse than the RSB GCD algorithm. I blame this on the fact that the left-shift binary algorithms must have two "if" tests and two subtractions, while the right-shift binary requires only one of each.

Challenges

1. Write a program that will produce a table showing the GCD for all numbers 1 through 20 versus all numbers 1 through 20.

2. Recursive algorithms often express more clearly the underlying nature of a method than do nonrecursive implementations. Euclid's algorithm is an example of an algorithm that can be written recursively, albeit less efficient. Write a version of the GCD function that is recursive. (Hint: $GCD(a, b) = GCD(b, a \% b)$.)

3. The left-shift binary GCD algorithm is based on the identity: $gcd(a, b) = gcd(b, a - 2^e \cdot b)$. Write a program to implement the left-shift binary GCD algorithm (Sorenson, 1994):

> INPUT: integers $a > 0, b > 0, a \ge b$
> OUTPUT: $gcd(a, b)$
> while $b > 0$
>   find $e \ge 0$ such that $2^e \cdot b \le a < 2^{e+1} \cdot b$
> $a = \min(a - 2^e \cdot b, 2^{e+1} \cdot b - a)$
>   if $a < b$ then swap (a, b)
> return(a)

4. Prove that $gcd(a^m - 1, a^n - 1) = a^{gcd(m, n)} - 1$. (Hint: Suppose that $m > n$ and $m = n \cdot q + r$.)

2.4 Least Common Multiple

The least common multiple or LCM is a direct counterpart to the GCD. It is the smallest number that can be evenly divided by two (or more) numbers. Using an earlier example, the LCM of 30 and 42 is 210. That is, 210 is the smallest number that can be evenly divided by both 30 and 42.

Like the GCD, the LCM function can be viewed from a set union point of view. Define two sets, A and B, as the set of prime divisors of 30 and 42, respectively. To find the LCM, we determine the union of A and B:

A = {2, 3, 5}, B = {2, 3, 7}, A ∪ B = {2, 3, 5, 7}.

The LCM is the product of all of factors in A ∪ B. For 30 and 42, the numbers are 2, 3, 5, and 7, whose collective product is 210.

When trying to find the LCM, we can factor the numbers and group the factors in a manner described above. Thankfully there is a simpler way that uses the GCD. The LCM can be computed by

$$lcm(a, b) = \frac{a \cdot b}{gcd(a, b)}$$

The following program determines the LCM for two integers. You will notice that we used the function called GCD() to compute the GCD of our two numbers. The returned value is the GCD.

Program 2_4_1.c. Find the least common multiple for two integers

```
/*
** Program 2_4_1 - Find the least common multiple for two numbers
**                 with help from Euclid's GCD algorithm
*/
#include "numtype.h"

void main(void)
{
  NAT a, b, l;

  printf("Least common multiple (Euclid's algorithm)\n\n");
  printf("a = ");
  scanf("%lu", &a);
  printf("b = ");
  scanf("%lu", &b);

  l = a * (b / GCD(a, b));

  printf("lcm = %lu", l);
}


/*
** GCD - find the greatest common divisor using Euclid's algorithm
*/
#include "numtype.h"

NAT GCD(NAT a, NAT b)
{
  NAT r;

  while (b > 0)
  {
    r = a % b;
    a = b;
    b = r;
  }

  return(a);
}
```

```
C> 2_4_1
Least common multiple (Euclid's algorithm)
a = 30
b = 42
lcm = 210
```

Overflow is the condition that arises when the result of an arithmetic operation exceeds the storage capacity of a particular variable. Because we are working with unsigned long integers in this program, any product that exceeds 4,294,967,295 (i.e., $2^{32} - 1$) will cause an overflow. To ensure that the division is performed before the multiplication, we use parentheses to group the operands in the line:

```
l = a * (b / GCD(a, b))
```

By performing the division first, we lower (but do not remove) the risk of overflow during the LCM computation. Also, since we computed the GCD from a and b, we know that the GCD will be a divisor of b and there will be no remainder. Of course, it is possible to know if an overflow will occur before the multiplication is performed. In the next program, we will return 0 as an error condition from the LCM() routine.

Like the greatest common divisor function, the LCM enjoys the property that $lcm(a, b, c) = lcm(lcm(a, b), c)$. In other words, one obtains the least common multiple by nesting the operations so that the computation is performed iteratively on two numbers at a time.

A few modifications to Program 2_3_1 give us the ability to enter any number of integers on the command line to obtain the LCM for all. Also, for this program we have created the function LCM that calls the function GCD.

Program 2_4_2.c. Find the LCM for any number of integers

```
/*
** Program 2_4_2 - Find the least common multiple for any number of
**                 inputs with help from Euclid's GCD algorithm
*/
#include "numtype.h"

main (int argc, char *argv[])
{
  NAT a, b;
  int i;

  printf("Least common multiple (Euclid's algorithm)\n\n");
  if (argc < 3)
  {
    printf("Usage: 2_4_2 number number [number ...]\n");
    exit(1);
  }

  a = atol(argv[1]);
  for (i = 2; i < argc; i++)
  {
    b = atol(argv[i]);
    if ((a = LCM(a, b)) == 0)
```

```
      {
        printf("overflow\n");
        exit(1);
      }
  }

  printf("lcm = %lu\n", a);
}

/*
** LCM - finds the least common multiple for two integers
**        LCM = a * b / GCD(a, b)
*/
#include "numtype.h"

NAT LCM(NAT a, NAT b)
{
  b /= GCD(a,b);

  if ( (0xffffffff / b) < a )
    return(0);                          /* overflow */
  else
    a *= b;                            /* ok to multiply */

  return(a);
}

/*
** GCD - find the greatest common divisor using Euclid's algorithm
*/
#include "numtype.h"

NAT GCD(NAT a, NAT b)
{
  NAT r;

  while (b > 0)
  {
    r = a % b;
    a = b;
    b = r;
  }

  return(a);
}
```

The LCMs of the numbers 30, 42, 56, and 80 are

```
C> 2_4_2 30 42 56 80
Least common multiple (Euclid's algorithm)

lcm = 1680
```

The need to find both the GCD and the LCM occurs often in computer science. Next is a simple application that uses both.

Application: A Simple Fraction Calculator

A common limitation of many all hand-held calculators and calculator programs today is that they work strictly with decimal numbers. There are times when it is more convenient to work with fractions, particularly when working with units of measure that are not decimal in nature. Familiar examples are stock prices, building construction measures, and kitchen recipes.

Fractions also can be more accurate than floating-point numbers. Consider the following code fragment using floating-point arithmetic:

```
for (i = 0, a = 0.0; i < 1000; i++) a += 0.1;
```

One might reasonably expect a to equal 100 after the execution of this statement; after all, 100 = 0.1 * 1000. But, alas, a is only equal to 99.9990. This is an instance of an internal representation error (round-off error) being accumulated over repeated additions. To be more specific, just as 1/3 cannot be represented in decimal form $(0.333\ldots_{10})$, 1/10 cannot be represented in binary (0.0001 1001 1001 1001 $\ldots_2$).

Using the GCD() and LCM() functions, a simple fraction calculator protected against overflow conditions can be written. The protection is provided during each arithmetic operations by reducing the numerators and denominators so that only the smallest whole numbers are used. Although the possibility of an overflow cannot be completely eliminated, these functions significantly reduce the risk.

The program uses a function called FractionCalc. Passed to it is the operator as a single character and the fractions. The fractions are actually a new data type called FRACTION. FRACTION is a simple structure that is composed of three integers: w for whole number, n for numerator, and d for denominator:

```
typedef struct fraction_ {
    unsigned long w, n, d;
} FRACTION;
```

To pass this structure efficiently and return values, the variables pass by address rather than by value. This is why we use the pointer notation in FractionCalc routine rather than the perhaps more familiar period notation.

Program 2_4_3.c. Perform arithmetic on fractions

```
/*
** Program 2_4_3 - A simple fraction calculator that uses
**                 command line input
*/
#include "numtype.h"

main (int argc, char *argv[])
{
  FRACTION a, b, c;
  char    op;

  printf("Fraction Calculator\n\n");
  if (argc < 4)
  {
    printf("Usage: 2_4_3 #,# oper #,#\n\n");
    printf("valid operators are: +, -, *, /\n");
    exit(1);
  }
  else
  {
    sscanf(argv[1], "%lu,%lu", &a.n, &a.d);
    op = *argv[2];
    sscanf(argv[3], "%lu,%lu", &b.n, &b.d);
  }

  FractionCalc(op, &a, &b, &c);

  printf(" %6lu    %6lu        %6lu\n", a.n, b.n, c.n);
  printf("-------- %c -------- = %6lu --------\n", *argv[2], c.w);
  printf(" %6lu    %6lu        %6lu\n", a.d, b.d, c.d);
}

/*
** FractionCalc - a simple fraction calculator function
*/
#include "numtype.h"

NAT FractionCalc(char op, FRACTION *a, FRACTION *b, FRACTION *c)
{
  switch ( op )                  /* do the arithmetic */
  {
  case '+':
    c->n = a->n * (LCM(a->d,b->d) / a->d) + b->n * (LCM(a->d,b->d) / b->d);
    if ((c->d = LCM(a->d, b->d)) == 0) { printf("overflow\n"); exit(1); }
    break;
  case '-':
    c->n = a->n * (LCM(a->d,b->d) / a->d) - b->n * (LCM(a->d,b->d) / b->d);
    if ((c->d = LCM(a->d, b->d)) == 0) { printf("overflow\n"); exit(1); }
    break;
  case '*':
    c->n = (a->n / GCD(a->n,b->d)) * (b->n / GCD(a->d,b->n));
    c->d = (a->d / GCD(a->d,b->n)) * (b->d / GCD(a->n,b->d));
    break;
  case '/':
    c->n = (a->n / GCD(a->n,b->n)) * (b->d / GCD(a->d,b->d));
    c->d = (a->d / GCD(a->d,b->d)) * (b->n / GCD(a->n,b->n));
```

```
    break;
  default:
    printf("unknown operator: %c\n", op);
    exit(1);
  }

  c->w = c->n / c->d;                      /* make a proper fraction */
  c->n %= c->d;

  return(0);
}

/*
** LCM -   finds the least common multiple for two integers
**         LCM = a * b / GCD(a, b)
*/
#include "numtype.h"

NAT LCM(NAT a, NAT b)
{
  b /= GCD(a,b);

  if ( (0xffffffff / b) < a )
    return(0);                             /* overflow */
  else
    a *= b;                                /* ok to multiply */

  return(a);
}

/*
** GCD - find the greatest common divisor using Euclid's algorithm
*/
#include "numtype.h"

NAT GCD(NAT a, NAT b)
{
  NAT r;

  while (b > 0)
  {
    r = a % b;
    a = b;
    b = r;
  }

  return(a);
}
```

To run Program 2_4_3, enter all data on the command line. The command format is somewhat rigid to simplify parsing the input:

```
    2_4_3  #,#  op  #,#,
```

where #,# is two integers, the first is the numerator and the second is the denominator, and op stands for arithmetic operator. The calculator recognizes the operators: +, −, *, and /, but cannot return a negative result. To separate the numerator from the denominator, we use a comma without any spaces.

An example where we add 5/6 and 7/8 looks like this:

```
C> 2_4_3   5,6 + 7,8
        5            7              17
   --------- + --------- =   1  ---------
        6            8              24
```

Application: Data Encryption Using XOR

Data security is a growing problem for computer users everywhere. Although it is safe to say that the exclusive-or (XOR) encryption algorithm is not suitable for security, it can be effective for privacy if several precautions are taken. Before discussing the security aspects, I will explain the why the exclusive-or can work for encryption purposes. The nature of this dyadic operator is to change unlike bits to 1 and like bits to 0 (Table 2.4):

Table 2.4. The XOR operator

XOR	0	1
0	0	1
1	1	0

If you take a bit and XOR twice with a second bit, you get the first bit's original value back regardless of the value of the second bit. Expressed as using the C/C++ operator "^" to mean the XOR operation,

$$x = x \wedge y \wedge y.$$

Of course we would not write a line of code like the one above. What we really want is to encrypt our message x, using key y, with a processing flow like the following:

$$e = x \wedge y$$

to encrypt the data. At this point, we have the encrypted message, e. Then, to get our original data back we apply the key a second time:

$$x = e \wedge y.$$

For the current application, binary data is encrypted a byte at a time using a key and the XOR operator. This form of encryption is a polyalphabetic cipher. It's as though you had a number of different substitution alphabets, one for each letter in your key, and applied them sequentially to your data, called *plaintext*. You can see at once that this procedure will engender a period in the encrypted data (or *ciphertext*) equal to the length of your key. It is worth noting that a cryptanalyst can (without too much effort) "crack" the XOR encryption simply by XORing the ciphertext against itself and counting coincidences (Schneier, 1996).

Longer keys mean fewer repetitions of the polyalphabet. Put another way, the longer the key the greater the difficulty in cracking the encryption. Suppose that the key is the four character text "MARK". The encryption would look like this:

```
This is the text to be encrypted.          (plaintext)
MARKMARKMARKMARKMARKMARKMARKMARKM          (key)
);8m(!k9)7k9$*?m5=k/$r.#" 2=57/c          (ciphertext)
```

In this case, the key MARK has an obvious period of 4. An experienced cryptanalyst could have the message and the key in a few minutes. However, if we XOR using two or more keys, then the total effective length of the key is the 'least common multiple of the keys' lengths. Using multiple keys amounts to using a greater number of alphabets for the substitution; the number of alphabets being equal to the total key period:

$$total\_key\_period = lcm(k1, k2, k3, ...),$$

where $k1$ is the length in bytes of key 1, $k2$ is the length in bytes of key 2, and so on. If we have 4 keys of lengths 5, 7, 9, and 11, the effective length of the (combined) key would be 3465 characters. If our key period equals or exceeds the length of our message, then we have, in effect, created an encryption where each character of the plaintext has its own substitution alphabet. Clearly, this would provide the greatest degree of security using the XOR cipher.

The table that follows shows how dramatically the effective key length increases as the number of keys increases:

Table 2.5. How Increasing the Number of Encryption Keys can Increase the Overall Key Length

Individual key lengths	Effective key length
5, 6	30
5, 6, 7	210
5, 6, 7, 11	2310
5, 6, 7, 11, 13	30,030
5, 6, 7, 11, 13, 17	510,510
5, 6, 7, 11, 13, 17, 19	9,699,690
5, 6, 7, 11, 13, 17, 19, 23	223,092,870

It's a simple matter to write a program to encrypt using XOR owing to the character-by-character nature of the operation. Although it is difficult to create a single long key, as noted above creating multiple keys that yield a long period is relatively easy. For Program 2_3_4, input is taken from the command line; the command format being

```
2_4_4 input_file output_file key1 [key2 ...],
```

where `input_file` is the name of the file to be encrypted (plaintext) and `output_file` is the file to contain the encrypted data (ciphertext). You may enter as many keys (i.e., `key1`, `key2`, ...) up to 10 to create a longer effective key.

Program 2_4_4.c. Encrypt/decrypt data using the XOR algorithm

```c
/*
** Program 2_4_4 - An XOR encryption program using multiple keys
*/
#include "numtype.h"

#define MAXKEY 13

main (int argc, char *argv[])
{
  NAT len;
  FILE *fpi, *fpo;
  int i;
  char c, *arg[MAXKEY];

/* check input, open files */

  printf("XOR encryption/decryption\n\n");
  if (argc < 4 || argc >= MAXKEY)
  {
    printf("Usage: 2_4_4 input_file output_file key1 [key2 ...]\n");
    exit(1);
  }

  if ((fpi = fopen(argv[1], "rb")) == NULL)
  {
    printf("error opening plaintext file: %s\n", argv[1]);
    exit(1);
  }

  if ((fpo = fopen(argv[2], "wb+")) == NULL)
  {
    printf("error opening ciphertext file: %s\n", argv[2]);
    exit(1);
  }

/* total length of the key */

  for ( i = 3, len = 1; i < argc; i++)
  {
    len = LCM(len, strlen(argv[i]));
```

```
      arg[i] = argv[i];
  }

  printf("Plaintext file          = %s\n", argv[1]);
  printf("Ciphertext file         = %s\n", argv[2]);
  printf("Effective key length    = %lu\n", len);

/* process the plaintext file */

  while ((c = getc(fpi)) != EOF)
  {
    for (i = 3; i < argc; i++)
    {
      if (*arg[i] == 0) arg[i] = argv[i];
      c ^= *arg[i]++;
    }
    putc(c, fpo);
  }

  fclose(fpi);
  fclose(fpo);
}

/*
** LCM - finds the least common multiple for two integers
**        LCM = a * b / GCD(a, b)
*/
#include "numtype.h"

NAT LCM(NAT a, NAT b)
{
  b /= GCD(a,b);

  if ( (0xffffffff / b) < a )
    return(0);                          /* overflow */
  else
    a *= b;                             /* ok to multiply */

  return(a);
}

/*
** GCD - find the greatest common divisor using Euclid's algorithm
*/
#include "numtype.h"

NAT GCD(NAT a, NAT b)
{
  NAT r;

  while (b > 0)
  {
    r = a % b;
    a = b;
    b = r;
  }

  return(a);
}
```

To encrypt a file using Program 2_3_4 using the encryption keys "mysteries," "redemption," and "problem," type the following at the command prompt:

```
C> 2_4_4 PLAIN.TXT CIPHER.TXT mysteries redemption problem
Plaintext file       = PLAIN.TXT
Ciphertext file      = CIPHER.TXT
Effective key length = 630
```

In the example shown above, the individual key lengths are 9, 10, and 7, respectively. This yields the effective key length of $lcm(9, 10, 7) = 630$. If the message is not significantly longer that 630 characters, then there would be no coincidences that a cryptanalyst could use to crack the message. Compare the XOR encryption methodology to the one-time pad encryption technique discussed in Chapter 4.

Challenges

1. Write a program that will produce a table showing the lcm for all numbers 1 through 20 versus all numbers 1 through 20.

2. Modify the FRACTION structure and fraction calculator to allow negative numbers.

3. Modify the fraction calculator to accept the input from standard input. You can put one fraction and one operator per line and use an equal sign to print the result. This would permit the fractions and operations to be written into a script file that might look something like this:

    5, 6
    +
    7, 8
    =

4. Continue with the modification made in Exercise 3. If you copy the result from variable c to a and read standard input repeatedly into variable b, a series of arithmetic operations can be performed. Make this change.

3

Number-Theoretic Functions

*Take all the strange and monstrous
transformations. These are all leveled by Tao.
Division is the same as creation; creation is the
same as destruction.*

CHUANG-TZU (CA. 300 B.C,)

3.0 Introduction to Multiplicative Arithmetic Functions

There exists a class of functions that operates on integers called *number-theoretic* or *arithmetic functions*. Number-theoretic functions are like mirror images of their real counterparts — if you are looking at the mirror edgewise. Any function whose domain of definition is the set of positive integers is said to be an arithmetic function. An arithmetic function $f(n)$ is said to be multiplicative if

$$f(m \cdot n) = f(m) \cdot f(n)$$

whenever $gcd(m, n) = 1$.

The functions that we will look at next are Euler's totient function $\phi(n)$, the Möbius function $\mu(n)$, the positive-divisors function $d(n)$, and the sum-of-the-divisors functions $\sigma(n)$. Each of these belongs to the family of multiplicative arithmetic functions. Additionally, we will look at the greatest integer function, also called the bracket function [ ]. For comparison purposes, below is a table (Table 3.1) of values computed for the number theoretic functions mentioned previously. Their definitions will be discussed in the sections that follow.

You might be impressed by the erratic nature of these functions. Although their values are strictly deterministic, the values they assume appear random. As we shall see, this "randomness" is due to the fact that the values of these functions are based on the prime factorization.

Table 3.1. Values of Euler's Totient Function, the Möbius Function, the Positive-Divisors Function, and the Sum-of-the-Divisors Function

n	Euler's totient function $\phi(n)$	Möbius function $\mu(n)$	Positive divisors $d(n)$	Sum of the divisors $\sigma(n)$
1	1	1	1	1
2	1	−1	2	3
3	2	−1	2	4
4	2	0	3	7
5	4	−1	2	6
6	2	1	4	12
7	6	−1	2	8
8	4	0	4	15
9	6	0	3	13
10	4	1	4	18
11	10	−1	2	12
12	4	0	6	28
13	12	−1	2	14
14	6	1	4	24
15	8	1	4	24
16	8	0	5	31
17	16	−1	2	18
18	6	0	6	39
19	18	−1	2	20
20	8	0	6	42

Factoring by Trial Division

Although factoring is considered in detail in Chapters 7 and 8, it is necessary to introduce it here so that some of the benefits of having a complete factorization for a given number can be explored. For now I will use only the simplest factoring methods: trial division.

Factoring by trial division, as its name would suggest, is a straightforward process. We simply try every number, divide out a factor, and try again. For extremely large numbers (>100 digits), this process is not practical. However, for unsigned long integers (numbers less than $2^{32} - 1$), this procedure is remarkably fast.

Why this procedure is acceptable for "small" integers is due to the fact that we only need to test numbers less than or equal to the square root of the number being factored. Once we factor out a divisor, we have a new number to factor and we can compute a new upper limit to test based on its square root. This can quickly reduce the amount of numbers to be tried as factors.

This can best be shown by a simple example. If we are factoring the number 2,000,000,003, we can count the total number of trial divisions. Here we are counting every number as a possible factor.

Table 3.2. Number of Trial Division to Factor the Number 2,000,000,003

Factor found	Number to factor	Upper limit to test	Total number of trial divisions
17	2,000,000,003	44,721	16
211	117,647,059	10,847	210
233	557,569	747	232
2393	(Exceeds upper limit to test so factorization is complete)		

Thus, the factors of 2,000,000,003 are 17, 211, 233, and 2393 and were discovered with merely 232 trial divisions. When we skip testing all even numbers greater than 2, the factors are discovered with only 117 trial divisions. In Program 3_0_1, we first test whether 2 is a divisor, and then we test the odd numbers beginning with 3. By eliminating all even numbers from the test first, roughly 50% of the testing will be avoided.

Consider a "worst" case for an unsigned long integer where we are trying to factor a prime, the total number of divisions is still tolerable. Using Program 3_0_1, if we try to factor 4,294,967,291, then 65,534 trials are necessary to determine that it is a prime number. Of course 65,534 divisions would seem like a lot if you are doing it by hand, but for your computer it's a snap!

Program 3_0_1.c. Factors a number using trial division

```
/*
** Program 3_0_1 - Factor a number by trial division
*/
#include "numtype.h"

main(int argc, char *argv[])
{
  NAT i, n, fk, f[32];

  if (argc < 2)
  {
    printf("n = ");
    scanf("%lu", &n);
  }
  else
    n = atol(argv[1]);

  fk = Factor(n, f);

  for (i = 0; i < fk; i++)
    printf("%lu\n", f[i]);
}
```

```
/*
** Factor - finds prime factorization by trial division
**          returned values: number of factors, array w/ factors
*/
#include "numtype.h"

NAT Factor(n, f)
NAT n, f[];
{
  NAT i, fk, s;

  fk = 0;                                /* factor count */

  while ((n % 2) == 0)                   /* factor all 2's */
  {
    f[fk++] = 2;
    n /= 2;
  }

  s = sqrt(n);
  for (i = 3; i <= s; i++, i++)          /* factor odd numbers */
    while ( (n % i) == 0L)
    {
      f[fk++] = i;
      n /= i;
      s = sqrt(n);
    }

  if (n > 1L) f[fk++] = n;

  for (i = fk; i < 32; i++) f[i] = 0;  /* zero out the rest */

  return(fk);
}
```

```
C> 3_0_1
Factor by trial division

n = 432345
3
5
19
37
41
```

Challenge

1. Determine the maximum number of factors an unsigned long integer can have.

3.1 Euler's Totient Function

A very useful function and one of the most important in number theory is Euler's totient function, usually represented by the Greek letter phi, ϕ. A type of counting function, $\phi(n)$ is defined as the number of positive integers not exceeding n that are relatively prime to n. For example, for $\phi(6) = 2$, since there are two numbers; that is, 1 and 5 that are relatively prime to 6. For any prime number p, $\phi(p) = p - 1$.

Euler's totient function can be expressed another way:

$$\phi(n) = n \cdot \prod_{p \mid n} (1 - 1/p),$$

where $\prod_{p \backslash n} (1 - 1/p)$ means the product of all numbers of the form $(1 - 1/p)$ with p being equal to *distinct prime divisors*. Expanding the products will make clear how these two definitions are equivalent. An interesting identity is associated with Euler's totient function.

Theorem 3.1. Given positive integer n,

$$n = \sum_{d \mid n} \phi(d),$$

where $\sum_{d \mid n} \phi(d)$ means the sum of the values of Euler's totient function for all of the proper divisors d of n. A combinatorial argument can be used to prove Theorem 3.1.

Since we know that two numbers are relatively prime if their greatest common divisor equals 1, we can compute values for $\phi(n)$ with the help of Euclid's GCD algorithm. For the program that follows, I will make use of the GCD function created above. To compute a value of ϕ for a given n, we simply count those numbers less than n where the $gcd(n, i) = 1$

Program 3_1_1.c. Evaluate Euler's totient function by counting GCDs equal to 1

```
/*
** Program 3_1_1 - Find the value of Euler's totient function
**                 by counting GCD's equal to 1
*/
#include "numtype.h"

void main(int argc, char *argv[])
{
  NAT n, et;
  int i;

  if (argc == 2) {
    n = atol(argv[1]);
  } else {
    printf("Euler's totient\n\n");
    printf("n = ");
    scanf("%lu", &n);
  }
```

```
    et = 0;
    for (i = 1; i <= n; i++)
      if (GCD(n, i) == 1) et++;

    printf("phi(%lu) = %lu\n", n, et);
}

/*
** GCD - find the greatest common divisor using Euclid's algorithm
*/
#include "numtype.h"

NAT GCD(NAT a, NAT b)
{
   NAT r;

   while (b > 0)
   {
     r = a % b;
     a = b;
     b = r;
   }

   return(a);
}
```

```
C> 3_1_1
Euler's totient

n = 280
phi(18) = 96
```

Although this function is easy to compute for small numbers, it becomes *extremely* slow for any n greater than 100,000. Do not try to compute $\phi(2,000,000,000)$ —you'll be rebooting your computer. We are lucky that ϕ has some properties that will permit us to create a more efficient algorithm.

Application: An Efficient Euler Totient Function

Because Euler's totient function is used extensively in trap-door cryptographic algorithms, improving its efficiency is a worthy goal. In Program 3_3_1, computing $\phi(100,000)$ is unbearably slow. We can improve significantly the speed by which we compute values for Euler's totient function, ϕ; if we make an algorithmic change. Using the factoring function, we can compute ϕ as quickly as we can factor a number.

Factoring works because of ϕ is a multiplicative arithmetic function and the greatest common divisor of any two primes is always 1. Also, ϕ exhibits some interesting behavior under certain conditions. Among the many interesting properties of the ϕ function are

$\phi(p) = p - 1$, whenever p is prime and

$\phi(p^n) = p^n - p^{n-1} = p^{n-1} \cdot (p - 1)$, whenever p is prime

Consider the value of $\phi(n)$ when $n = 280$:

$$\phi(280) = \phi(2^3 \cdot 5 \cdot 7)$$
$$= \phi(2^3) \cdot \phi(5) \cdot \phi(7)$$
$$= (2^3 - 2^2) \cdot (5 - 1) \cdot (7 - 1)$$
$$= 96$$

Clearly the greater the value n we are trying to evaluate the greater the savings. Using a factorization and these facts together we can create a much, much more efficient algorithm for computing the value of $\phi(n)$.

Program 3_1_2.c. Computes values for Euler's totient function using a prime factorization

```
/*
** Program 3_1_2 - Compute values for Euler's totient function using
**                 a prime factorization by trial division
*/
#include "numtype.h"

main (int argc, char *argv[])
{
  NAT n, et;

  if (argc == 2) {
    n = atol(argv[1]);
  } else {
    printf("Euler's totient (factorization)\n\n");
    printf("n = ");
    scanf("%lu", &n);
  }

  et = EulerTotient(n);

  printf("phi(%lu) = %lu\n", n, et);
}

/*
** EulerTotient - compute values for Euler's totient function using
** a prime factorization by trial division
*/
#include "numtype.h"

NAT EulerTotient(NAT n)
{
  NAT i, j, et, pk, fk, f[32];
```

```
  et = 1;                               /* totient value */
  pk = 0;                               /* power count */

  if (n == 1) return(1);

  fk = Factor(n, f);                    /* factor count, factors */
  if (fk > 32) exit(1);

  for (i = 0; i < fk; i++)
  {
    if (f[i] == f[i+1])
    {
      pk++;
    }
    else
    {
      for (j = 0; j < pk; j++) et *= f[i];
      et *= f[i] - 1;
      pk = 0;
    }
  }

  return(et);
}

/*
** Factor — finds prime factorization by trial division
**          returned values: number of factors, array w/ factors
*/
#include "numtype.h"

NAT Factor(NAT n, NAT f[])
{
  NAT i, fk, s;

  fk = 0;                               /* factor count */

  while (!(n & 1))                      /* even - factor all 2's */
  {
    f[fk++] = 2;
    n /= 2;
  }

  s = sqrt(n);
  for (i = 3; i <= s; i++, i++)    /* factor odd numbers */
    while ( (n % i) == 0L)
    {
      f[fk++] = i;
      n /= i;
      s = sqrt(n);
    }

  if (n > 1L) f[fk++] = n;

  for (i = fk; i < 32; i++) f[i] = 0;  /* zero out the rest */

  return(fk);
}
```

```
C> 3_1_2
Euler's totient (factorization)

n = 280
phi(280) = 96
```

Application: Making Yourself a Star

Would you have believed when you woke up this morning that you would be spending time thinking about stellated polygons? Yet, the question of how many ways an n-pointed star can be drawn (without lifting the pen) has been around at least since the time of Euler, and probably well before. The answer to this artistic conundrum can be obtained using Euler's totient function.

Let's begin by describing the process of star-making. Starting with an n-sided polygon, we draw consecutive diagonals from one vertex to another until returning to the starting vertex. Note the following rules: We must skip at least one vertex or else we would obtain the polygon itself. For the same reason, we must skip less than $n - 1$ vertices.

Consider a five-pointed star, the star with the fewest number of points. Number the vertices from 0 to 4 with 0 at the top and proceed clockwise. If we draw a line to every second vertex (skipping one), we connect: 0, 2, 4, 1, 3, 0. Going to every third vertex yields the star: 0, 3, 1, 4, 3, 0: the sequence in reverse. Connecting every fourth (skipping three) only draws the polygon itself: 0, 4, 3, 2, 1, 0. Ditto for drawing sequentially (skipping none), but reversed. There are two values between and not including 1 and $n - 1$ that will generate stars: 2 and 3.

The six-pointed star has a little problem. Only the numbers 2, 3, and 4 are between 1 and 5 (i.e., $n - 1$). However, all these numbers will return us to the starting vertex before reaching all corners. In fact, there is no way to draw a six-pointed star without lifting your pen — I wonder how many people have tried before coming to this conclusion.

Let $k - 1$ be the number of vertices skipped: that is, we connect to the kth vertex. One can deduce that to be able to create a star without lifting the pen, k must be relatively prime to n, the number of vertices. If k is not relatively prime to n then we return to the starting vertex before visiting all the vertices.

Because of symmetry when $k > n/2$, we get a star that is identical to a star generated when $k < n/2$. As noted above, $k > 1$ and $k < n-1$, and $gcd(k, n) = 1$, the number of unique n-pointed stars that can be drawn without lifting one's pen can be expressed as

$$\text{unique stars} = \frac{\phi(n) - 2}{2}.$$

At last we have the answer to our original question. Clearly the points to be made in making stars rests more in the totient than in the quotient.

Based on our formula, we can construct the following table of stellated polygonal possibilities:

Table 3.3. Number of Unique n-Pointed Stars

n	$\phi(n)$	Unique stars
5	4	1
6	2	0
7	6	2
8	4	1
9	6	2
10	4	1
11	10	4
12	4	1
13	12	5
14	6	2
15	8	3
16	8	3

Using Microsoft C/C++'s graphics routines, we can create a program to draw various types of stars. In this and several other applications in this book, I will use this graphics library, although I will limit the use of these routines to only the very basic drawing functions. Program 3_1_3 uses the four functions: _setvideomode, _getvideoconfig, _moveto, and _lineto.

Please refer to Appendix B for a description of these graphics functions and information on their use. Also, you may want to consult a run-time library reference manual for more detailed information regarding the various graphics commands. If you are compiling these programs on a different platform, you should be able to find equivalent functions.

To draw a star, you will need two pieces of information. The first is the number of vertices n. The second is the vertex connection rate k. Not placing any constraints on the input parameters makes the program fairly simple and allows you to try different star formats. Some choices of input will, however, make only polygons or single points.

Program 3_1_3.c. Draw stars having n points

```
/*
** Program 3_1_3 - Draw stars having a certain number of vertices
**                 using Microsoft C graphics functions
*/
#include "numtype.h"

void main(void)
{
  struct videoconfig vc;
  short int            i, j, k, n, x, y, xs, ys;
  float                a;         /* aspect ratio */

/* get number of points, connect factor */

  printf("Draw n-point stars\n\n");
  printf("n, k = ");
  scanf("%d,%d:", &n, &k);

/* initialize graphics, compute center of the graph */

  _setvideomode( _MRESNOCOLOR);
  _getvideoconfig( &vc);

  x = vc.numxpixels / 2 - 1;      /* x center */
  y = vc.numypixels / 2 - 1;      /* y center */
  a = (float) y / (float) x;      /* aspect ratio */

/* create spiral graphic function */

  printf("%d,%d", n, k);

  for (i = 0, j = 0; i <= n; i++)
  {
    j = (j + k) % n;
    xs = x + sin(((float) j) * 6.28329 / (float) n)*x*a;
    ys = y - cos(((float) j) * 6.28329 / (float) n)*y;
    if (i == 0) _moveto(xs, ys); else _lineto(xs, ys);
  }

  while(!kbhit());
  _setvideomode( _DEFAULTMODE);
}
```

You can test that there are indeed only two 7-pointed stars. When you try different stars and vary the connection rate, you will notice some interesting behavior. For example, a 16-pointed star with a connection rate of 7 yields the stellated polygon shown in Figure 3.1.

However, if you graph a 16-pointed star with a connection rate of 6, it will not actually have 16 points (Figure 3.2).

Knowing what you do about the greatest common divisor and Euler's totient function and their relationship to drawing stars, you should have no trouble explaining this phenomenon.

Figure 3.1. Sixteen-pointed star with a connection rate of 7.

Figure 3.2. Sixteen-pointed star with a connection rate of 6.

Challenge

1. Is there any star other than the six-pointed star that is impossible to draw without lifting one's pencil? Explain.

3.2 Möbius Function

The difficult problem of inverting number theoretic relationships was first explored by Gauss and was further developed by his student A. F. Möbius (1790–1860) of one-sided paper strip fame. The Möbius function, usually denoted by the Greek letter μ, is important because it allows us to invert some relationships involving divisors. As Schroeder (1997) explained, summation over divisors is to integration as the Möbius function is to differentiation. The Möbius function is defined as:

$$\phi(n) = \begin{cases} 1 & \text{if } n = 1, \\ 0 & \text{if } p^2 \mid n \text{ for some prime } p \\ (-1)^k & \text{if } n = p_1 p_2 p_3, \ldots p_k, \text{ where } p_i \text{ are distinct primes} \end{cases}$$

Look back at Table 3.1 and consider the factorizations when studying the behavior of the Möbius function. Like the other functions presented in this section, its values appear quite unpredictable.

Using the Möbius function, we can create inversions of arithmetic functions. For example, another way to express Euler's totient function is

$$\phi(n) = \sum_{d \mid n} \mu(d) \frac{n}{d}.$$

This formula and the related identity

$$n = \sum_{d \mid n} \phi(d)$$

represent an instance of an application of a more general theorem on the Möbius function.

Möbius Inversion Formula — Theorem 3.2. If two arithmetic functions $f(n)$ and $g(n)$ satisfy one of the following two conditions:

$$f(n) = \sum_{d \mid n} g(d)$$

and

$$g(n) = \sum_{d \mid n} \mu(d) \cdot f(n/d)$$

for each n, then they satisfy both conditions. (For a proof of Theorem 3.2 see Andrews, 1971.)

By way of definition, any two arithmetic functions f and g that are related by the Möbius inversion formula are called *Möbius pairs* and denote it by { $f(n), g(n)$ }.

Application: An Efficient Möbius Function

The definition of the Möbius function lends itself in a very natural way to the conscription of a number's factors into service. Using the factorization we know if the number is square free and the number of distinct primes it is the product of.

Program 3_2_1.c. Compute values of the Möbius function using a prime factorization

```
/*
** Program 3_2_1 - Compute values for the Möbius function
**               using a prime factorization by trial division
*/
#include "numtype.h"

main (int argc, char *argv[])
{
  NAT n, mf;

  if (argc == 2) {
    n = atol(argv[1]);
  } else {
    printf("Möbius function (factorization)\n\n");
    printf("n = ");
    scanf("%lu", &n);
  }

  mf = Mobius(n);

  printf("µ(%lu) = %ld\n", n, mf);
}

/*
** Möbius - compute values for the Möbius function
**          using a prime factorization by trial division
*/
#include "numtype.h"

INT Möbius(NAT n)
{
  NAT i, j, mf, fk, f[32];

  if (n == 1) return(1L);

  fk = Factor(n, f);                    /* factor count, factors */
```

```
   if (fk > 32) exit(1);

   for (i = 0; i < fk; i++)                /* check if square free */
     if (f[i] == f[i+1]) return(0L);

   if (fk & 1)
     return(-1L);                      _    /* odd distinct primes */
   else
     return(1L);                           /* even distinct primes */
}

/*
** Factor - finds prime factorization by trial division
**             returned values: number of factors, array w/ factors
*/
#include "numtype.h"

NAT Factor(NAT n, NAT f[])
{
  NAT i, fk, s;

  fk = 0;                                 /* factor count */

  while (!(n & 1))                        /* even - factor all 2's */
  {
    f[fk++] = 2;
    n /= 2;
  }

  s = sqrt(n);
  for (i = 3; i <= s; i++, i++)           /* factor odd numbers */
    while ( (n % i) == 0L)
    {
      f[fk++] = i;
      n /= i;
      s = sqrt(n);
    }

  if (n > 1L) f[fk++] = n;

  for (i = fk; i < 32; i++) f[i] = 0;  /* zero out the rest */

  return(fk);
}
```

```
C> 3_2_1
Möbius function (factorization)

n = 19
μ(19) = -1

C> 3_2_1
Möbius function

n = 280
μ(280) = 0
```

3.3 Positive-Divisors Function

The positive-divisors function is another type of counting function, represented by the italic letter d (and sometimes by the Greek letter τ). $d(n)$ (or $\tau(n)$) is defined as the number of positive divisors of n (including n itself):

$$d(n) = \sum_{d \mid n} 1$$

It has the interesting property that

$$d(n) \leq 2 \cdot \sqrt{n}$$

for $n > 1$. This is easily discerned if you consider the divisors of n. Let a be a divisor of n such that $a \geq n$. Then there corresponds another divisor $a' = n / a$ and $1 \leq a' \leq \sqrt{n}$. The number of these a cannot exceed $\sqrt{n}$, so $d(n) \leq 2 \cdot \sqrt{n}$. The next program counts divisors with the help of this property.

Program 3_3_1.c. Evaluate the positive-divisors function

```
/*
** Program 3_3_1 - Find the value of the positive-divisors function
*/
#include "numtype.h"

void main(int argc, char *argv[])
{
  NAT i, n, pd;

  if (argc == 2) {
    n = atol(argv[1]);
  } else {
    printf("Number of positive divisors\n\n");
    printf("n = ");
    scanf("%lu", &n);
  }

  pd = 0;
  for (i = 1; i*i < n; i++)          /* divisors < n */
    if ((n % i) == 0) pd += 2;

  if ((n % i) == 0) pd++;            /* perfect square */

  printf("d(%lu) = %lu\n", n, pd);
}
```

```
C> 3_3_1
Number of positive divisors

n = 18
d(18) = 6
```

Like the totient function, the positive-divisors function, $d(n)$, works fine for small n but becomes slow for large n. But also like the totient function, the algorithm can be improved dramatically once the prime factorization of n is in hand.

Application: An Efficient Positive-Divisors Function

To improve our algorithm for the positive-divisors function, we must first know the following:

If $n = p_1^{al} \cdot p_2^{a2} \cdot \ldots \cdot p_m^{am}$, where the ps are prime factors of n, then:

$$d(n) = (al + 1) \cdot (a^2 + 1) \cdot \ldots \cdot (am + 1),$$

Consider the value of $d(n)$ when $n = 280$:

$$d(280) = d(2^3 \cdot 5^1 \cdot 7^1)$$

$$= (3 + 1) \cdot (1 + 1) \cdot (1 + 1)$$

$$= 16$$

Once again, we obtain greater savings the larger n is. Now our algorithm can be significantly improved:

Program 3_3_2.c. Compute values of the positive-divisors function using a prime factorization

```
/*
** Program 3_3_2 - Compute the number of positive-divisors for n
**                 using a prime factorization by trial division
*/
#include "numtype.h"

main (int argc, char *argv[])
{
  NAT n, pd;

  if (argc == 2) {
    n = atol(argv[1]);
  } else {
    printf("Positive divisors (factorization)\n\n");
    printf("n = ");
    scanf("%lu", &n);
  }

  pd = PositiveDivisors(n);

  printf("d(%lu) = %lu\n", n, pd);
}

/*
```

```
** PositiveDivisors - compute values for the positive-divisors
** function using a prime factorization by trial division
*/
#include "numtype.h"

NAT PositiveDivisors(n)
NAT n;
{
  NAT i, j, pd, pk, fk, f[32];

  pd = 1;                          /* positive divisors */
  pk = 1;                          /* power count */

  if (n == 1) return(1L);

  fk = Factor(n, f);               /* factor count, factors */
  if (fk > 32) exit(1);

  for (i = 0; i < fk; i++)
  {
    if (f[i] == f[i+1])
    {
      pk++;
    }
    else
    {
      pd *= pk + 1;
      pk = 1;
    }
  }

  return(pd);
}

/*
** Factor - finds prime factorization by trial division
**          returned values: number of factors, array w/ factors
*/
#include "numtype.h"

NAT Factor(NAT n, NAT f[])
{
  NAT i, fk, s;

  fk = 0;                          /* factor count */

  while (!(n & 1))                 /* even - factor all 2's */
  {
    f[fk++] = 2;
    n /= 2;
  }

  s = sqrt(n);
  for (i = 3; i <= s; i++, i++)    /* factor odd numbers */
    while ( (n % i) == 0L)
    {
      f[fk++] = i;
      n /= i;
```

```
        s = sqrt(n);
    }

    if (n > 1L) f[fk++] = n;

    for (i = fk; i < 32; i++) f[i] = 0;   /* zero out the rest */

    return(fk);
}
```

C> *3_3_2*
Positive divisors (factorization)

n = *280*
d(280) = 16

3.4 Sum-of-the-Divisors Function

The last multiplicative arithmetic function I will look at is the sum-of-the-divisors function, usually indicated by the Greek letter sigma, σ. $\sigma(n)$ is the sum of the divisors of n (including n):

$$\sigma(n) = \sum_{d \mid n} d$$

For example, if $n = 18$, its divisors are 1, 2, 3, 6, 9, and 18; the sum of which in 39. Because I have not used any factoring methodologies, you may find this program almost too simple. It tests every number up to and equal to n.

Program 3_4_1.c. Evaluate the sum-of-the-divisors function

```
/*
** Program 3_4_1 - Find the value of the sum-of-the-divisors
**                 function
*/
#include "numtype.h"

void main(int argc, char *argv[])
{
    NAT i, n, sd;

    if (argc == 2) {
        n = atol(argv[1]);
    } else {
        printf("Sum of the divisors\n\n");
        printf("n = ");
        scanf("%lu", &n);
    }

    sd = 0;
    for (i = 1; i <= n; i++)
```

```
    if ((n % i) == 0) sd += i;

  printf("sigma(%lu) = %lu\n", n, sd);
}
```

```
C> 3_4_1
Sum of the divisors

n = 18
sigma(18) = 39
```

Evaluation of this function is also substantially benefited by first performing a prime factorization.

Application: An Efficient Sum-of-the-Divisors Function

If we have a prime factorization of an integer, then we can use those prime factors in a useful formula.

If $n = p_1^{a1} \cdot p_2^{a2} \cdot \ldots \cdot p_m^{am}$, where the ps are prime factors of n, then:

$$\sigma(n) = \frac{(p_1^{(a1 + 1)} - 1)}{(p_1 - 1)} \cdot \frac{(p_2^{(a2 + 1)} - 1)}{(p_2 - 1)} \cdot \ldots \cdot \frac{(p_m^{(am + 1)} - 1)}{(p_{am} - 1)}$$

Consider the value of $d(n)$ when $n = 280$:

$$\sigma(280) = \sigma(2^3 \cdot 5^1 \cdot 7^1)$$

$$= (2^{3+1} - 1)/(2 - 1) \cdot (5^{1+1} - 1)/(5 - 1) \cdot (7^{1+1} - 1)/(7 - 1)$$

$$= 15 \qquad \cdot 6 \qquad \cdot 8$$

$$= 720$$

For the fourth time, we obtain much greater savings the larger n is. The new algorithm based on trial factorization is dramatically faster:

Program 3_4_2.c. Computes values of the sum-of-the-divisors function using a prime factorization

```
/*
** Program 3_4_2 - Compute the sum-of-the-divisors of n
**                 using a prime factorization by trial division
*/
#include "numtype.h"

main (int argc, char *argv[])
{
```

```
  NAT n, sd;

  if (argc == 2) {
    n = atol(argv[1]);
  } else {
    printf("Sum of the divisors (factorization)\n\n");
    printf("n = ");
    scanf("%lu", &n);
  }

  sd = SumDivisors(n);

  printf("sigma(%lu) = %lu\n", n, sd);
}

/*
** SumDivisors - compute values for the sum-of-the-divisors function
**               using a prime factorization by trial division
*/
#include "numtype.h"

NAT SumDivisors(n)
NAT n;
{
  NAT i, j, sd, pk, fp, fk, f[32];

  sd = 1;                       /* sum of the divisors */
  pk = 1;                       /* power count */

  if (n == 1) return(1);

  fk = Factor(n, f);            /* factor count, factors */
  if (fk > 32) exit(1);

  for (i = 0; i < fk; i++)
  {
    if (f[i] == f[i+1])
    {
      pk++;
    }
    else
    {
      fp = f[i];
      for (j = 0; j < pk; j++) fp *= f[i];
      sd *= (fp - 1) / (f[i] - 1);
      pk = 1;
    }
  }

  return(sd);
}

/*
** Factor - finds prime factorization by trial division
** returned values: number of factors, array w/ factors
*/
#include "numtype.h"
```

```
NAT Factor(n, f)
NAT n, f[];
{
  NAT i, fk, s;

  fk = 0; /* factor count */

    while ((n % 2) == 0)                   /* factor all 2's */
    {
  f[fk++] = 2;
    n /= 2;
    }

  s = sqrt(n);
  for (i = 3; i <= s; i++, i++)            /* factor odd numbers */
  while ( (n % i) == 0L)
    {
      f[fk++] = i;
      n /= i;
      s = sqrt(n);
    }

  if (n > 1L) f[fk++] = n;

  for (i = fk; i < 32; i++) f[i] = 0;  /* zero out the rest */

  return(fk);
}
```

```
C> 3_4_2
Sum of the divisors (factorization)

n = 280
sigma(280) = 720
```

3.5 The Greatest Integer Function

Although not considered a number-theoretic function because its
domain includes real numbers, the greatest integer function or "bracket" function has some properties that make it useful to examine. It is
represented formulaically by the square brackets [ ]. Often though, we
take its existence for granted since it is nothing more than the largest
integer after any fraction has been removed. It pops up all the time as
a natural consequence of integer arithmetic on a computer.

The [ ] yields the largest integer satisfying

$$x - 1 < [\,x\,] \le x.$$

To illustrate its application, consider the following examples using
real numbers

$$[\,3\,] = 3, [\,9/4\,] = 2, [\sqrt{2}\,] = 1, [\,-1.5\,] = -2, [\,-\pi\,] = -4.$$

The greatest integer function is implemented in C/C++ as the runtime function floor().

If x and y are real numbers then the following identities are true:

1. $x - 1 < [x] \leq x, [x] \leq x < [x] + 1, 0 \leq x - [x] < 1$.
2. $[x] + [-x] = \{ 0$ if x is an integer, -1 otherwise $\}$.
3. $x - [x]$ is the fractional part of x.
4. $[x + m] = [x] + m$ if m is a positive integer.
5. $[[x]/m] = [x/m]$ if m is a positive integer.
6. $[x] + [y] \leq [x + y] \leq [x] + [y] + 1$.
7. $[x] = \sum_{1 \leq i \leq x} 1$ if $x \geq 0$.

One of the greatest integer function more interesting properties is how it can be used to count how many times a particular prime p can divide $n!$ (factorial). This can be done using the following theorem:

Theorem 3.3. If n is a positive integer and p is a prime, then the exponent of the highest power of p that divides $n!$ is

$$e = \sum_{i=1}^{\infty} [n/p^i].$$

(Note that this is not an infinite series since at some point $p^i > n$.)

Suppose $n = 15$ and $p = 2$, then the largest divisor having the form 2^e of 15! is determined by finding the exponent e using Theorem 3.3. To simplify computations, we can use identity 5) to help:

$$[15/2] = 7, [7/2] = 3, [3/2] = 1, [1/2] = 0$$

Adding the values $7 + 3 + 1$ gives us the exponent of the largest divisor having the form 2^e. Therefore 2^{11} is the largest divisor of 15! (and 2^{12} is not).

The relationship of the greatest integer function to other number theoretic functions is expressed in the following theorem:

Theorem 3.4. Let f and F be number-theoretic functions such that

$$F(n) = \sum_{d \mid n} f(d),$$

then for any positive integer N,

$$\sum_{n=1}^{N} F(n) = \sum_{k=1}^{N} f(d) \cdot [N/k]$$

Proof of this theorem can be found in most books on elementary number theory.

Challenges

1. Write a program to find the largest exponent e of prime divisors p for $n!$. Use this program to find the number of trailing 0s (zeroes) in 1000!

2. Given integers a and $m > 0$, prove that $a - m \cdot [\, a \,/\, m]$ is the least nonnegative residue of a modulo m.

3. Prove the identity: $n = \sum_{d \mid n} \phi(d)$.

4. Show that $\{\, n,\, \phi(n) \,\}$, $\{d(n),\, 1\, \}$, and $\{\, \sigma(n),\, n\, \}$ are all Möbius pairs.

5. Write a program that, given a positive integer n, finds the smallest integer x such that $\phi(x) = n$, $d(x) = n$, $\sigma(x) = n$.

4

Congruence Arithmetic

Arithmetic is where numbers fly like pigeons in and out of your head.

CARL SANDBURG, *ARITHMETIC* 1950

4.0 Introduction

Congruence arithmetic is one of the fundamental calculating devices in number theory. It is the mathematics of remainders — that which remains after division has been performed. There are many times when we are interested only in remainders. From time-keeping to cryptography to periodic physical phenomena, remainders often contain very useful information.

Suppose we want to know what day it will be 15 days from Wednesday (the third day of the week); we might think 3 + 15 = (3 + 1) + (2 · 7) ≡ 4 or Thursday (the fourth day of the week). Here we are only interested in the remaining days after all of the 7s (whole weeks) have been removed. As an arithmetic convenience, the days are numbered starting at 0 for Sunday.

The name for the type of mathematics we will be using is called *modular* or *modulo arithmetic*. The nature of modular arithmetic is grounded in the concept of divisibility. Recall Euclid's Division Lemma (Chapter 2):

$$a = x \cdot m + b \qquad 0 \leq b < m$$

After rewriting this expression:

$$(a - b) = x \cdot m$$

it becomes clear that m is a divisor of $(a - b)$; that is, $m \mid (a - b)$. Gauss, in his great work *Disquisitiones Arithmeticae* (1801) realized that this type of expression has some limitations with respect to manipulating the remainder. To study the divisibility properties of the integers he developed the notation:

$$a \equiv b \pmod m$$

This is read "a is congruent to b mod (or modulo) m." From the above statement, we may also say "b is a residue of a mod (or modulo) m." Congruence, indicated by the symbol "$\equiv$", is called an equivalence relation. It is not the same as equality, "$=$", but it has striking similarities. How it is different will become apparent when we look at residual classes.

Congruences have properties that permit us to manipulate the quantities and unknowns in much the same way we manipulate equations. Consider the basic properties of modular arithmetic congruences having the same modulus:

If $a \equiv b \pmod m$ and $c \equiv d \pmod m$, then:

$a \equiv a \pmod m$,

$b \equiv a \pmod m$,

$a + c \equiv b + d \pmod m$,

$a \cdot (b + c) \equiv a \cdot b + a \cdot c \pmod m$,

$a \cdot c \equiv b \cdot d \pmod m$,

$k \cdot a \equiv k \cdot b \pmod m$, for all k, and

$a^k \equiv b^k \pmod m$, for all k

Congruences also have the following transitive property:

If $a \equiv b \pmod m$ and $b \equiv c \pmod m$, then $a \equiv c \pmod m$.

We can add the following statement with respect to the GCD:

If $k \cdot a \equiv k \cdot b \pmod m$ and $gcd(k,m) = d$, then $a \equiv b \pmod{m / d}$

Another useful fact relating powers and congruences, based on the divisibility of binomial coefficients, is

$(a + b)^n \equiv a^n + b^n \pmod n$, if n is prime or absolute pseudoprime

The concept of the pseudoprime will be discussed in Chapter 4 where we concentrate on finding prime numbers and testing primality.

Challenges

1. Interpret each of the following: ≡ (mod 0), ≡ (mod 1), and ≡ (mod –1).

2. Congruence and modular arithmetic can sometimes be found in "magic" tricks. In a recent example, a magician appeared on television and said that he could "telepathically" guess the letter (A, B, C, D, E) pointed to by the viewing audience at home. The setup was similar to this:

He makes his secret guess before we begin. Next, we, the audience, are presented with a large list of animal names (e.g., tiger, elephant, parrot, cat, etc.) are and instructed to choose one. We may use any letter appearing in that name as our starting position. To indicate our choice, we touch the letter on the television screen showing five cards. We are then told to move our hands three card places to the left or the right in any combination, moving one card at a time. The televised magician removes the two end cards A and E. We are then told to move again three places to the left or right as before. From coast to coast, people are touching the letter C on their televisions as he removes the cards B and D. As for his telepathic guess — the letter C, of course! How does this trick work? (Hint: Consider what must be a common characteristic of the animal names in the list.)

4.1 Modular Arithmetic and Residue Classes

When a number is divided by a modulus, m, there are only m possible remainders, r. From Euclid's Division Lemma, we know that $0 \le r < m$. Every allowable arithmetic operation between any two operands can be reduced to a table of values where the number of row equals the number of columns equals m.

If we make our days of the week (modulo 7) example more abstract by numbering the days 0 through 6, we can create the following table for modular addition (Table 4.1):

Table 4.1. Modulo 7 Arithmetic Table

+	0	1	2	3	4	5	6
0	0	1	2	3	4	5	6
1	1	2	3	4	5	6	0
2	2	3	4	5	6	0	1
3	3	4	5	6	0	1	2
4	4	5	6	0	1	2	3
5	5	6	0	1	2	3	4
6	6	0	1	2	3	4	5
7	0	1	2	3	4	5	6
8	1	2	3	4	5	6	0
9	2	3	4	5	6	0	1
0	3	4	5	6	0	1	2
11	4	5	6	0	1	2	3
12	5	6	0	1	2	3	4
13	6	0	1	2	3	4	5
14	0	1	2	3	4	5	6
15	1	2	3	4	5	6	0
16	2	3	4	5	6	0	1
17	3	4	5	6	0	1	2
18	4	5	6	0	1	2	3
19	5	6	0	1	2	3	4
20	6	0	1	2	3	4	5

You can see in Table 4.1 what we determined before: $3 + 15 = 18 \equiv 4$ (mod 7). Selecting only those numbers that satisfy $a \equiv 4$ (mod 7) gives the set: { ..., –10, –3, 4, 11, 18, ... }. This set is one of seven possible *residue classes*. In Table 4.2 each of the seven different residue classes are shown with their related algebraic form. If we select one member from each class, we would have a complete residue system modulo 7.

Table 4.2. Modulo 7 Algebraic Forms and Residue Classes

Algebraic form, for all b				Residue class			
$a = 7 \cdot b$	... ,	–14,	–7,	0,	7,	14,	...
$a = 7 \cdot b + 1$	... ,	–13,	–6,	1,	8,	15,	...
$a = 7 \cdot b + 2$	... ,	–12,	–5,	2,	9,	16,	...
$a = 7 \cdot b + 3$	... ,	–11,	–4,	3,	10,	17,	...
$a = 7 \cdot b + 4$	... ,	–10,	–3,	4,	11,	18,	...
$a = 7 \cdot b + 5$	... ,	–9,	–2,	5,	12,	19,	...
$a = 7 \cdot b + 6$	... ,	–8,	–1,	6,	11,	20,	...

Any two elements from the same residue class are said to be congruent. For example, 1 and 8 are from the same residue class and so they are congruent. Conversely, any two elements from different residue classes are incongruent: 2 and –2 are from two different residue classes and so they are incongruent. This is why we use congruence notation as distinguished from equality; if any one member from a particular class satisfies a relation, then *all* of the members from that residual class satisfy the relation.

In general, we prefer to restrict ourselves to a particular residue system (mod m). Our usual preference is called the "arithmetic residue system (mod m)": { 0, 1, 2, ..., $m – 1$ }. These are also called the "least positive remainders." You probably recognize this as the system produced from the C/C++ modulus operator, %, when it is applied to unsigned operands.

Subtraction (mod m) behaves much the same as addition. For example, $3 – 15 = –12 \equiv –5 \equiv 2$ (mod 7). Generally, we will want to ensure that we obtain the arithmetic residue system in operations between signed operands. For this reason, the modulus to the reduced result.

Modular multiplication is performed in a manner similar to addition. For example, $3 \cdot 5 = 15 \equiv 1$ (mod 7). As noted in the initial discussion above, the distributive law is respected in congruence arithmetic: $3 \cdot (4 + 8) = 3 \cdot 4 + 3 \cdot 8 = 36 \equiv 1$ (mod 7). What may not be apparent is how division might work, if it works at all.

What would be the result if you divided 1 by 2 (mod 7)? This is the same as asking what number times 2 is equal to 1. The value 4 can be found and verified with little effort. Of course members of the same residue class are also valid answers (e.g., 11, 18, –3). Similarly, $5 \cdot 3 \equiv 1 \pmod 7$, or put another way, $5 \equiv 1 / 3 \pmod 7$. Because the product of 5 and 3 is congruent to 1, 5 is called the inverse of 3 (mod 7), and 3 is the inverse of 5 (mod 7). Inverses (mod m) find their way into may number theoretic analyses. Consider the Table 4.3, a table of inverses (mod 7).

Table 4.3. Mod 7 Inverses

a	$1/a$
0	Undefined
1	1
2	4
3	5
4	2
5	3
6	6

The general case for finding inverses (mod m) whenever m is relatively prime to a is

$$1/a = a^{(m-2)}$$

from Fermat's Little Theorem.

Among the many uses of this expression is its ability to solve equations of the form

$$a \cdot x \equiv b \pmod m,$$

which we shall see later on in this chapter.

Use the next program to create tables of modular addition and multiplication:

Program 4_1_1.c. Creates a table of modular addition and multiplication

```
/*
** Program 4_1_1 — Print tables of modular arithmetic
*/
#include "numtype.h"

void main(int argc, char *argv[])
{
   INT i, j, m, n;

   if (argc == 2) {
   m = atol(argv[1]);
   } else {
```

```
printf("Modular arithmetic tables\n\n");
printf("m = ");
scanf("%ld", &m);
}

printf("\n\nAddition (mod %ld)\n  ", m);
for ( j = 0; j < m; j++) printf("__%2ld", j);
for ( i = 0; i < m; i++)
{
printf("\n%3ld  |", i);
for ( j = 0; j < m; j++)
{
  n = (i + j) % m;
  printf(" %2ld", n);
  }
}

printf("\n\nMultiplication (mod %ld)\n  ", m);
for ( j = 0; j < m; j++) printf("__%2ld", j);
for ( i = 0; i < m; i++)
{
printf("\n%3ld  |", i);
for ( j = 0; j < m; j++)
{
  n = (i * j) % m;
  printf(" %2ld", n);
  }
}
}
```

Output from Program 4_4_1 looks like

```
C> 4_1_1
Modular arithmetic tables

m = 5

Addition (mod 5)
    __ 0__ 1__ 2__ 3__ 4
 0 |  0   1   2   3   4
 1 |  1   2   3   4   0
 2 |  2   3   4   0   1
 3 |  3   4   0   1   2
 4 |  4   0   1   2   3

Multiplication (mod 5)
    __ 0__ 1__ 2__ 3__ 4
 0 |  0   0   0   0   0
 1 |  0   1   2   3   4
 2 |  0   2   4   1   3
 3 |  0   3   1   4   2
 4 |  0   4   3   2   1
```

Another important residue system is derived from a complete residue system (arithmetic or otherwise); it is called the *reduced residue system*. Simply stated, the reduced residue system is a set of numbers, one from each residue class, that is relatively prime to the modulus.

Definition. If p is a prime, the ring Z_p is a *finite field* having p elements.

By way of an example, we observe that the complete arithmetic residue system (mod 10) is { 0, 1, 2, 3, 4, 5, 6, 7, 8, 9 } and that the reduced residue system (mod 10) is { 1, 3, 7, 9 }. In general, the reduced residue system for a prime modulus p is { 1, 2, 3, ... , $p - 1$ }. By definition, the number of members in the reduced residue system is equal to the value of Euler's totient function $\phi(p)$. Reduced residue systems form a cyclic group under multiplication.

Program 4_1_2.c. Determines the reduced residue system for a given modulus

```
/*
** Program 4_1_2 - Determine the reduced residue system for
**                 a given modulus
*/
#include "numtype.h"

void main(int argc, char *argv[])
{
  NAT i, g, m;

  if (argc == 2) {
    m = atol(argv[1]);
  } else {
    printf("Reduced residue system\n\n");
    printf("m = ");
    scanf("%lu", &m);
  }

  printf("reduced residue class (mod %lu) = { 1", m);
  for (i = 2; i < m; i++)
    if ((g = GCD(m, i)) == 1) printf(", %lu", i);

  printf(" }\n");
}

/*
** GCD - find the greatest common divisor using Euclid's algorithm
*/
#include "numtype.h"

NAT GCD(NAT a, NAT b)
{
  NAT r;

  while (b > 0)
  {
    r = a % b;
    a = b;
    b = r;
  }

  return(a);
}
```

```
C> 4_1_2
Reduced residue system

m = 10
reduced residue class (mod 10) = { 1, 3, 7, 9 }
```

We will return to a discussion of reduced residue classes when we look at primitive roots and powers of integers.

Application: The Perpetual Calendar

Certainly one of the most everyday applications of congruence arithmetic must be embodied in the calendar. Along with the printed form, the calendar is ubiquitous in software. In this application, we will make a calendar that will compute the day of the week for any day of any year. Each day of the week is assigned a number from the arithmetic residue system: Sunday = 0, Monday = 1, Tuesday = 2, Wednesday = 3, Thursday = 4, Friday = 5, and Saturday = 6.

Our modern calendar is steeped in human history. The first modern calendar was attributed to Julius Caesar, who in 46 B.C. corrected the earlier less accurate Egyptian calendar of exactly 365 days. The Julian calendar was based on a year having 365 days and every fourth (leap) year had 366 days. This means that each calendar year had 365 + 1/4 = 365.25 days. As time passed, the ability of astronomers to measure the length of a day improved and it was calculated that the true length of a year was closer to 365.2422 days. This meant that there was an extra 0.0078 days per year in the calendar.

Over the centuries, that extra 0.0078 accrued to 10 extra days and in 1582 Pope Gregory modified the calendar again. On October 5th, 1582, the calendar was advanced 10 days to October 15th thereby skipping October 6th through the 14th. When this calendar was adopted in 1752 by the American colonies and Great Britain 11 days had to be added. You can well imagine the consternation felt at that time by both creditors and debtors alike when they cried out "Give us back our fortnight!" and rioted in the streets.

To correct the problem on an ongoing basis, a decision was made that leap years would be those years evenly divisible by 4, *except* those that are evenly divisible by 100, the century years. Further, those years that mark centuries are only leap years when divisible by 400, the quadricentennial years. Therefore the years 1600 and 2000 are leap years but 1700, 1800, 1900, and 2100 are not. Such considerations have important implications when refinancing a home or planning vacations into the distant future.

After all of this jiggery-pokery, we now have a calendar with a four-hundred-year period and a calendar day with an average length of

365 + 1/4 − 1/100 + 1/400 = 365.2425 days. Future generations will need to grapple with the time still remaining to account for an even tinier discrepancy of 0.0003 days per year or 3 days per 10,000 years. Please let me be the first to suggest that we not have a leap year in years divisible by 3332 (i.e., 3332, 6664, 9996, etc.).

The computation of the day of the week is accomplished by computing the number of days since an arbitrary date, say March 1, 1600. Why March 1st? This gives us a convenient base to work from so our leap year days will be added as the very last day of the year. Our months are therefore numbered March = 0, April = 1, ..., January = 10, February = 11. Using this convention, we begin by computing the day of the week that March 1st falls on for a given year y.

The first thing we should note is that 365 (mod 7) $\equiv$ 1, so each year advances the day by 1. Next we will account for the "oddball" years that have passed from 1600 to year y:

number of quadrennial years	$= [\,(y - 1600)\,/\,4\,]$
number of centennial years	$= [\,(y - 1600)\,/\,100\,]$
number of quadricentennial years	$= [\,(y - 1600)\,/\,400\,]$

The "[ ]" notation means the greatest integer. The day of the week w_y is computed by

$$w_y = b + y - 1600 + [\,(y-1600)\,/\,4\,] - [\,(y-1600)\,/\,100\,] + [\,(y-1600)\,/\,400\,],$$

where b is the day of the week on March 1st, 1600 (which is easily determined later). Simplifying this expression yields:

$$w_y = b + y + [\,y\,/\,4\,] - [\,y\,/\,100\,] + [\,y\,/\,400\,] - 1988$$

or

$$w_y \equiv b + y + [\,y\,/\,4\,] - [\,y\,/\,100\,] + [\,y\,/\,400\,] \;(\text{mod } 7).$$

March 1st, 1996 was a Friday (w_{1996} = 5). Substituting back into the formula and solving for b gives

$$5 \equiv b + 1996 + [\,1996\,/\,4\,] - [\,1996\,/\,100\,] + [\,1996\,/\,400\,] \;(\text{mod } 7);$$

therefore,

$$b = 3$$

Because the year starts on March 1st, January and February of the current year should actually belong to the previous year. To correct the year, we subtract 1 from the year count if we are in the months of January or February. This correction is implemented using C's conditional operator ? when the year is computed:

```
((m < 3) ? -1 : 0)
```

Next we compute the monthly increment. The months with 30 days will shift the first day of the next month up 2 days and months with 31 days will shift the first day of the next month up 3. What we need is a function that will compute the number of days that a given month shifts the day of the week (Table 4.4):

Table 4.4. Months of the Year – Days (Mod 7)

Month	Days (mod 7)	Sum
February	0	0
March	3	3
April	2	5
May	3	8
June	2	10
July	3	13
August	3	16
September	2	18
October	3	21
November	2	23
December	3	26
January	3	29

Ignoring February, there are 11 increments with an average of 2.6 days. With a little bit of trial and error, the following formula can be derived, which will yield the desired increments for the day offset:

$$w_m = [2.6 \cdot m + 0.5]$$

or

$$w_m \equiv [2.6 \cdot m + 0.5] \pmod 7,$$

where w_m is the day of the week for the first day of month. Because we start our year on March 1st, we must apply a correction to align the date correctly with the variables within the formula. The month numbers are shifted circularly: $m = (month + 9) \% 12$.

Lastly the day of the week relative to the first day of the month for day d is $w_d = (d - 1)$. Putting it all together, the day of the week, given the month m, the day d, and the year y:

$$w \equiv (w_y + w_m + w_d) \pmod 7.$$

Substituting in the respective parts gives

$$w \equiv 3 + y + [y / 4] - [y / 100] + [y / 400] + [2.6 \cdot m + 0.5] + (d - 1) \pmod 7$$

and combining the constant terms

$$w \equiv y + [y / 4] - [y / 100] + [y / 400] + [2.6 \cdot m + 2.5] + d \pmod 7.$$

After all that work, I'm sure you're glad that the coding is relatively simple. Putting it all together, we get the following program:

Program 4_1_3.c. A perpetual calendar

```
/*
** Program 4_1_3 - A perpetual calendar
*/
#include "numtype.h"

void main(void)
{
   char *day[] = { "Sunday",
                   "Monday",
                   "Tuesday",
                   "Wednesday",
                   "Thursday",
                   "Friday",
                   "Saturday"};

   int m, d, y, wd, wm, wy, w;

   printf("Perpetual calendar\n\n");
   printf("month-day-year: ");
   scanf("%d-%d-%d", &m, &d, &y);

   wy = 3 + y + (y / 4) - (y / 100) + (y / 400) + ((m < 3) ? -1 : 0);
   wm = 2.6 * ((m + 9) % 12) + 0.5;
   wd = d - 1;

   w = (wy + wm + wd) % 7;    /* day of the week */

   printf("%s %d ≡ (%d + %d + %d) (mod 7)\n", day[w], w, wy, wm, wd);
}
```

```
C> 4_1_3
Perpetual calendar

month-day-year: 11-13-1996
Wednesday 3 ≡ (2483 + 21 + 12) (mod 7)
```

For other interesting topics in time keeping, visit the website for the Time Service Dept., U.S. Naval Observatory, the official source of time used in the United States, at http://tycho.usno.navy.mil/.

Application: Linear Congruential Random Number Generator

Programming for games and simulations frequently leads to a need for sequences of random numbers. Of course one cannot generate truly random numbers using arithmetic procedures, but so-called pseudo-randomness is easy to achieve. Pseudorandomness refers to the fact that these numbers have the statistical qualities of honest-to-goodness random numbers, even though they can be generated identically any number of times. In the discussion that follows, we will use the term random numbers but keep in mind their pseudorandom nature.

Congruence relationships are a simple means of generating random numbers. The following linear recursive equation

$$x_{n+1} = (a \cdot x_n + b) \% m$$

with three parameters is commonly used. Many nonlinear and compound variations are possible (Knuth, 1981; Niederreiter, 1992; Press et al, 1992; Eichenauer-Herrmann, 1995; Eichenauer-Herrmann and Emmerich, 1996). An informative taxonomy of pseudorandom number generators can be found in Lagarias (1990).

Also we use the arithmetic modulus operator %, meaning that we want the residues from the arithmetic residue system (mod m). Clearly if a, b, and m are selected correctly, then the maximal period for x_n can be obtained. If the period is the longest possible, then all the values between 0 and $m - 1$ will occur at some point and the choice of the initial x_0 (called the seed) only changes the offset into the series.

If we let $m = 2^{16}$ or 2^{32}, the maximum integer range, then several benefits are immediately realized. First, the modulus operation is free since any overflow condition will be handled automatically by the computer hardware with the arithmetic residue being retained. Second, the rules for selecting the remaining parameters so that we achieve the maximal period are

$$a \equiv 1 \ (\text{mod } 4)$$

and

$$gcd(b, m) = 1.$$

The selection of a can be done so as to make for a very efficient algorithm. If $a = 2^k + 1$, then instead of performing multiplication we can substitute a binary left shift and addition. Since m's only factor is 2, b only needs to be odd.

For our program, we will let $a = 5$, $b = 1$, and $m = 65,536$ (2^{16}). By letting $a = 5$, we can multiply by 5 by performing a right bit shift 2 places and adding the value back. Since $b = 1$, we use the increment operator. Selecting $m = 2^{16}$ is just a matter of convenience and speed. Using the program provided here and these parameters, a 90 MHz Pentium PC will generate 500,000 random numbers per second.

If more diversity in the random numbers is required, change the declarations from unsigned int to unsigned long int. You should also increase the value of a and b to lessen the likelihood of having relatively long sequences of ascending numbers.

The implementation of the above algorithm with the seed optionally provided on the command line follows:

Program 4_1_4.c. \A linear congruential pseudorandom number generator

```
/*
** Program 4_1_4 - Test a congruential random number generator
**                 seeded from a command line argument
*/
#include "numtype.h"

void main(int argc, char *argv[])
{
  unsigned int x, x0, random();

  printf("Congruential random number generator\n\n");
  if (argc == 2)                  /* seed random number generator */
    x0 = x = atol(argv[1]);
  else
    x0 = x = 0;                   /* default = 0 */

  while (!kbhit())                /* check for key pressed */
  {
    x = random(&x);              /* generate RN */
    printf("%u\n", x);
    if (x == x0) break;          /* check for duplicate */
  }
}


/*
** random - a multiplicative congruential random number generator
**           x = (a * x + b) % m, where a = 5, b = 1, m = 2^16
*/

unsigned int random(x)
unsigned int *x;
{
  *x += (*x << 2);               /* a: multiply by 5 */
  (*x)++;                        /* b: add 1 */
  return(*x);
}
```

Our random number generator yields integer values between 0 and 65,535. The program will print numbers until the keyboard is hit or a duplicate is found (after which the sequence would necessarily repeat):

```
C> 4_1_4
Congruential random number generator

1
6
31
156
781
3906
19531
32120
29529
16574
17335
21140
40165
4218
. . .
```

It is worth pointing out that it is generally not a good practice to extract and use the lowest bits in the random numbers. These bits tend to be less random than their high order counterparts. Thus, if numbers are desired in the range of [0, 16), the following statement should *not* be used:

```
r = random(&x) % 16;
```

The preferred coding would be

```
r = random(&x) / 4096
```

because it extracts the high-order bits. (For speed you could use the right-shift operator instead of division.) Similarly, random numbers should not be broken into pieces to produce more than one random number.

Application: One-Time Pad Data Encryption

Congruence arithmetic lends itself in many ways to applications relating to data encryption. Using the random number generator just created, we can develop a "one-time pad" encryption program. A one-time pad is a large nonrepeating set of values, historically random characters (Schneier, 1996). To encrypt the data we simply add the text message to the one-time pad character (mod n).

The name "one-time pad" comes from the fact that the encrypting set of values is used only once and then discarded. Clearly both the sender and receiver must each have a copy of the encrypting values. Also, the pad itself must be destroyed after the message is decrypted or the secrecy may be compromised at some future date.

There are two problems in using a one-time pad generated with a pseudorandom number generator. The first concern you should have when using a computer-generated set of one-time pad values is that they are not truly random. In fact, the general problem of encryption schemes relying on pseudorandom number generators have been studied extensively (Reeds, 1294). It may strike you as remarkable that such schemes can be broken even if the encrypting parameters are unknown (Stearn, 1987). The second consideration is that the one-time pad values can be reconstructed at any time in the future if the algorithm that was used to construct them is discovered.

The one-time pad encryption technique using a random number generator may be suitable for privacy. However, like the XOR method described in Chapter 2, it is not recommended for any application where the encrypted data is subject to serious attack. Even with the aforementioned limitations, there are times when a simple encryption scheme may be satisfactory, especially for short messages. Keep in mind that the encrypting program is itself the one-time pad. Of course you are probably not going to destroy the program after only one use. You should always change the offset into the one-time pad or vary parameters associated with the random number generator to produce another one-time pad sequence.

As noted in the previous section, we may select a, b, m, and x_0 to generate a sequence of pseudorandom digits using the following recurrence relation:

$$x_{n+1} = (a \cdot x_n + b) \% m$$

that has the maximum period. That means that

$$a \equiv 1 \;(\text{mod } 4)$$

and

$$gcd(b, m) = 1.$$

Using the sample random number generator from the last section, let $a = 5$, $b = 1$, and $m = 65{,}536$ (2^{16}) $\cdot$ $x_0 = 1996$ will be used to seed the random number generator; this is the offset into the one-time pad. Also, we will let $n = 26$ and we will work only with uppercase letters where A = 0, B = 1, ... Z = 25.

Now consider the following text, called plaintext:

```
CRYPTOGRAPHY.
```

The one-time pad encrypting string is derived from the random number generator:

```
EYZBGEWSCNEY.
```

The resulting ciphertext obtained by the following addition one character at a time

```
(C + E) % 26 = (67 + 69) % 26 = 6  =  G
(R + Y) % 26 = (82 + 89) % 26 = 15 =  P
(Y + Z) % 26 = (89 + 90) % 26 = 23 =  X
. . .
```

to yield the ciphertext

```
GPXQZSCJCCLW.
```

The great strength of the one-time pad is that the above encrypted text can be obtained from any plaintext message with equal likelihood. Why equal? Because the one-time pad is a random sequence of encrypting characters. Since any one encrypting character has a uniform probability of occurring, so do the characters in the resulting ciphertext. Our plaintext could have read:

```
NOWISTHETIME
```

or:

```
TOOMANYUSERS.
```

Without the one-time pad used to encrypt, who could know? Only with the one-time pad can the message be correctly decrypted. To decrypt the ciphertext subtract the one-time pad encrypting string:

```
(G - E) % 26 = (71 - 69) % 26 = 2  =  C
(P - Y) % 26 = (80 - 89) % 26 = 17 =  R
(X - Z) % 26 = (88 - 90) % 26 = 24 =  Y
. . .,
```

which gives back the original plaintext.

The next program implements the one-time pad encryption and decryption algorithm for character data using a pseudorandom number generator. Its uses to the function random(). To encrypt data using a random number seed of 1996, enter the following command:

```
C> 3_5_1 /D 1996.
```

The message is typed after the command is given. To end input, type
^Z (Control-Z). To decrypt data, type

```
C> 3_5_1 /E 1996
```

Instead of typing the message you can use a file and direct it into
the program using "<". The following program is used to create an
encrypted file called XDATA. As you would expect, in the example the
file is then decrypted using the same random number generator key.

Program 4_1_5.c. One-time pad encryption/decryption using a pseudorandom number generator

```
/*
** Program 4_1_5 - One-time pad data encryption program using
**                 a multiplicative congruential random number
**                 generator seeded from the command line
*/
#include "numtype.h"

void main(int argc, char *argv[])
{
   unsigned int x, seed, random();
   char   c, e, k, text[128];
   int    i, j, mode;

   if (argc < 3)                      /* check command line */
   {
     printf("One-time pad encryption\n\n");
     printf("command format: 4_1_5 /D|/E seed\n");
     exit(1);
   }
   if (toupper(argv[1][1]) == 'E')
     mode = 1; else mode = 0;         /* encrypt/decrypt */
     x = atol(argv[2]);               /* seed RN generator */

   while (gets(text) != NULL)
   {
     for (i = 0, j = 0; i < strlen(text); i++)
     {
       c = toupper(text[i]);
       if (isalpha(c))
       {
       x = random(&x);
       k = (x / 2048) % 26 + 'A';       /* one-time pad char */
       if (mode)
         text[j++] = (c + k) % 26 + 'A';      /* encrypted text */
       else
         text[j++] = (26 + c - k) % 26 + 'A'; /* decrypted text */
     }
       }
     text[i] = '\0';
     printf("%s\n", text);
   }
}
```

```
/*
** random - a multiplicative congruential random number generator
**             x = (a * x + b) % m, where a = 5, b = 1, m = 2^16
*/

unsigned int random(x)
unsigned int *x;
{
  *x += (*x << 2);                    /* a: multiply by 5 */
  (*x)++;                             /* b: add 1 */
  return(*x);
}
```

```
C> 4_1_5 /E 1996 > XDATA
CRYPTOGRAPHY
^Z

C> TYPE XDATA
GPXQZSCJCCLW

C> 4_1_5 /D 1996 < XDATA
CRYPTOGRAPHY
```

Challenges

1. Modify Program 4_1_1 to perform modular subtraction. Print the results from the arithmetic residue system.

2. Modify Program 4_1_1 to perform modular division. Print the results from the arithmetic residue system.

3. Program 4_1_4 found random numbers in the range of 0 to 65,535. It is often desirable to have random numbers over the interval [0, 1). Modify the program to produce floating-point values in this range.

4. Devise or use a standard statistical test (e.g., chi-square) to measure if the sequence of numbers produced in Program 4_1_4 is random. Look for patterns in neighbors having a constant offset. Modify the program to shuffle the random number selections and see if you can improve the statistical randomness.

5. Modify Program 4_1_5 to operate on binary digits in the range from 0 to 255.

4.2 Multiplication over a Modulus

You may have detected a potential limitation when using fixed-precision arithmetic —the risk of overflow while multiplying and exponentiating. Obviously, nothing can be done for the usual operations, but is it possible to reduce this risk when operating over a modulus? Yes, and we shall see how. In the next two sections, I present two utility functions called MulMod and PowMod that will be helpful in this regard. MulMod computes r in $a \cdot b \equiv r$ (mod m), given a, b, and m. PowMod computes r in $a^n \equiv r$ (mod m), given a, n, and m. These functions should be used whenever the possibility of overflow exists.

The multiplication method presented here is based on an age-old principle of halving and doubling. Consider the following multiplication of 41 times 26 (without reduction by the modulus). The first column (multiplicand) has successive halving, and the second column (multiplier) has successive doubling. If the number in the first column is odd (after discarding any fractional part), then the number in the second column is added to the product (Table 4.5):

Table 4.5. Multiplication by Halving and Doubling: 41 · 26

Multiplicand	Odd?	Multiplier	Product
41	Yes	26	26
20	No	52 = 26 + 26	26
10	No	104 = 52 + 52	26
5	Yes	208 = 104 + 104	208 + 26 = 234
2	No	416 = 208 + 208	234
1	Yes	832 = 416 + 416	832 + 234 = 1066

The situation is the same when reducing over the modulus. Look at the same example where multiplying 41 times 26 (mod 11). At each step, test if the multiplicand is odd, and if it is, then we add it to the product (mod m). Next we halve the multiplicand, discarding any fractional part. Then double the multiplier (mod m) and return to the test (Table 4.6):

Table 4.6. Multiplication over a Modulus by Halving and Doubling: 41 · 26 (mod 11)

Multiplicand	Odd?	Multiplier	Product
41	Yes	$4 \equiv 26$ (mod 11)	4
20	No	$8 \equiv 4 + 4$ (mod 11)	4
10	No	$5 \equiv 8 + 8$ (mod 11)	4
5	Yes	$10 \equiv 5 + 5$ (mod 11)	$10 + 4 \equiv 3$ (mod 11)
2	No	$9 \equiv 10 + 10$ (mod 11)	3
1	Yes	$7 \equiv 9 + 9$ (mod 11)	$7 + 3 \equiv 10$ (mod 11)

So $41 \cdot 26 \equiv 10$ (mod 11). You can probably see the advantage to this methodology as opposed to writing in our code,

```
r = (a * b) % m
```

or even

```
r = ((a % m) * (b % m)) %m.
```

The problem is that the product $(a \cdot b)$, depending on the values of a, b, and m, can be very close to m^2 in value. If you consider the situation for unsigned long integers ($2^{32} - 1$ max) when multiplying two numbers that are each greater than $2^{16}-1$, an overflow condition will arise. Using the MulMod method above, you can multiply two integers (mod m) up to $2^{27} - 1$ without overflow and even some larger with limited risk.

Another interesting positive feature of this methodology is that we never actually do any division or multiplication. We halve by right shifting and double by left-shifting. On the down side, the process can perform a modulus operation after the sum in accumulated. Some of these operations are avoided because the remainder is computed only when the sum exceeds the modulus.

Use this little utility function to extend the useful range for your applications as I have in this book. Program 4_2_1 compares the results of performing natural multiplication (without overflow protection) with MulMod.

Program 4_2_1.c. Computes products over a modulus using the halving/doubling method

```c
/*
** Program 4_2_1 - Compute r in the relation a * b ≡ r (mod m)
*/
#include "numtype.h"

void main(void)
{
  NAT a, b, m, r;

  printf("Compute r in: a*b ≡ r (mod m)\n\n");
  printf("a, b, m = ");
  scanf("%lu,%lu,%lu", &a, &b, &m);

  r = (a * b) % m;
  printf("without MulMod: %lu*%lu ≡ %lu (mod %lu)\n", a, b, r, m);

  r = MulMod(a, b, m);
  printf("using MulMod : %lu*%lu ≡ %lu (mod %lu)\n", a, b, r, m);
}

/*
** MulMod - computes r for a * b ≡ r (mod m) given a, b, and m
*/
#include "numtype.h"

NAT MulMod(NAT a, NAT b, NAT m)
{
  NAT r;
```

```
    if (m == 0) return(a * b);           /* (mod 0) */

    r = 0;
    while (a > 0)
    {
      if (a & 1)                          /* test lowest bit */
        if ((r += b) > m) r %= m;         /* add (mod m) */
      a >>= 1;                            /* divided by 2 */
        if ((b <<= 1) > m) b %= m;        /* times 2 (mod m) */
    }

    return(r);
}
```

```
C> 4_2_1
Compare r in: a*b ≡ r (mod m)

a, b, m = 100000, 100000, 9
without MulMod: 100000*100000 ≡ 2 (mod 9)
using MulMod:   100000*100000 ≡ 1 (mod 9)
```

100,000 * 100,000 = 10,000,000,000 = 9,999,999,999 + 1. 1 is clearly the correct remainder. However, the natural multiplication (without MulMod), followed by the modulus operation, produced the wrong answer. Under the constraint of fixed precision, only by halving and doubling can the correct result be obtained.

Challenges

1. Describe the MulMod algorithm in terms of binary arithmetic.

2. Explain why 100,000 * 100,000 (mod 9) gives the answer 2 when using long integers (as in program 4_2_1).

3. Modify MulMod() to handle (mod 1) and (mod 2) more efficiently.

4.3 Exponentiation over a Modulus

Exponentiating a number over a modulus, like multiplying over a modulus, often occurs in number theoretic applications. Like the MulMod technique, the algorithm for computing an exponentiated value is fairly well known but certainly worth examining.

In this method, if the exponent is odd, then we keep the partial product for multiplication later; if it is even, we ignore it. Next the exponent is halved, the partial product is squared, and the test is performed again. In Table 4.7 we compute 3^{10} using the halving and doubling method for exponentiation.

Table 4.7. Exponentiation by Halving and Doubling: 3^{10}

Exponent	Odd?	Partial Product	Product
10	No	3	1
5	Yes	$9 = 3 \cdot 3$	$1 \cdot 9 = 9$
2	No	$81 = 9 \cdot 9$	9
1	Yes	$6561 = 81 \cdot 81$	$9 \cdot 6561 = 59{,}049$

The final product is the answer, so $3^{10} = 59{,}049$.

If we think of the exponent in terms of its binary representation, then each binary digit i that is equal to 1 represents the partial product a raised to the 2^i power: $a^1, a^2, a^4, a^8, \ldots$. Multiplying the partial products together gives the complete product (a^n). Repeating the example, if $a = 3$ and $n = 10$ (1010 in binary notation), then $a^n = 3^{10} = 1 \cdot 3^8 \cdot 0 \cdot 3^4 \cdot 1 \cdot 3^2 \cdot 0 \cdot 3^1 = 59{,}049$. To get the residue when $m > 1$ in $a^n \equiv r \pmod{m}$, we simply reduce both the partial product r and the complete product a by the modulus at each step.

Program 4_3_1.c. Computes exponentiated values over a modulus using the halving/doubling method

```
/*
** Program 4_3_1 - Compute r in the relation: a^n ≡ r (mod m)
*/
#include "numtype.h"

void main(void)
{
  NAT a, i, m, n, r;

  printf("Compare r in: a^n ≡ r (mod m)\n\n");
  printf("a, n, m = ");
  if (scanf("%lu,%lu,%lu", &a, &n, &m) != 3) exit(1);

  r = 1;
  for (i = 0; i < n; i++)
    r *= a;
  r %= m;
  printf("without PowMod: %lu^%lu ≡ %lu (mod %lu)\n", a, n, r, m);

  r = PowMod(a, n, m);
  printf("using PowMod : %lu^%lu ≡ %lu (mod %lu)\n", a, n, r, m);
}

/*
** PowMod - computes r for a^n ≡ r (mod m) given a, n, and m
*/
#include "numtype.h"

NAT PowMod(NAT a, NAT n, NAT m)
{
  NAT r;
```

```
  r = 1;
  while (n > 0)
  {
     if (n & 1)                        /* test lowest bit */
        r = MulMod(r, a, m);           /* multiply (mod m) */
     a = MulMod(a, a, m);              /* square */
     n >>= 1;                          /* divided by 2 */
     }

  return(r);
}

/*
** MulMod - computes r for a * b ≡ r (mod m) given a, b, and m
*/
#include "numtype.h"

NAT MulMod(NAT a, NAT b, NAT m)
{
  NAT r;

  if (m == 0) return(a * b);           /* (mod 0) */

  r = 0;
  while (a > 0)
  {
     if (a & 1)
        if ((r += b) > m) r %= m;       /* add (mod m) */
     a >>= 1;                           /* divided by 2 */
     if ((b <<= 1) > m) b %= m;         /* times 2 (mod m) */
     }

  return(r);
}
```

```
C> 4_3_1
Compare r in: a^n ≡ r (mod m)

a, n, m = 100, 100, 9
without PowMod: 100^100 ≡ 0 (mod 9)
using PowMod  : 100^100 ≡ 1 (mod 9)
```

Challenges

1. Make a table showing each step of the PowMod algorithm when the modulus is applied.

2. Modify PowMod() to handle (mod 1) and (mod 2) more efficiently.

4.4 Primitive Roots and Powers of Integers

Let us begin our discussion of primitive roots with an example and return to the definition a little later. Consider the following congruence where $gcd(a, m) = 1$:

$a^h \equiv b \pmod{m}$.

If we let $a = 4$, $m = 19$, and $h = 1, 2, 3, \ldots$, we can create the following sequence of congruences (mod 19):

$$
\begin{aligned}
4^1 &\equiv 4 & &\equiv 4 \\
4^2 &\equiv 4 \cdot 4 = 16 & &\equiv 16 \\
4^3 &\equiv 4 \cdot 16 = 64 & &\equiv 7 \\
4^4 &\equiv 4 \cdot 7 = 28 & &\equiv 9 \\
4^5 &\equiv 4 \cdot 9 = 36 & &\equiv 17 \\
4^6 &\equiv 4 \cdot 17 = 68 & &\equiv 11 \\
4^7 &\equiv 4 \cdot 11 = 44 & &\equiv 6 \\
4^8 &\equiv 4 \cdot 6 = 24 & &\equiv 5 \\
4^9 &\equiv 4 \cdot 5 = 20 & &\equiv 1 \\
4^{10} &\equiv 4 \cdot 1 = 4 & &\equiv 4 \\
4^{11} &\equiv 4 \cdot 4 = 16 & &\equiv 16 \\
4^{12} &\equiv 4 \cdot 16 = 64 & &\equiv 7 \\
4^{13} &\equiv 4 \cdot 7 = 28 & &\equiv 9 \\
4^{14} &\equiv 4 \cdot 9 = 36 & &\equiv 17 \\
\end{aligned}
$$

...

You can see that the sequence begins to repeat after $4^9 \equiv 1 \pmod{19}$ and in fact has a period of 9. This means that $4^{9+i} \equiv 4^9 \cdot 4^i \equiv 4^i \pmod{19}$. Earlier in this chapter we exploited this behavior to create our random number generator that iteratively incremented the exponent.

The period exhibited shall be the focus of our attention for now. We call the length of the period *the order of a (mod m)* , written as $\text{ord}_m a = h$. It is sometimes said, particularly in older texts, *that a belongs to the exponent h mod m*. I prefer the former wording because it is less verbose and agrees nicely with the written notation.

The order also can be defined as the smallest positive integer h such that

$$a^h \equiv 1 \pmod{m}$$

If follows that if $\text{ord}_m a = h$ and $a^x \equiv 1 \pmod{m}$, then $h \mid x$.

In the next program, we will compute the order of all members of the reduced residue system given the modulus as an input value:

Program 4_4_1.c. Generates a table showing the order of each member of a reduced residue class

```
/*
** Program 4_4_1 - Compute the order of each member of a residue class
*/
#include "numtype.h"
```

```
void main(int argc, char *argv[])
{
  NAT a, g, h, m, r;

  if (argc == 2) {
    m = atol(argv[1]);
  } else {
    printf("Order of each member of a residue class\n\n");
    printf("m = ");
    scanf("%ld", &m);
  }

/* column headers */

  for (h = 1; h < m; h++)
    printf(" %-2lu ", h);
  printf("\n");

    for (h = 1; h < m; h++)
  printf("a ");
  printf("\n");

  for (h = 1; h < m; h++)
    printf("--- ");
  printf("\n");

/* compute the values */

  for (a = 1; a < m; a++)
    if ((g = GCD(a, m)) == 1)        /* reduced residue system */
      {
      for (h = 1; h < m; h++)        /* each exponent */
      {
      r = PowMod(a, h, m);
      printf("%3lu ", r);
      if (r == 1) break;
      }
      printf("\n");
      }
}

/*
** PowMod - computes r for a^n ≡ r (mod m) given a, n, and m
*/
#include "numtype.h"

NAT PowMod(NAT a, NAT n, NAT m)
{
  NAT r;

  r = 1;
  while (n > 0)
  {
    if (n & 1)                       /* test lowest bit */
      r = MulMod(r, a, m);           /* multiply (mod m) */
    a = MulMod(a, a, m);             /* square */
    n >>= 1;                         /* divided by 2 */
  }
```

```
    return(r);
}

/*
** MulMod - computes r for a * b ≡ r (mod m) given a, b, and m
*/
#include "numtype.h"

NAT MulMod(NAT a, NAT b, NAT m)
{
  NAT r;

  if (m == 0) return(a * b);          /* (mod 0) */

  r = 0;
  while (a > 0)
  {
    if (a & 1)                        /* test lowest bit */
      if ((r += b) > m) r %= m;       /* add (mod m) */
    a >>= 1;                          /* divided by 2 */
    if ((b <<= 1) > m) b %= m;        /* times 2 (mod m) */
  }

  return(r);
}

/*
** GCD - find the greatest common divisor using Euclid's algorithm
*/
#include "numtype.h"

NAT GCD(NAT a, NAT b)
{
  NAT r;

  while (b > 0)
  {
    r = a % b;
    a = b;
    b = r;
  }

  return(a);
}
```

Across the top of the columns is the exponent that a is raised to. Continuing with our example modulo 19, you can see that the lengths of the periods of the numbers in the reduced residue system are 1, 3, 6, 9, and 18.

```
C> 4_4_1
Order of each member of a residue class

m = 19
   1   2   3   4   5   6   7   8   9  10  11  12  13  14  15  16  17  18
   a   a   a   a   a   a   a   a   a   a   a   a   a   a   a   a   a   a
  --- --- --- --- --- --- --- --- --- --- --- --- --- --- --- --- --- ---
   1
```

(continued)

```
 2   4   8  16  13   7  14   9  18  17  15  11   3   6  12   5  10   1
 3   9   8   5  15   7   2   6  18  16  10  11  14   4  12  17  13   1
 4  16   7   9  17  11   6   5   1
 5   6  11  17   9   7  16   4   1
 6  17   7   4   5  11   9  16   1
 7  11   1
 8   7  18  11  12   1
 9   5   7   6  16  11   4  17   1
10   5  12   6   3  11  15  17  18   9  14   7  13  16   8   4   2   1
11   7   1
12  11  18   7   8   1
13  17  12   4  14  11  10  16  18   6   2   7  15   5   8   9   3   1
14   6   8  17  10   7   3   4  18   5  13  11   2   9  12  16  15   1
15  16  12   9   2  11  13   5  18   4   3   7  10  17   8   6  14   1
16   9  11   5   4   7  17   6   1
17   4  11  16   6   7   5   9   1
18   1
```

We are now ready for our definition of a primitive root. If a is an integer belonging to exponent $\equiv \phi(m)$ modulo m, then a is called a *primitive root* modulo m (Andrews, 1971). To put it another way, if the order of a (mod m) = $\phi(m)$, then a is called a *primitive root* modulo m. Regardless of whether or not a is a primitive root, it is always true that $\text{ord}_m a \mid \phi(m)$.

Of our example above we can say that the order of 4 mod 19 equals 9; that is, $\text{ord}_{19}4 = 9$. Although $\text{ord}_{19}4 \mid \phi(19)$, since $\phi(19) = 18$, 4 is not a primitive root mod 19. On the other hand, 3 is a primitive root mod 19 because $\text{ord}_{19}3 = \phi(19) = 18$.

The relationship between primitive roots and reduced residue classes is direct. If a is a primitive root (mod m), then $\{ a, a^2, a^3, \ldots, a^{\phi(m)} \}$ are mutually incongruent and form a reduced residue system (mod m). For this reason a primitive root is also called a *generator* or a generating element because it completely generates a permutation of a reduced residue system.

Not all moduli have primitive roots. There are no primitive roots modulo 8. If you run Program 4_4_1, you can see how this fact expresses itself. There are no numbers in the reduced residue system of order $\phi(8) = 4$; therefore, 8 has no primitive roots.

```
C> 4_4_1
Order of each member of a residue class

m = 8
1   2   3   4   5   6   7
a   a   a   a   a   a   a
--- --- --- --- --- ---

1
3   1
5   1
7   1
```

This fact may lead us to ask: For which moduli do primitive roots exist? Although a proof can be found in many books on number theory, the following theorem tells us which numbers have primitive roots.

Theorem 4.1. Those numbers having primitive roots are only those that have the form 2, 4, p^n, $2p^n$, where p is any odd prime. From this statement we can conclude that every prime has a primitive root.

If primitive roots exist for the particular modulus m, then the number of mutually incongruent primitive roots is equal to $\phi(\phi(m))$. From the definition of $\phi(m)$ we can conclude that if m is prime, then there are exactly $\phi(m - 1)$ primitive roots.

In our example, $m = 19$, so $\phi(m) = 18$. How many primitive roots should we expect? $\phi(18) = 6$; therefore, there are six primitive roots for the prime 19. The primitive roots are those six with a period of 18: namely, 2, 3, 10, 13, 14, and 15.

Fermat's Little Theorem and Euler's Theorem

A very useful theorem, called Fermat's Little Theorem, developed by P. Fermat (1601–1665) states the following:

Fermat's Little Theorem (4.2). If p is prime and a is a positive integer, then $p \mid a^p - a$. This important theorem can be restated as a congruence:

If p is prime and a is a positive integer, then $a^p \equiv a \pmod{p}$, or as it often appears:

If p is prime and a is a positive integer, then $a^{p-1} \equiv 1 \pmod{p}$. This result is proved below for Euler's generalization of Fermat's Little Theorem.

It is important to note that you cannot prove primality with Fermat's Little Theorem, as we shall see in Chapter 8. In fact it is fairly easy to find an a and nonprime p for which $a^{p-1} \equiv 1 \pmod{p}$. For example, $2^{340} \equiv 1 \pmod{341}$.

Fermat's Little Theorem also has application with respect to polynomial congruences. If p is prime and $d \mid p - 1$, then there exactly d roots of the congruence

$$x^d \equiv 1 \pmod{p}.$$

This theorem can be proved with help from Lagrange Factor Theorem (Section 5.5). Since $d \mid p - 1$, we can have

$$x^{p-1} - 1 = (x^d - 1) \cdot q(x),$$

where q is a polynomial of degree $(p - 1 - d)$. By the Lagrange Factor Theorem, the congruence $q(x) \equiv 0 \pmod{p}$ has $(p - 1 - d)$ solutions. Similarly, since $x^{p-1} \equiv 1 \pmod{p}$ has $(p - 1)$ solutions, $x^d \equiv 1 \pmod{p}$ must have $(p - 1) - (p - 1 - d) = d$ solutions.

You may think that Fermat's Little Theorem looks familiar to the discussion of Euler's totient function introduced in Chapter 2. Because $\phi(p) = p - 1$ for prime p, Fermat's Little Theorem is actually a special case of Euler's Theorem, which applies to all moduli:

Euler's Theorem (4.3). If $gcd(a,m) = 1$, then $a^{\phi(m)} \equiv 1 \pmod{m}$.

To prove Euler's Theorem, let $\{ c_1, c_2, ..., c_{\phi(m)} \}$ be a reduced residue system $\pmod{m}$. Let $gcd(a,m) = 1$. Then $\{ a \cdot c_1, a \cdot c_2, ..., a \cdot c_{\phi(m)} \}$ also is a reduced residue system $\pmod{m}$. Now:

$$\prod_{i=1}^{\phi(m)} a \cdot c_i \;=\; \prod_{i=1}^{\phi(m)} c_i \pmod{m}.$$

Moving a outside the multiplication gives:

$$a^{\phi(m)} \cdot \prod_{i=1}^{\phi(m)} c_i \;=\; \prod_{i=1}^{\phi(m)} c_i \pmod{m}.$$

Because $gcd(m, \prod c_i)$ by definition of reduced residue systems, $a^{\phi(m)} \equiv 1 \pmod{m}$.

An interesting mental experiment involving Fermat's Little Theorem is in the analysis of riffling cards. In an ordinary deck of cards, there are 52 cards. To perfect faro shuffle, the deck is cut into two equal 26-card decks and shuffled by alternating cards from each deck. Obviously this is a vast simplification over real-life shuffling, which is much less predictable.

Each time the deck is perfect faro shuffled, card x is moved to position y according to

$$2 \cdot x \equiv y \pmod{53}.$$

After n shuffles, the card will be in position w: $2^n \cdot x \equiv w \pmod{53}$. It's natural to ask how many times must the deck be shuffled so that the cards are returned to their original order. In this case we must solve the congruence:

$$2^n \cdot x \equiv x \pmod{53}.$$

From Fermat's Little Theorem we know that

$$2^{53} \equiv 2 \pmod{53},$$

which gives

$2^{52} \equiv 1 \pmod{53}$.

Therefore, the cards will be returned to their original order after 52 shuffles. As it turns out, 52 is the least number of shuffles required. In general, if one starts with a deck of m cards, then it will take only n shuffles to return the deck to its original order (Andrews, 1971), providing that $2^n \equiv 1 \pmod{m + 1}$.

Challenges

1. Prove that if p is an odd prime and g is a primitive root modulo p, then g belongs to h modulo p^m, where $h = (p - 1)p^e$, for some exponent e.

4.5 The Discrete Logarithm

Suppose a number m has primitive roots and a is one of them. From our definition, we know that $\{ a, a^2, a^3, ..., a^{\phi(m)} \}$ are mutually incongruent and form a reduced residue system (mod m). For any number b, where $gcd(b,m) = 1$, we can find a smallest exponent of a power of a that is congruent to b. That is to say,

$a^h \equiv b \pmod{m}$.

The exponent h is called the index (or discrete logarithm) of b to the base a and is written as

$h = \text{ind}_a b$.

Thus, the preceding expression is the discrete antilogarithm of b to base a.

The behavior of the index is strikingly similar to the logarithm of a real number. Let m be any integer that admits a primitive root a and $gcd(b, m) = 1$, and where $a^h \equiv b \pmod{m}$, then the following laws hold:

$\text{ind}_a b \equiv \text{ind}_a c \pmod{\phi(m)}$ if $b \equiv c \pmod{m}$

$\text{ind}_a (b \cdot c) \equiv \text{ind}_a b + \text{ind}_a c \pmod{\phi(m)}$

$\text{ind}_a b^n \equiv n \cdot \text{ind}_a b \pmod{\phi(m)}$

$\text{ind}_a b \cdot \text{ind}_b a = 1 \pmod{\phi(m)}$

Finding indices depends on knowing a primitive root — an important problem in its own right (see Section 4.4). If a primitive root a is known, then the exponent is the index of that number. For modulus 19, we know that 2 is a primitive root. Recalling the output from

Program 4_4_1 presented above:

1	2	3	4	5	6	7	8	9	10	11	12	13	14	15	16	17	18
a	a	a	a	a	a	a	a	a	a	a	a	a	a	a	a	a	a
2	4	8	16	13	7	14	9	18	17	15	11	3	6	12	5	10	1

It can be seen that 2 generates the complete reduced residue class for the modulus 19. This is a table of discrete logarithms to the base 2. To use it, look at the bottom row and find the number whose index is sought; the index is the exponent on a. For example: $\text{ind}_2 2 = 1$, $\text{ind}_2 7 = 6$, and $\text{ind}_2 10 = 17$. The anti-index is read in the other direction (mod 19): $2^1 = 2$, $2^6 = 7$, $2^{17} = 10$.

The simplest method of finding the index for general groups is to create a table with the same flavor as Program 4_4_1. Clearly, to build the table from a known generator would take n operations and we would need to store n indices. To find the index, we can use the function IndMod() presented here in Program 4_5_1. As input, we need a number to take the index of the modulus and its primitive root.

Program 4_5_1. Compute the discrete logarithm of a number

```
/*
** Program 4_5_1 - Compute the index x in: a^x ≡ b mod m
*/
#include "numtype.h"

void main(int argc, char *argv[])
{
  NAT g, b, m, x;

  if (argc == 4) {
    g = atol(argv[1]);
    b = atol(argv[2]);
    m = atol(argv[3]);
  } else {
    printf("Find x in: ind(g) b ≡ x (mod m)\n\n");
    printf("g, b, m = ");
    scanf("%lu,%lu,%lu", &g, &b, &m);
  }

  x = IndMod(g, b, m);

  if (x)
    printf("ind(%lu) %lu ≡ %lu (mod %lu)\n", g, b, x, m);
  else
    printf("no index\n");
}

/*
** IndMod - find the index (discrete logarithm): ind(g) b mod m
**
** input: b = number to find index for
```

```
**          g = primitive root for m
**          m = modulus w/ primitive root g
*/
#include "numtype.h"

NAT IndMod(NAT g, NAT b, NAT m)
{
  NAT i, x;

  i = 1;
  x = g;

  if (GCD(b, m) != 1) return(0);

  while ((x != b) && (i < m))
  {
    x = MulMod(x, g, m);
    i++;
  }

  if (i == m)
    return(0);                      /* index not found */
  else
    return(i);                      /* got it */
}

/*
** MulMod - computes r for a * b ≡ r (mod m) given a, b, and m
*/
#include "numtype.h"

NAT MulMod(NAT a, NAT b, NAT m)
{
  NAT r;

  if (m == 0) return(a * b);        /* (mod 0) */

  r = 0;
  while (a > 0)
  {
    if (a & 1)                      /* test lowest bit */
      if ((r += b) > m) r %= m;     /* add (mod m) */
    a >>= 1;                        /* divided by 2 */
    if ((b <<= 1) > m) b %= m;      /* times 2 (mod m) */
  }

  return(r);
}

/*
** GCD - find the greatest common divisor using Euclid's algorithm
*/
#include "numtype.h"

NAT GCD(NAT a, NAT b)
{
  NAT r;

  while (b > 0)
  {
    r = a % b;
```

```
   a = b;
   b = r;
 }

 return(a);
}
```

Finding a primitive root for $\phi(m)$ is the first problem. Recall that the numbers having primitive roots are only those that have the form 2, 4, p^n, $2p^n$, where p is any odd prime and n is a positive integer. To program this as an index function, it is generally necessary to have the totient function's value of the modulus being used. If it turns out that m is prime, then $\phi(m) = m - 1$ is an important simplification of the problem.

Further, finding a number a that is a primitive root is greatly facilitated by the prime factorization of $m - 1$ (a major limitation of this method). Let $p_1, p_2, p_3, \ldots, p_k$ be the distinct prime factors of $m - 1$. Then if for any a in the reduced residue system (mod m)

$$a^{(m-1)/p} \equiv 1 \ (\text{mod } m)$$

for any p_i, then a is *not* a primitive root. The usual procedure is to start at $a = 2, 3, \ldots$ and test the prime factors until a primitive root is found (Cohen, 1995).

Luckily for the primes, there are plenty of primitive roots and one will quickly be discovered. Table 4.8 shows strictly increasing small-

Table 4.8. Table of Strictly Increasing Smallest Primitive Roots for Primes Less than 1,000,000

Prime	Smallest Primitive Root
2	1
3	2
7	3
23	5
41	6
71	7
191	19
409	21
2,161	23
5,881	31
36,721	37
55,441	38
71,761	44
110,881	69
760,321	73

est primitive roots for primes less than 1,000,000. For any prime less than 1,000,000, the primitive root will be 73 or less. Primitive roots for composites are a bit more scarce (see Challenge 1).

The next program finds primitive roots using the strategy outlined above. Given an *m*, and a factorization, it finds the first primitive root.

Program 4_5_2.c. Find primitive roots

```
/*
** Program 4_5_2 - Find a primitive root for m
*/
#include "numtype.h"

void main(int argc, char *argv[])
{
  NAT g, m;

  if (argc == 2) {
    m = atol(argv[1]);
  } else {
    printf("Find the smallest primitive root\n\n");
    printf("m = ");
    scanf("%lu", &m);
  }

  g = PrimitiveRoot(m);

  if (g > 0)
    printf("g = %lu\n", g);
  else
    printf("no primitive roots\n");
}

/*
** PrimitiveRoot - finds the smallest primitive root for m
*/
#include "numtype.h"

NAT PrimitiveRoot(m)
NAT m;
{
  NAT g, i, t, fk, f[32];

/* find a primitive root for m */

  if (m == 2)
    g = 1;
  else
  {
    t = EulerTotient(m);
    fk = Factor(t, f);
    for (g = 2; g < m; g++)
    {
      if (GCD(g, m) == 1)
      {
```

```
            for (i = 0; i < fk; i++)
              if (PowMod(g, t / f[i], m) == 1) break;
            if (i == fk) break;          /* primitive root found */
      }
    }
  }

  if (g == m)
    return(0);                          /* none */
  else
    return(g);
}

/*
** EulerTotient - compute values for Euler's totient function using a
**                prime factorization by trial division
*/
#include "numtype.h"

NAT EulerTotient(NAT n)
{
  NAT i, j, et, pk, fk, f[32];

  et = 1;                               /* totient value */
  pk = 0;                               /* power count */

  if (n == 1) return(1);

  fk = Factor(n, f);                    /* factor count, factors */
  if (fk > 32) exit(1);

  for (i = 0; i < fk; i++)
  {
    if (f[i] == f[i+1])
      {
        pk++;
      }
    else
    {
      for (j = 0; j < pk; j++) et *= f[i];
      et *= f[i] - 1;
      pk = 0;
    }
  }

  return(et);
}

/*
** Factor - finds prime factorization by trial division
**          returned values: number of factors, array w/ factors
*/
#include "numtype.h"

NAT Factor(NAT n, NAT f[])
{
  NAT i, fk, s;
```

```
   fk = 0;                                    /* factor count */

   while (!(n & 1))                           /* even - factor all 2's */
   {
     f[fk++] = 2;
     n /= 2;
   }

   s = sqrt(n);
   for (i = 3; i <= s; i++, i++)              /* factor odd numbers */
     while ( (n % i) == 0L)
     {
       f[fk++] = i;
       n /= i;
       s = sqrt(n);
     }

   if (n > 1L) f[fk++] = n;

     for (i = fk; i < 32; i++) f[i] = 0;  /* zero out the rest */

   return(fk);
}

/*
** PowMod - computes r for aⁿ ≡ r (mod m) given a, n, and m
*/
#include "numtype.h"

NAT PowMod(NAT a, NAT n, NAT m)
{
  NAT r;

  r = 1;
  while (n > 0)
  {
    if (n & 1)                          /* test lowest bit */
      r = MulMod(r, a, m);              /* multiply (mod m) */
    a = MulMod(a, a, m);                /* square */
    n >>= 1;                            /* divided by 2 */
  }
  return(r);
}

/*
** MulMod - computes r for a * b ≡ r (mod m) given a, b, and m
*/
#include "numtype.h"

NAT MulMod(NAT a, NAT b, NAT m)
{
  NAT r;

  if (m == 0) return(a * b);            /* (mod 0) */

  r = 0;
  while (a > 0)
```

```
{
    if (a & 1)                          /* test lowest bit */
        if ((r += b) > m) r %= m;       /* add (mod m) */
    a >>= 1;                            /* divided by 2 */
    if ((b <<= 1) > m) b %= m;          /* times 2 (mod m) */
}

return(r);
}

/*
** GCD - find the greatest common divisor using Euclid's algorithm
*/
#include "numtype.h"

NAT GCD(NAT a, NAT b)
{
    NAT r;

    while (b > 0)
    {
        r = a % b;
        a = b;
        b = r;
    }

    return(a);
}
```

```
C> 4_5_2
Find the smallest primitive root

m = 760321
g = 73
```

Challenge

1. Write a program that will find primitive roots for composites as
well as prime. Use it to make a table of strictly increasing smallest
primitive roots for numbers less than 1000 (in the same vein as in
Table 4.8).

5

Solving Congruences

Find strength in what remains behind.

WILLIAM WORDSWORTH, *INTIMATIONS OF IMMORTALITY*

5.0 Introduction

As when you solve equations, it is often necessary to solve specific congruences. The procedure for solving congruences that are linear, quadratic, or exponential are well known in the literature. Also, the criteria for solvability are well known. In this chapter, I will discuss a number of methods for solving these types of congruences.

5.1 Linear Congruences with One Unknown

Congruences of the form

$$a \cdot x \equiv b \ (\text{mod } m)$$

have solutions if and only if $gcd(a, m) \mid b$. Additionally, if solutions exist, then there are $gcd(a, m)$ unique solutions (that is, solutions that are incongruent modulo m). While this may seem obvious, the easiest way to find solutions to $a \cdot x \equiv b \ (\text{mod } m)$ is to try every possible $x = 0$, 1, ..., $m - 1$. This inelegant technique does not provide much insight into the nature of these types of congruences, but it does find all the incongruent solutions in the arithmetic residue system.

The following program will prompt for a, b, and m to solve for x in

$$a \cdot x \equiv b \ (\text{mod } m)$$

if a, b, and m are not entered on the command line. The program accepts negative values for a and b and adjusts them so that they are both greater than 0. Whether or not solutions exist is determined first by the GCD function. If they do, then the program proceeds to try every possible x in the range of 0 to $m - 1$.

Program 5_1_1.c. Solve the congruence $a \cdot x \equiv b \pmod{m}$ by trial

```
/*
** Program 5_5_1 - Solve for x in the relation: a * x ≡ b mod m
**                 using the trial method
*/
#include "numtype.h"

void main(int argc, char *argv[])
{
    INT a, b, m, x;

    if (argc == 4) {
        a = atol(argv[1]);
        b = atol(argv[2]);
        m = atol(argv[3]);
    } else {
        printf("Solve for x in: a * x ≡ b (mod m)\n\n");
        printf("a, b, m = ");
        scanf("%ld,%ld,%ld", &a, &b, &m);
    }

    if ((a %= m) < 0) { a = -a; b = -b; } /* transfer sign to b */
    if ((b %= m) < 0) b += m;             /* make b > 0 */

    if ((b % GCD(a, m)) == 0)
    {
        for ( x = 0; x < m; x++)
            if (MulMod(a, x, m) == b)
                printf("solution = %lu\n", x);
    }
    else
        printf("no solutions\n");
}

/*
** MulMod - computes r for a * b ≡ r (mod m) given a, b, and m
*/
#include "numtype.h"

NAT MulMod(NAT a, NAT b, NAT m)
{
    NAT r;

    if (m == 0) return(a * b);                /* (mod 0) */

    r = 0;
    while (a > 0)
    {
        if (a & 1)                            /* test lowest bit */
```

```
      if ((r += b) > m) r %= m;          /* add (mod m) */
    a >>= 1;                             /* divided by 2 */
    if ((b <<= 1) > m) b %= m;           /* times 2 (mod m) */
  }

  return(r);
}

/*
** GCD - find the greatest common divisor using Euclid's algorithm
*/
#include "numtype.h"

NAT GCD(NAT a, NAT b)
{
  NAT r;

  while (b > 0)
  {
    r = a % b;
    a = b;
    b = r;
  }

  return(a);
}
```

For the equation $21 \cdot x \equiv 11 \pmod{33}$, we can ascertain that it has no solutions using the GCD criterion: $gcd(21,33) = 3$ and 3 is not a divisor of 11. Program 5_1_1 makes a similar determination.

```
C> 5_1_1
Solve for x in: a*x ≡ b (mod m)


a, b, m = 21,11,33
no solutions
```

On the other hand, the equation $15 \cdot x \equiv 9 \pmod{12}$ has three incongruent solutions because $gcd(15,12) = 3$ and $3 \mid 9$:

```
C> 5_1_1
Solve for x in: a*x ≡ b (mod m)


a, b, m = 15,9,12
solution = 3
solution = 7
solution = 11
```

The risk in using this program is that of overflow when the product a*x exceeds the precision of the declared data type; in this case, unsigned long integers ($2^{32} - 1$). To reduce this risk, the application uses the function MulMod() (see Chapter 4) for multiplication of numbers that are to be reduced by a modulus.

A method developed by Gauss for the case in which $gcd(a, m) = 1$ (one unique solution) has the advantages of (usually) being much faster for a large m while at the same time reducing the risk of exceeding the precision of the variables is lessened. To find the unique solution to the congruence $a \cdot x \equiv b$ (mod m), we rewrite it as

$$x \equiv b \ / \ a \ (\text{mod } m).$$

Next, we add multiples of m to b and divide out the greatest common divisor. When a has been reduced to 1, we have found the solution, x. Look at the following example:

$$20 \cdot x \equiv 3 \ (\text{mod } 31)$$

$$x \equiv 3/20 \equiv 34/20 \equiv 17/10 \equiv 48/10 \equiv 24/5 \equiv 55/5 \equiv 11 \ (\text{mod } 31).$$

It is easy to verify that $20 \cdot 11 = 220 \ \% \ 31 = 3$. It is worth noting, however, that the solution obtained using this method may not be from the arithmetic residual system. Even so, the solution determined can always be reduced easily to the arithmetic residual system by the modulus.

Overflow is generally not a problem because the values are reduced by the greatest common divisor whenever possible. Program 5_1_2 implements Gauss's method for solving simple linear congruences having one solution. It is worth noting that congruences having more than one incongruent solution cannot be solved completely using this method:

Program 5_1_2.c. Solve linear congruences using Gauss' method

```
/*
** Program 5_1_2 - Solve for x in the relation: a*x ≡ b mod m
**                 using Gauss' method
*/
#include "numtype.h"

void main(int argc, char *argv[])
{
  NAT a, b, g, m;

  if (argc == 4) {
    a = atol(argv[1]);
    b = atol(argv[2]);
    m = atol(argv[3]);
  } else {
    printf("Solve for x in: a*x ≡ b (mod m)\n\n");
    printf("a, b, m = ");
    scanf("%ld,%ld,%ld", &a, &b, &m);
  }
```

```
if (GCD(a,m) == 1)
{
   if ((g = GCD(a,b)) > 1) { a /= g; b /= g; }
   while (a > 1)
   {
      b += m;
      if ((g = GCD(a,b)) > 1) { a /= g; b /= g; }
   }
   printf("solution = %lu\n", b);
}
else
{
   if ((b % GCD(a,m)) == 0)
      printf("congruence cannot be solved with Gauss' method\n");
   else
      printf("no solutions\n");
}
}

/*
** GCD - find the greatest common divisor using Euclid's algorithm
*/
#include "numtype.h"

NAT GCD(NAT a, NAT b)
{
   NAT r;

   while (b > 0)
   {
      r = a % b;
      a = b;
      b = r;
   }

   return(a);
}
```

```
C> 3_4_2
Solve for x in: a*x ≡ b (mod m)

a, b, m = 20,3,31
solution = 11
```

Consider the following congruence: $a \cdot z \equiv 1 \pmod{m}$. This is a special instance of the linear congruence $a \cdot x \equiv b \pmod{m}$. In this case, z is said to be the inverse of a, mod m, and is sometimes expressed as

$$z \equiv 1 / a \pmod{m}.$$

It is possible for certain numbers not to have inverses with respect to a particular modulus. If a and m are relatively prime, that is, $gcd(a,m) = 1$, then inverses can be computed. Otherwise the inverse does not exist. It is sometimes convenient to compute a table of inverses for reference when performing a great number of calculations using the same modulus. Program 5_1_3 creates a table of inverses for a given modulus.

Program 5_1_3.c. Create a table of inverses for a given modulus

```
/*
** Program 5_1_3 - Create a table of inverses for a particular
**                 modulus by solving for x in the relation:
**                 a*x ≡ 1 mod m
*/
#include "numtype.h"

void main(int argc, char *argv[])
{
  NAT a, m, z;

  if (argc == 2) {
    m = atol(argv[1]);
  } else {
    printf("Table of Inverses\n\n");
    printf("m = ");
    scanf("%lu", &m);
  }

  printf("Table of Inverses for Modulus %lu\n---------------------
- \n", m);

  for (a = 1; a < m; a++)
  {
    z = ComputeInverse(a, m);
    if (z)
      printf("%lu ≡ 1/%lu (mod %lu)\n", z, a, m);
    else
      printf("       %lu has no inverse (mod %lu)\n", a, m);
  }
}

/*
** ComputeInverse -  find the inverse of the congruence a * z ≡ 1
**                   (mod p) using Gauss' method
*/
#include "numtype.h"

NAT ComputeInverse(NAT a, NAT p)
{
  NAT g, z;

  if (GCD(a,p) > 1) return(0); /* no inverse */

  z = 1;

  while (a > 1)
  {
    z += p;
    if ((g = GCD(a,z)) > 1)
    {
      a /= g;
      z /= g;
```

```
      }
   }

   return(z);
}

/*
** GCD - find the greatest common divisor using Euclid's algorithm
*/
#include "numtype.h"

NAT GCD(NAT a, NAT b)
{
   NAT r;

   while (b > 0)
   {
      r = a % b;
      a = b;
      b = r;
   }

   return(a);
}
```

```
C> 5_1_3
Table of Inverses

m = 7
Table of Inverses for Modulus 7
----------------------------------
1 ≡ 1/1  (mod 7)
4 ≡ 1/2  (mod 7)
5 ≡ 1/3  (mod 7)
2 ≡ 1/4  (mod 7)
3 ≡ 1/5  (mod 7)
6 ≡ 1/6  (mod 7)
```

Based on the previous discussion you would expect that prime moduli will have inverses for all numbers other than 0. For a composite modulus the situation is markedly different. Compare the previous table to the table generated using a composite modulus 9:

```
C> 5_1_3
Table of Inverses

m = 9
Table of Inverses for Modulus 9
----------------------------------
1 ≡ 1/1  (mod 9)
5 ≡ 1/2  (mod 9)
      3 has no inverse (mod 9)
7 ≡ 1/4  (mod 9)
2 ≡ 1/5  (mod 9)
      6 has no inverse (mod 9)
4 ≡ 1/7  (mod 9)
8 ≡ 1/8  (mod 9)
```

You may have noticed that the count of multiplicative inverses for a given modulus is equal to the value of Euler's totient function, $\phi(m)$, described in Chapter 2. This makes sense when one considers the fact that an inverse is obtainable only when a and m are relatively prime: the definition of $\phi(m)$.

It would be nice to have a general formula for computing inverses rather than solving for them as general congruences. As noted in Section 3.1, if m is relatively prime to a, then

$$z = 1/a = a^{(m-2)}.$$

Using the inverse, we can solve the congruence $a \cdot x \equiv b \pmod{m}$. The solution follows from the definition of the inverse. If m is prime then

$$a \equiv b / a = b \cdot a^{(m-2)} = z \cdot b \pmod{m}.$$

Interestingly, certain linear congruences can be solved using Euler's totient function. The relation we need is Euler's theorem:

If $gcd(a,m) = 1$, then $a^{\phi(m)} \equiv 1 \pmod{m}$.

From Euler's theorem we can see that $a \cdot a^{\phi(m)-1} \equiv 1 \pmod{m}$, or $a^{\phi(m)-1} \equiv 1/a \pmod{m}$. It follows then that $a^{\phi(m)-1}$ is the inverse of a, mod m. We can then write a program that will solve $a \cdot x \equiv b \pmod{m}$ by setting $a^{\phi(m)-1} \cdot b \equiv x \pmod{m}$. The advantage to using the totient function to derive the inverse is that its application is less restricted than inverse $a(m-1)$. The obvious disadvantage is that a factorization of m must be known. Note that program needs to be linked to `EulerTotient()`, `Factor()`, `GCD()` as defined in Chapter 2.

Program 5_1_4.c. Solve linear congruences using Euler's theorem and totient function

```
/*
** Program 5_1_4 - Solve for x in the relation: a*x ≡ b mod m
**                 using Euler's theorem and the totient function
*/
#include "numtype.h"

void main(int argc, char *argv[])
{
   NAT a, b, e, i, m, x;

   if (argc == 4) {
      a = atol(argv[1]);
      b = atol(argv[2]);
      m = atol(argv[3]);
   } else {
```

```
      printf("Solve for x in: a*x ≡ b mod m\n\n");
      printf("a, b, m = ");
      scanf("%lu,%lu,%lu", &a, &b, &m);
   }

   if (GCD(a,m) == 1)
   {
      e = 1;
      for (i = 0; i < EulerTotient(m) - 1; i++)
      {
         e *= a;
         e %= m;
      }
      x = e * b % m;
      printf("solution = %lu\n", x);
   }
   else
   {
      if ((b % GCD(a,m)) == 0)
         printf("congruence cannot be solved with the totient function\n");
      else
         printf("no solutions\n");
   }
}

/*
** EulerTotient - compute values for Euler's totient function using a
**                prime factorization by trial division
*/
#include "numtype.h"

NAT EulerTotient(n)
NAT n;
{
   NAT i, j, et, pk, fk, f[32];

   et = 1;                              /* totient value */
   pk = 0;                              /* power count */

   if (n == 1) return(1);

   fk = Factor(n, f);                   /* factor count, factors */
   if (fk > 32) exit(1);

   for (i = 0; i < fk; i++)
   {
      if (f[i] == f[i+1])
      {
         pk++;
      }
      else
      {
         for (j = 0; j < pk; j++) et *= f[i];
         et *= f[i] - 1;
         pk = 0;
      }
   }
```

```
  return(et);
}
/*
** Factor - finds prime factorization by trial division
**             returned values: number of factors, array w/ factors
*/
#include "numtype.h"

NAT Factor(n, f)
NAT n, f[];
{
  NAT i, fk, s;

  fk = 0;                              /* factor count */

  while ((n % 2) == 0)                 /* factor all 2's */
  {
    f[fk++] = 2;
    n /= 2;
  }

  s = sqrt(n);
  for (i = 3; i <= s; i++, i++)        /* factor odd numbers */
    while ( (n % i) == 0L)
    {
      f[fk++] = i;
      n /= i;
      s = sqrt(n);
    }

  if (n > 1L) f[fk++] = n;

  for (i = fk; i < 32; i++) f[i] = 0; /* zero out the rest */

  return(fk);
}


/*
** GCD - find the greatest common divisor using Euclid's algorithm
*/
#include "numtype.h"

NAT GCD(NAT a, NAT b)
{
  NAT r;

  while (b > 0)
  {
    r = a % b;
    a = b;
    b = r;
  }

  return(a);
}
```

```
C> 5_1_4
Solve for x in: a*x ≡ b mod m

a, b, m = 20,3,31
solution = 11
```

Challenge

1. Modify Program 5_1_2 so it will find all solutions if $gcd(a,m) > 1$. Make a similar modification for Program 5_1_4.

5.2 Simultaneous Linear Congruences

Having explored single linear congruences, we can now turn our attention to systems of linear congruences. The solution to the following system of congruences

$$a_1 \cdot x \equiv b_1 \ (\text{mod } m_1)$$

$$a_2 \cdot x \equiv b_2 \ (\text{mod } m_2)$$

$$a_3 \cdot x \equiv b_3 \ (\text{mod } m_3)$$

...

$$a_n \cdot x \equiv b_n \ (\text{mod } m_n)$$

is the single integer, x, which satisfies each and every congruence in the system.

It is important to know when the solution exists. The answer to the question is provided by the *Chinese Remainder Theorem*, which states the following.

Chinese Remainder Theorem (5.1). Given the system of n linear congruences: $a_1 \cdot x \equiv b_1 \ (\text{mod } m_1)$, $a_2 \cdot x \equiv b_2 \ (\text{mod } m_2)$, $a_3 \cdot x \equiv b_3 \ (\text{mod } m_3)$, ..., $a_n \cdot x \equiv b_n \ (\text{mod } m_n)$, where $gcd(m_i,m_j) = 1$ for all $i \neq j$ and $gcd(a_i,m_i) = 1$ for each i, then the system has a solution that is unique to $m_1 \cdot m_2 \cdot m_3 \cdot \ldots \cdot m_n$.

The proof to the Chinese Remainder Theorem provides the algorithm for determining the solution. To find the solution, we will seek a solution for each of the particular congruences, and from those we will construct one that is common to the entire system. First, find integers $c_1, c_2, c_3, \ldots, c_n$, such that $a_i \cdot c_i \equiv b_i \ (\text{mod } m_i)$. Next we will need the product of the moduli $M = m_1 \cdot m_2 \cdot m_3 \cdot \ldots \cdot m_n$.

From the product of the moduli we obtain the product "complement": $q_i = M / m_i$, for each i, $i = 1$ to n. Lastly we need the inverse of the moduli product complement z_i, such that $q_i \cdot z_i \equiv 1 \ (\text{mod } m_i)$, for each i, $i = 1$ to n.

With these quantities, we can now define the particular solution as

$$x = c_1 \cdot q_1 \cdot z_1 + c_2 \cdot q_2 \cdot z_2 + c_3 \cdot q_3 \cdot z_3 + \ldots + c_n \cdot q_n \cdot z_n.$$

This is the solution because x is constructed in such a way to be a solution to each congruence in the system. Take any congruence as the general case:

$$a_i \cdot x = a_i \cdot c_1 \cdot q_1 \cdot z_1 + a_i \cdot c_2 \cdot q_2 \cdot z_2 + a_i \cdot c_3 \cdot q_3 \cdot z_3 + \ldots + a_i \cdot c_n \cdot q_n \cdot z_n.$$

Each m_i divides q_i, except when $i = j$. Since the remainders are zero, we can cancel all the terms except:

$$a_i \cdot x \equiv a_i \cdot c_i \cdot q_i \cdot z_i \ (\text{mod } m_i),$$

Since $q_i \cdot z_i \equiv 1 \ (\text{mod } m_i)$, this leaves:

$$a_i \cdot x \equiv a_i \cdot c_i \ (\text{mod } m_i),$$

and by definition,

$$a_i \cdot c_i \equiv b_i \ (\text{mod } m_i).$$

Therefore x is the solution we seek. Lastly, we will reduce x to be the least positive remainder using the modulus M:

$$x \equiv c_1 \cdot q_1 \cdot z_1 + c_2 \cdot q_2 \cdot z_2 + c_3 \cdot q_3 \cdot z_3 + \ldots + c_n \cdot q_n \cdot z_n \ (\text{mod } M)$$

Implementation in a program requires arrays for holding the as, bs, and ms. The maximum number of congruences input should not exceed this number. Input is taken from standard input and ends when $\land Z$ (control + Z) is entered.

Program 5_2_1.c. Solve linear congruences using the Chinese remainder theorem

```
/*
** Program 5_2_1 - Solve simultaneous congruences: a*x ≡ b mod m
**                 using the Chinese Remainder theorem and
**                 Gauss' method
*/
#include "numtype.h"

void main(void)
{
    INT a[32], b[32], m[32];
    INT i, j, n, x, M, g;
    char str[32], *io;

/* get the system of simultaneous congruences from standard input */

    printf("Solve for x in: a[i]*x ≡ b[i] (mod m[i])\n\n");

    n = 0;
    do
```

```
   {
     printf("[%ld]: a, b, m = ", n);
     if ((io = gets(str)) != NULL)
     {
       if (sscanf(str, "%ld,%ld,%ld", &a[n], &b[n], &m[n]) != 3)
       exit(1); n++;
     }
   } while ( io != NULL);

/* echo input */

   printf("\nSolution\n--------\n");
   for ( i = 0; i < n; i++)
     printf("%ld * x ≡ %ld (mod %ld)\n", a[i], b[i], m[i]);

/*
** Chinese Remainder Theorem criteria
**   1) if gcd(a,m) of any pair > 1, then exit
**   2) if gcd(m,m) of any pair > 1, then exit
*/

   M = 1;
   for (i = 0; i < n; i++)
   {
     if (GCD(a[i],m[i]) > 1)
     {
       printf("No solution: gcd(a[%ld],m[%ld]) > 1\n", i, i);
       exit(1);
     }
     for (j = i+1; j < n; j++)
     {
       if (GCD(m[i],m[j]) > 1)
       {
         printf("No solution: gcd(m[%ld],m[%ld]) > 1\n", i, j);
         exit(1);
       }
     }
     M *= m[i];                      /* product of moduli */
   }

/*
** solve the congruences:
**   1) a[i] * c[i] ≡ b[i] (mod m[i])      <= intermediate congruence
**   2) (M / m[i]) * z[i] ≡ 1 (mod m[i])   <= inverse complement moduli
** build sum for solution
*/

   x = 0;
   for (i = 0; i < n; i++)
     x += SolveLinearCongruence(a[i], b[i], m[i]) *
          SolveLinearCongruence(M/m[i], 1L, m[i]) *
          M / m[i];

   x %= M;                 /* least positive solution */

   printf("x ≡ %ld (mod %ld)\n", x, M);
}
/*
```

```
** SolveLinearCongruence - solve a linear congruence: ax ≡ b mod m
**                         using Gauss' method
*/
#include "numtype.h"

INT SolveLinearCongruence(INT a, INT b, INT m)
{
  NAT g;

  if ((a %= m) < 0) { a = -a; b = -b; }  /* transfer sign to b */
  if ((b %= m) < 0) b += m;              /* make b > 0 */

  if ((b % GCD(a,m)) == 0)
  {
    if ((g = GCD(a,b)) > 1) { a /= g; b /= g; }
    while (a > 1)
    {
      b += m;
      if ((g - GCD(a,b)) > 1) { a /= g; b /= g; }
    }
    return(b);
  }

  return(-1);                            /* solution not possible */
}

/*
** GCD - find the greatest common divisor using Euclid's algorithm
*/
#include "numtype.h"

NAT GCD(NAT a, NAT b)
{
  NAT r;

  while (b > 0)
  {
    r = a % b;
    a = b;
    b = r;
  }

  return(a);
}
```

For convenience, Gauss's method for solving simple linear congruences has been made into a function SolveLinearCongruence. Rather than using the ComputeInverse() function, I've used SolveLinearCongruence() to find the inverse of the moduli product complement by setting the second argument to 1. This comes directly from the definition of the modular inverse.

```
C> 5_2_1
Solve for x in: a[i]*x ≡ b[i] (mod m[i])
[0]: a, b, m = 2,1,5
```

```
[1]:  a,  b,  m = 1,3,2
[2]:  a,  b,  m = 4,1,7
[3]:  a,  b,  m = 5,9,11
[4]:  a,  b,  m = ^z

Solution
--------

2 * x 4 1  (mod 5)
1 * x 4 3  (mod 2)
4 * x 4 1  (mod 7)
5 * x 4 9  (mod 11)
x ≡ 653  (mod 770)
```

5.3 Linear Congruences with More than One Unknown

Linear congruences having more than one unknown present a more challenging problem, particularly with respect to the number of possible solutions, if any exist at all. A solution to a linear congruence with more than one unknown

$$a_1 \cdot x_1 + a_2 \cdot x_2 + a_3 \cdot x_3 + \ldots + a_n \cdot x_n \equiv b \ (\text{mod } m)$$

will exist if and only if $gcd(a_1, a_2, a_3, \ldots, a_n, m) \mid b$, just as in the case of single linear congruence. If solutions exist, then the number of solutions is $gcd(a_1, a_2, a_3, \ldots, a_n, m) \cdot m^{n-1}$. Clearly, the number of solutions can be a very large number even for a modest modulus.

Linear congruences having more than one unknown can be attacked by successively solving a number of congruences having a single unknown. This is the usual procedure for finding solutions by hand, but it becomes a complex task for a computer program.

Solving a linear congruence with more than one unknown by reducing the number of unknowns is solved recursively in several steps. We will assume that $gcd(a_1, a_2, a_3, \ldots, a_n, m) = 1$ so the number of solutions equals m^{n-1}. If $gcd(a_2, a_3, a_4, \ldots, a_n, m) = d > 1$, we must have

$$a_1 \cdot x_1 \equiv b \ (\text{mod } d).$$

Since $gcd(a_1, d) = 1$, we know that this congruence with one unknown has a unique solution. There exist m/d solutions (mod m), and these are easily generated by successively adding d to the least positive solution. Substituting each of these solutions in turn give us m/d congruences in $n - 1$ unknowns. We repeat this process until the last congruence has only one unknown and has been solved.

Consider the following congruence (mod 15) with two unknowns. We would expect to have to compute $15^{2-1} = 15$ solutions:

$$3 \cdot x_1 + 5 \cdot x_2 \equiv 7 \ (\text{mod } 15).$$

Since $gcd(5,15) = 5$, we have

$$3 \cdot x_1 \equiv 7 \pmod 5.$$

This gives $x_1 \equiv 4 \pmod 5$, and therefore $x_1 \equiv 4, 9, 14 \pmod{15}$. Substituting these solutions into the original equation gives

$$3 \cdot 4 + 5 \cdot x_2 \equiv 7 \pmod{15}$$

$$5 \cdot x_2 \equiv 10 \pmod{15}$$

$$x_2 \equiv 2 \pmod 3$$

$$x_2 \equiv 2, 5, 8, 11, 14 \pmod{15}$$

The complete set of solutions is

$$(x_1, x_2) \equiv \quad (4,2), \ (4,5), \ (4,8), \ (4,11), \ (4, 14),$$

$$(9,2), \ (9,5), \ (9,8), \ (9,11), \ (9, 14),$$

$$(14,2), (14,5), (14,8), (14,11), (14,14).$$

You can see how this can become a bookkeeping nightmare for relations that have more than two congruences with different greatest common divisors in each coefficient subset. Also, from a software development point of view, memory management and array indexing become a complex problem.

For these reasons, the easiest way to solve congruences with more than one unknown is we must use the trial method with recursion. To use this technique, we must take steps to combat the risk of an overflow if the left side of our congruence exceeds the precision of our data type. We use the basic properties of congruences to reduce the size of the product and thereby the sum.

First we test to see if solutions exist. If they are possible, we construct each possible solution from the arithmetic residue system. When we evaluate the congruence and the result is zero, we have discovered a solution. This is an obvious case of number theory versus muscle programming.

Program 5_3_1.c. Solve linear congruences with morethan one unknown

```
/*
** Program 5_3_1 - Solve a linear congruence with more than one
**                 unknown using the recursive trial method:
**                 a[0]*x[0] + a[1]*x[1] + ... ≡ b mod m
*/
```

```
#include "numtype.h"
void main(void)
{
  INT a[32], x[32];
  INT b, i, j, k, m, n, g;
  char str[128], *begptr, *endptr;

/* get the linear congruence from standard input */

  printf("Solve for x's in: a0*x0 + a1*x1 + ... ≡ b (mod m)\n\n");
  printf("a0, a1, ..., an, b, m = ");
  if ((begptr = gets(str)) == NULL) exit(1);

  n = 0;
  while (n < 32)
  {
    a[n++] = strtol( begptr, &endptr, 10);
    if (*endptr)
      begptr = endptr + 1;
    else
      break;
  }

  if (n < 3) exit(1);
  m = a[--n];
  b = a[--n];

/* echo input */

  printf("\nSolution\n--------\n");
  printf("%ld*x[0] ", a[0]);
  for ( i = 1; i < n; i++)
    printf("+ %ld*x[%ld] ", a[i], i);
  printf("≡ %ld (mod %ld)\n\n", b, m);

/* test if solution exists */

  g = m;
  for (i = 0; i < n; i++) g = GCD(g,a[i]);

  if ((b % g) != 0)
  {
    printf("no solution\n");
    exit(1);
  }

/* find solutions to congruence */

  SolveLinearCongruenceN(a, x, b, n, m, n);
}

/*
** SolveLinearCongruenceN - solve a linear congruence with more than
**                          one unknown using the recursive trial
**                          method:a[1]*x[1] + a[2]*x[2] + ...
**                          a[n]* x[n] **    ≡ b mod m
*/
#include "numtype.h"
```

```
INT SolveLinearCongruenceN(INT a[], INT x[], INT b, INT l, INT m,
INT n)
{
  INT i, s;

  l--;                                    /* level of recursion */

  if (l < 0)
  {
    s = 0;
    for (i = 0; i < n; i++)
      s = ((s + ((a[i] % m) * x[i]) % m) % m);
    if (((s - b) % m) == 0)
    {
      printf("(%ld", x[0]);
      for (i = 1; i < n; i++) printf(", %ld", x[i]);
      printf(")\n");
    }
  }
  else
  {
    for (i = 0; i < m; i++)
    {
      x[l] = i;
      SolveLinearCongruenceN(a, x, b, l, m, n);
    }
  }

  return(l);
}

/*
** SolveLinearCongruence - solve a linear congruence: ax ≡ b mod m
**                         using Gauss' method
*/
#include "numtype.h"

INT SolveLinearCongruence(INT a, INT b, INT m)
{
  NAT g;

  if ((a %= m) < 0) { a = -a; b = -b; } /* transfer sign to b */
  if ((b %= m) < 0) b += m;            /* make b > 0 */

  if ((b % GCD(a,m)) == 0)
  {
    if ((g = GCD(a,b)) > 1) { a /= g; b /= g; }
    while (a > 1)
    {
      b += m;
      if ((g = GCD(a,b)) > 1) { a /= g; b /= g; }
    }
    return(b);
  }

  return(-1);                          /* solution not possible */
}
```

```
/*
** MulMod - computes r for a * b ≡ r (mod m) given a, b, and m
*/
#include "numtype.h"

NAT MulMod(NAT a, NAT b, NAT m)
{
  NAT r;

  if (m == 0) return(a * b);                /* (mod 0) */

  r = 0;
  while (a > 0)
  {
    if (a & 1)                              /* test lowest bit */
      if ((r += b) > m) r %= m;             /* add (mod m) */
    a >>= 1;                                /* divided by 2 */
    if ((b <<= 1) > m) b %= m;              /* times 2 (mod m) */
  }

  return(r);
}


/*
** GCD - find the greatest common divisor using Euclid's algorithm
*/
#include "numtype.h"

NAT GCD(NAT a, NAT b)
{
  NAT r;

  while (b > 0)
  {
    r = a % b;
    a = b;
    b = r;
  }

  return(a);
}
```

```
C> 5_3_1
Solve for x's in: a0*x0 + a1*x1 + ... ≡ b (mod m)

a0, a1, ..., an, b, m = 3,5,7,15

Solution
--------
3*x[0] + 5*x[1] ≡ 7 (mod 15)

(4, 2)
(9, 2)
(14, 2)
(4, 5)
(9, 5)
(14, 5)
(4, 8)
```

```
(9, 8)
(14, 8)
(4, 11)
(9, 11)
(14, 11)
(4, 14)
(9, 14)
(14, 14)
```

Challenges

1. Solve the following congruence by hand: $3 \cdot x_1 + 5 \cdot x_2 + 9 \cdot x_3 \equiv 7$ (mod 15). Check your work using Program 5_3_1.

2. Modify Program 5_3_1 to preprocess the coefficients, a[], so that none exceeds the modulus. How will this help protect against overflow?

3. Modify Program 5_3_1 to solve a system of equations having more than one unknown.

5.4 Quadratic Congruences

In the preceding sections, we completely solved linear congruences. We now will focus on congruences involving quadratic polynomials. The simplest of these is also the most crucial congruence to solve:

$$x_2 \equiv a \pmod p.$$

If a solution exists, we could say that a is a square mod p. However, for historical reasons the customary phrase is that a is a *quadratic residue* of p. If the congruence cannot be solved, then a is a *quadratic nonresidue*. Similar nomenclature is used with respect to higher power residues and nonresidues.

Consider the following example where $x = 1$ to 15 and $p = 7$. From Table 5.1, we see that the numbers 1, 2, and 4 are the quadratic residues of 7. We do not consider 0 to be a quadratic residue because it only occurs when p is a proper divisor of x. It is also worth noting that the quadratic residue pattern repeats indefinitely.

Our first order of business is to identify those numbers that are quadratic residues of primes.

Euler's Criterion, the Legendre Symbol, and the Jacobi Symbol

Euler developed a means for determining if a solution exists for quadratic congruences of the form $x^2 \equiv a \pmod p$. Knowing if a particular

Table 5.1. Table of Values for x, x^2, x^2 mod 7

x	x^2	x^2 mod 7
1	1	1
2	4	4
3	9	2
4	16	2
5	25	4
6	36	1
7	49	0
8	64	1
9	81	4
10	100	2
11	121	2
12	144	4
13	169	1
14	196	0
15	225	1

quadratic congruence has a solution is not the same as finding the solution, but it's a first important step that we shall use.

Definition (Euler's criterion). Given that p is an odd prime and $gcd(a, p) = 1$, then a number a is a quadratic residue modulo p if and only if

$$a^{(p-1)/2} \equiv 1 \ (\text{mod } p)$$

This congruence is called Euler's criterion. It is worth noting that a quadratic congruence can always be solved if $p = 2$ since every odd $x^2 \equiv 1 \ (\text{mod } 2)$.

It is a relatively simple matter to write a program that will give us some insight into the distribution of solvable quadratic residues. The following program will generate a table of Euler's criterion values for a versus p.

Program 5_4_1.c. Prints a table of Euler's criterion identifying quadratic residues for primes less than 32

```
/*
** Program 5_4_1 - Print table of Euler's criterion for quadratic
**                 congruences
```

```
*/
#include "numtype.h"

void main(void)
{
  NAT a, i;
  NAT p[] = { 3, 5, 7, 11, 13, 17, 19, 23, 29, 31};   /* primes */

  printf("Table of Euler's criterion\n\n");
  printf(" a\\p");
  for (i = 0; i < 10; i++) printf("__%2ld", p[i]);
  for (a = 1; a < 31; a++)
  {
    printf("\n%3ld |", a);
    for (i = 0; i < 10; i++)
    {
      if ( PowMod(a, (p[i] - 1)/2, p[i]) == 1)
        printf("  X");
      else
        printf("   ");
    }
  }
}

/*
** PowMod - computes r for a^n ≡ r (mod m) given a, n, and m
*/
#include "numtype.h"

NAT PowMod(a, n, m)
NAT a, n, m;
{
  NAT r;

  r = 1;
  while (n > 0)
  {
    if (n & 1)                      /* test lowest bit */
      r = MulMod(r, a, m);          /* multiply (mod m) */
    a = MulMod(a, a, m);            /* square */
    n >>= 1;                        /* divided by 2 */
  }

  return(r);
}

/*
** MulMod - computes r for a * b ≡ r (mod m) given a, b, and m
*/
#include "numtype.h"

NAT MulMod(a, b, m)
NAT a, b, m;
{
  NAT r;

  if (m == 0) return(a * b);        /* (mod 0) */
```

```
    r = 0;
    while (a > 0)
    {
        if (a & 1)                      /* test lowest bit */
            if ((r += b) > m) r %= m;   /* add (mod m) */
        a >>= 1;                        /* divided by 2 */
        if ((b <<= 1) > m) b %= m;      /* times 2 (mod m) */
    }

    return(r);
}
```

An "X" in the table output from Program 5_4_1 indicates that $x^2 \equiv a$ (mod p) is a solvable quadratic congruence. Otherwise, the congruence is not solvable. For example, the quadratic congruence $x^2 \equiv 7$ (mod 19) is solvable whereas $x^2 \equiv 13$ (mod 19) is not.

C> **5_4_1**
Table of Euler's criterion

a\p	3	5	7	11	13	17	19	23	29	31
1	X	X	X	X	X	X	X	X	X	X
2			X			X		X		X
3				X	X			X		
4	X	X	X	X	X	X	X	X	X	X
5				X			X		X	X
6		X					X	X	X	
7	X						X		X	X
8			X			X		X		X
9		X	X	X	X	X	X	X	X	X
10	X				X					X
11		X	X				X			
12				X	X			X		
13	X					X		X	X	
14		X		X	X					X
15			X	X		X				
16	X	X	X	X	X	X	X	X	X	X
17					X		X			
18			X			X		X		X
19	X	X				X				X
20				X			X		X	X
21		X				X				
22	X		X		X				X	
23			X	X	X		X		X	
24		X					X	X	X	
25	X		X	X	X	X	X	X	X	X
26		X		X		X	X	X		
27				X	X			X		
28	X						X		X	X
29		X	X		X			X		
30			X		X	X	X		X	

Having determined which quadratic congruences are solvable, we will now develop a procedure for finding the solution. Before proceeding, we need to introduce the *Legendre symbol*.

The Legendre symbol is a type of classifying function. In this book, I will use the notation $(a \mid p)$ to indicate the Legendre symbol. Other books may use $\left(\frac{a}{p}\right)$ or the somewhat ambiguous notation of (a/p).

Definition (Legendre symbol). If p is an odd prime, then

$$(a \mid p) = \begin{cases} 1 & \text{if } a \text{ is a quadratic residue modulo } p \\ 0 & \text{if } p \mid a \\ -1 & \text{otherwise.} \end{cases}$$

We can use Euler's criterion to determine if a is a quadratic residue mod p after ensuring that $gcd(a, p) = 1$. The Legendre symbol has many interesting properties summarized in the theorem that follows.

Theorem 5.2. Let p, q, be distinct odd primes and a be any integer. Then

1. $(a \mid p) \equiv a^{(p-1)/2} \pmod{p}$ and in the following cases

$$(-1 \mid p) = \begin{cases} 1 & \text{if } p \equiv 1 \pmod 4 \\ -1 & \text{if } p \equiv 3 \pmod 4 \end{cases}$$

$$(2 \mid p) = \begin{cases} 1 & \text{if } p \equiv 1, 7 \pmod 8 \\ -1 & \text{if } p \equiv 3, 5 \pmod 4 \end{cases}$$

2. $(a^2 \mid p) = 1$ if $p \mid a$

3. $(a \mid p)(b \mid p) = (ab \mid p)$

4. $(a \mid p) = (b \mid p)$ if $a \equiv b \pmod{p}$

5. if $p \neq q$, then $(p \mid q) = (q \mid p)$, unless $p \equiv q \equiv 3 \pmod 4$, in which case $(p \mid q) = -(q \mid p)$.

(This property is known as the *law of quadratic reciprocity*.)

The Prussian mathematician Carl Gustav Jacobi (1804–1851), known for his research in elliptic functions, also made contributions to number theory. Among these are his generalization of the Legendre symbol $(a \mid n)$ for any integer a and n.

Definition (Jacobi symbol). Let n be odd with the prime factorization, $p_1^{e_1} p_2^{e_2} p_3^{e_3} \cdots p_k^{e_k}$, then

$$(a \mid n) = (a \mid p_1)^{e_1} (a \mid p_2)^{e_2} (a \mid p_3)^{e_3} \ldots (a \mid p_k)^{e_k}$$

Like the Legendre symbol, the Jacobi symbol has noteworthy properties as well.

Theorem 5.3. Let m, n be positive odd integers and let a, b be any integer. Then the Jacobi symbol has the properties of the Legendre symbol and additionally,

1. $(a \mid n)(b \mid n) = (ab \mid n)$

2. $(a \mid m)(a \mid n) = (a \mid mn)$

The Jacobi symbol's use is limited when using a composite n because even if $(a \mid n) = 1$, a is not necessarily a quadratic residue modulo n. For example, $(8 \mid 15) = 1$ but 8 is not a quadratic residue of 15. If n is prime then the Jacobi symbol returns the same result as the Legendre symbol.

Computing the Jacobi symbol can be performed efficiently if we use the law of quadratic reciprocity to invert the symbol. This permits us to diminish the size of the variables continually and ensures that the algorithm will terminate.

Program 5_4_2.c. Evaluate the Jacobi symbol

```
/*
** Program 5_4_2 - Evaluate the Jacobi symbol
*/
#include "numtype.h"

void main(int argc, char *argv[])
{
  INT a, j, n;

  if (argc == 3) {
    a = atol(argv[1]);
    n = atol(argv[2]);
  } else {
    printf("Jacobi symbol (a | n)\n\n");
    printf("a, n = ");
    scanf("%ld,%ld", &a, &n);
  }

  j = JacobiSymbol(a, n);

  printf("(%ld | %ld) = %ld\n", a, n, j);
}

/*
** JacobiSymbol - evaluates the Jacobi Symbol (generalized Legendre
**                Symbol) using Euler's criterion and the Law of
**                Quadratic Reciprocity
*/
#include "numtype.h"

int JacobiSymbol(INT a, INT n)
{
  INT e, t;
```

```
  e = 1;
  while (a != 0)
  {
    while (!(a & 1))                   /* a is even */
    {
      a >>= 1;
      if ((n % 8) == 3 || (n % 8) == 5) e = -e;
    }

    t = a; a = n; n = t;               /* swap */

    if ((a % 4) == 3 && (n % 4) == 3) e = -e;

    a %= n;
  }

  if (n == 1)
    return(e);
  else
    return(0);
}
```

```
C> 5_4_2
Jacobi symbol (a | n)

a, n = 184, 37
(184 | 37) = 1
```

With these basic tools in place, we can now look at solving quadratic congruences.

Solving Quadratic Congruences
of the Form $x^2 \equiv a$ (mod p)

In this section we will be solving simple quadratic congruences of the form

$$x^2 \equiv a \pmod{p}.$$

To solve this congruence we need to identify quadratic residues and, for that purpose, a function that computes the value of the Legendre symbol using Euler's criterion is employed. For finding the quadratic residue, we will use the RESSOL algorithm (Shanks, 1972) owing to its algorithmic simplicity and relative efficiency. Unfortunately, although the algorithm is simple, the explanation of it is not. As will be seen in the description that follows, the simplicity develops as one cancels intermediate quantities.

Before beginning to find quadratic residues using the RESSOL algorithm, it is useful to introduce some definitions relating to cyclic groups.

Definition. An additive group G is cyclic if and only if there is an $x \in G$ such that each $y \in G$ is a integral multiple of x. The element x is a generator of G.

Definition. A multiplicative group G is cyclic if and only if there is an $x \in G$ such that each $y \in G$ is an integral power of x. The element x is a generator of G.

Although cyclic groups may be finite or infinite, our interest is in finite cyclic groups. In the current application, the finite cyclic group is the set $\{0, 1, 2, ..., p - 1\}$, the arithmetic residue class.

Given $x^2 \equiv a \pmod{p}$, p being an odd prime and $(a \mid p) = 1$, we first find s and k such that

$$p = 2^s \cdot (2 \cdot k + 1) + 1$$

You can easily verify that s and k can be found for any odd number and therefore for any odd prime. In our program, this is done by dividing s 2s out of $(p - 1)$, then solving for k.

Also, we need another number z, such that $(z \mid p) = -1$. This is easily found by starting with $z = 2$ and testing it with the function LegendreSymbol(). If z is not a quadratic nonresidue, we simply increment by 1 and test again.

Next compute the quantities

$$r \equiv a^{k+1} \pmod{p} \quad \text{and} \quad n \equiv a^{2 \cdot k+1} \pmod{p}$$

and have

$$r^2 \equiv a \cdot n \pmod{p}.$$

If $n \equiv 1$, then r is a solution to $x^2 \equiv a \pmod{p}$. If not, then we iterate until we find an n, such that $n \equiv 1$. The new values of r and n are computed by

$$r_{i+1} \equiv r_i \cdot b_i \pmod{p} \quad \text{and} \quad n_{i+1} \equiv n_i \cdot b_i^2 \pmod{p}$$

for any b_i. The key here is to choose b_i so that the order s of n_0, n_1, n_2, ... strictly decreases. Thus, we will eventually arrive at the solution r_i; that is, when $n_i \equiv 1 \pmod{p}$.

We know that if p is prime, there will be $p - 1$ (i.e., $2^s \cdot (2 \cdot k + 1)$) residue classes prime to p. These classes are cyclic under multiplication $\pmod{p}$.

Since z is a quadratic nonresidue, if we take

$$c \equiv z^{(2 \cdot k + 1)} \pmod{p},$$

then c is of order 2^s.   For some u,

$$n \equiv c^u \pmod{p},$$

but this still does not help us find the bs that will create the decreasing sequence of ns. To do this takes three steps (Shanks, 1972):

$$s_{i+1}$$

1. ni is squared until $n_i^2 \equiv 1 \pmod{p}$;

2. si + 1 is saved and c_i is substituted for n_i;

3. ci is squared to obtain $b_i \equiv c_i^u \pmod{p}$, where $u \equiv 2^{t+1} \pmod{p}$

The program that follows for finding quadratic residues includes the functions RESSOL(), PowMod(), MulMod() and LegendreSymbol(). Each of these perform their specific task described above.

Program 5_4_3.c. Solve quadratic congruences using Shanks' RESSOL algorithm

```
/*
** Program 5_4_3 - Solve for x in the relation x^2 ≡ a (mod p)
**                 using the RESSOL algorithm
*/
#include "numtype.h"

void main(int argc, char *argv[])
{
  INT a, p, x;
  NAT b, c, i, l, k, n, q, r, s, z;

  if (argc == 3) {
    a = atol(argv[1]);
    p = atol(argv[2]);
  } else {
    printf("Solve for x in: x^2 ≡ a (mod p)\n\n");
    printf("a, p = ");
    scanf("%lu,%lu", &a, &p);
  }

  if (p <= 0) exit(1);
  if (a == 0) exit(1);
  a = ((a % p) + p) % p;            /* make a > 0 and a < p */

  x = QuadraticResidue(a, p);

  printf("%lu^2 ≡ %lu (mod %lu)\n", x, a, p);
}

/*
** QuadraticResidue - find x such that x² ≡ a (mod p) using the
**                    RESSOL algorithm
*/
#include "numtype.h"

NAT QuadraticResidue(NAT a, NAT p)
{
```

```
    NAT b, c, i, l, k, n, q, r, s, x, z;

    if ( LegendreSymbol(a, p) != 1)
    {
      printf("%ld is not a quadratic residue of %ld\n", a, p);
      exit(1);
    }

/* find initial value for s, k, such that p = 2^s * (2*k + 1) - 1
*/

    s = 0;
    k = p - 1;
    while ( !(k & 1) ) { s++; k /= 2; }     /* reduce to 2*k + 1 */
    k = (k - 1) / 2;                         /* solve for k */

/* find z such that z is not a quadratic residue of p */

    z = 2;
    while ( LegendreSymbol(z, p) == 1) z++;

/* find initial values for r, n, and c */

    r = PowMod(a, k + 1, p);
    n = PowMod(a, 2*k + 1, p);
    c = PowMod(z, 2*k + 1, p);

/* derive the solution */

    while (n > 1)
    {
      l = s;
      b = n;
      for ( i = 0; i < l; i++)
        if ( b == 1)
        {
          b = c;
          s = i;
        }
        else
        {
          b = (b * b) % p;
        }

      c = (b * b) % p;
      r = (b * r) % p;
      n = (c * n) % p;
    }

    return(r);
}

/*
** LegendreSymbol -  evaluates the Legendre Symbol using modulus
**                   and Euler's criterion for quadratic congruences
**
** returns: 0 if p divides a
**          1 if a is a quadratic residue modulo p
**          -1 otherwise
```

```
*/
#include "numtype.h"

int LegendreSymbol(NAT a, NAT p)
{
  if ( (a % p) == 0) return(0);
  if ( PowMod(a, (p - 1)/2, p) == 1) return(1);
  return(-1);
}

/*
** PowMod - computes r for a^n ≡ r (mod m) given a, n, and m
*/
#include "numtype.h"

NAT PowMod(NAT a, NAT n, NAT m)
{
  NAT r;

  r = 1;
  while (n > 0)
  {
    if (n & 1)                    /* test lowest bit */
      r = MulMod(r, a, m);        /* multiply (mod m) */
    a = MulMod(a, a, m);          /* square */
    n >>= 1;                      /* divided by 2 */
  }

  return(r);
}

/*
** MulMod - computes r for a * b ≡ r (mod m) given a, b, and m
*/
#include "numtype.h"

NAT MulMod(NAT a, NAT b, NAT m)
{
  NAT r;

  if (m == 0) return(a * b);          /* (mod 0) */

  r = 0;
  while (a > 0)
  {
    if (a & 1)                        /* test lowest bit */
      if ((r += b) > m) r %= m;       /* add (mod m) */
    a >>= 1;                          /* divided by 2 */
    if ((b <<= 1) > m) b %= m;        /* times 2 (mod m) */
  }

  return(r);
}
```

```
C> 5_4_3
Solve for x in: x^2 ≡ a (mod p)

a, p = 5, 11
4^2 ≡ 5 (mod 11)
```

Generalized Quadratic Congruences

We are now prepared to find solutions to the general quadratic congruence:

$$a \cdot x^2 + b \cdot x + c \equiv 0 \ (\text{mod } p).$$

Several approaches are possible to solving this congruence. The most straight-forward is to find the *discriminant d*:

$$d = b^2 - 4 \cdot a \cdot c$$

If $(d \mid p) = 1$, then we can solve the general quadratic congruence. If the congruence is soluble, then we need to solve the congruence:

$$u^2 \equiv d \ (\text{mod } p)$$

for u. As noted in Section 3.4, we can find a multiplicative inverse z for a (mod p) by solving the linear congruence $a \cdot z \equiv 1 \ (\text{mod } p)$. Once we have found z and u, we can find x explicitly by computing the following quantities:

$$x = z \cdot (-b + u) / 2 \quad \text{and} \quad x = z \cdot (-b - u) / 2$$

If $(b \pm u)$ is odd, then we use $(b \pm u + p)$.
Let's look at an example where we solve the congruence

$$5 \cdot x^2 + 11 \cdot x + 3 \equiv 0 \ (\text{mod } 47).$$

then

$$d = 121 - 4 \cdot 5 \cdot 3 = 61.$$

Since $(61 \mid 47) = 1$, we can solve the quadratic congruence. Using Program 5_4_3, we find that

$$u_2 \equiv 61 \ (\text{mod } 47), \text{ so } u = 25.$$

To find z, the inverse of a, use Program 3_4_2 that solves linear congruences. In this case $a = 5$, $b = 1$, $m = 47$:

$$5 \cdot z \equiv 1 \ (\text{mod } 47), \text{ so } z = 19.$$

With u and z numbers in hand, we can compute the solutions to the quadratic congruence

$$x = 19 \cdot (-11 + 25) / 2 = 133 \ \% \ 47 = 39$$

and

$$x = 19 \cdot (-11 - 25) / 2 = -133 \% 47 = 34.$$

Checking our work,

$$5 \cdot 39^2 + 11 \cdot 39 + 3 = 8037 \equiv 0 \pmod{47} \checkmark$$

and

$$5 \cdot 34^2 + 11 \cdot 34 + 3 = 6157 \equiv 0 \pmod{47} \checkmark.$$

The solutions to the quadratic congruence are $x = 34$ and 39. And lastly, here is a program that implements the methodology described above.

Program 5_4_4.c. Solve general quadratics using RESSOL and Gauss' method

```
/*
** Program 5_4_4 - Solve for x in the relation a*x² + b*x + c ≡ 0
**                 (mod p) using the RESSOL algorithm and Gauss' &&
                   method
*/
#include "numtype.h"

void main(int argc, char *argv[])
{
  INT a, b, c, d, p, u, z;
  INT x1, x2;

  if (argc == 5) {
    a = atol(argv[1]);
    b = atol(argv[2]);
    c = atol(argv[3]);
    p = atol(argv[4]);
  } else {
    printf("Solve for x in: a*x² + b*x + c ≡ 0 (mod p)\n\n");
    printf("a, b, c, p = ");
    scanf("%ld,%ld,%ld,%ld", &a, &b, &c, &p);
  }

  SolveQuadraticCongruence(a, b, c, p, &x1, &x2);

  printf("x = %ld, %ld\n", x1, x2);
}

/*
** SolveQuadraticCongruence - find x such that a*x² + b*x + c ≡ 0
**                            (mod p) by the RESSOL algorithm
**                            and Gauss' method
*/
#include <stdio.h>
#include <stdlib.h>

#include "numtype.h"
```

```
INT SolveQuadraticCongruence(INT a, INT b, INT c, INT p, INT *x1,
INT *x2)
{
  INT d, u, z;

/* first find the discriminant and the quadratic residue */

  d = (b * b - 4 * a * c) % p;
  if (d < 0) d += p;                       /* make d > 0 */
  u = QuadraticResidue(d, p);

/* find the z, inverse of a */

  z = PowMod(a, p - 2, p);

/* now find the solutions */

  if ((b + u) & 1)
  {
    *x1 = (z * ( - b + u + p) / 2) % p;
    *x2 = (z * ( - b - u + p) / 2) % p;
  }
  else
  {
    *x1 = (z * ( - b + u) / 2) % p;
    *x2 = (z * ( - b - u) / 2) % p;
  }
  if (*x1 < 0) *x1 += p;
  if (*x2 < 0) *x2 += p;

  return(0);
}

/*
** QuadraticResidue -  find x such that x² ≡ a (mod p) using the
**                     RESSOL algorithm (Shanks, 1972)
*/
#include "numtype.h"

NAT QuadraticResidue(NAT a, NAT p)
{
  NAT b, c, i, l, k, n, q, r, s, x, z;

  if ( LegendreSymbol(a, p) != 1)
  {
    printf("%ld is not a quadratic residue of %ld\n", a, p);
    exit(1);
  }

/* find initial value for s, k, such that p = 2^s * (2*k + 1) - 1
*/

  s = 0;
  k = p - 1;
  while ( !(k & 1) ) { s++; k /= 2; }    /* reduce to 2*k + 1 */
  k = (k - 1) / 2;                       /* solve for k */

/* find z such that z is not a quadratic residue of p */
```

```
  z = 2;
  while ( LegendreSymbol(z, p) == 1) z++;

/* find initial values for r, n, and c */

  r = PowMod(a, k + 1, p);
  n = PowMod(a, 2*k + 1, p);
  c = PowMod(z, 2*k + 1, p);

/* derive the solution */

  while (n > 1)
  {
    l = s;
    b = n;
    for ( i = 0; i < l; i++)
      if ( b == 1)
      {
        b = c;
        s = i;
      }
      else
      {
        b = (b * b) % p;
      }

    c = (b * b) % p;
    r = (b * r) % p;
    n = (c * n) % p;
  }

  return(r);
}

/*
** LegendreSymbol — evaluates the Legendre Symbol using modulus
**                  and Euler's criterion for quadratic congru-
ences
**
** returns:  0 if p divides a
**           1 if a is a quadratic residue modulo p
**          -1 otherwise
*/
#include "numtype.h"

int LegendreSymbol(NAT a, NAT p)
{
  if ( (a % p) == 0) return(0);
  if ( PowMod(a, (p - 1)/2, p) == 1) return(1);
  return(-1);
}

/*
** PowMod - computes r for a^n ≡ r (mod m) given a, n, and m
*/
#include "numtype.h"

NAT PowMod(NAT a, NAT n, NAT m)
{
```

```
    NAT r;
    r = 1;
    while (n > 0)
    {
      if (n & 1)                          /* test lowest bit */
        r = MulMod(r, a, m);              /* multiply (mod m) */
      a = MulMod(a, a, m);                /* square */
      n >>= 1;                            /* divided by 2 */
    }

    return(r);
}

/*
** MulMod - computes r for a * b ≡ r (mod m) given a, b, and m
*/
#include "numtype.h"

NAT MulMod(NAT a, NAT b, NAT m)
{
    NAT r;

    if (m == 0) return(a * b);            /* (mod 0) */

    r = 0;
    while (a > 0)
    {
      if (a & 1)                          /* test lowest bit */
        if ((r += b) > m) r %= m;         /* add (mod m) */
      a >>= 1;                            /* divided by 2 */
      if ((b <<= 1) > m) b %= m;          /* times 2 (mod m) */
    }

    return(r);
}
```

```
C> 5_4_4
Solve for x in:  a*x² + b*x + c ≡ 0 (mod p)

a, b, c, p = 5,11,3,47
x = 39, 34
```

Challenges

1. Is the congruence $x^2 \equiv 4 \pmod{p}$ solvable for all odd prime p? Explain.

2. Write a program to solve linear congruences using the PowMod function.

5.5 Polynomial Congruences

Using the properties of modular arithmetic described above, we can make the following statement with respect to polynomials:

If $f(x)$ is a polynomial with integer coefficients and $a \equiv b \pmod{m}$, then $f(a) \equiv f(b) \pmod{m}$.

This is a powerful result that has wide application. You probably know that a number is divisible by 9 if the sum of the digits is divisible by 9. The basis for the ancient "asting out 9s" rule can be illustrated by an example. Let $n = 327,987$

$$n = 3 \cdot 10^5 + 2 \cdot 10^4 + 7 \cdot 10^3 + 9 \cdot 10^2 + 8 \cdot 10 + 7.$$

If you think of n in terms of a polynomial, then $n = f(x)$, where $x = 10$ and

$$f(x) = 3 \cdot x^5 + 2 \cdot x^4 + 7 \cdot x^3 + 9 \cdot x^2 + 8 \cdot x + 7.$$

We know that $10 \equiv 1 \pmod{9}$, so using our statement above, we can conclude that $f(10) \equiv f(1) \pmod{9}$. Written out, $f(1)$ is

$$f(1) = 3 \cdot 1^5 + 2 \cdot 1^4 + 7 \cdot 1^3 + 9 \cdot 1^2 + 8 \cdot 1 + 7$$

$$= 3 + 2 + 7 + 9 + 8 + 7$$

$$= 36.$$

Since 36 is divisible by 9, we conclude that 327,987 is also divisible by 9. In fact, any permutation of the digits of 327,987 is also divisible by 9.

The principle behind casting out 9s is not limited to base 10. When using the procedure with another base m, we cast out all $(m - 1)$s.

This technique can be extended to serve as a divisibility test for any two numbers using a procedure called "radix shifting." To test if n is divisible by m, we convert n to a new radix of $m + 1$. Add the digits of the new representation, and if the sum of the digits is divisible by m, then n is divisible by m.

Consider the following example. Is 7 a proper factor of 555,555? To find out we let $n = 555,555$ and $m = 7$.

$$n_{10} = 555,555_{10} \quad \text{converts to} \quad n_8 = 2,075,043_8$$

using Program 5_5_1. Adding the octal digits, we get

$$2_8 + 0_8 + 7_8 + 4_8 + 0_8 + 5_8 + 3_8 = 25_8 = 21_{10}.$$

Because 21 is divisible by 7, we know that 555,555 is divisible by 7.

This method can be implemented efficiently for any divisor d of the form $(2^i - 1)$ because the extraction of the digit is accomplished by masking i bits and the division is performed by shifting i bits. The next program performs radix shifting to test if a number is divisible by 7:

Program 5_5_1.c. Test if a number is divisible by 7 using radix shifting

```
/*
** Program 5_5_1 - Test if a number is divisible by 7 by radix
**                 shifting
*/
#include "numtype.h"

void main(int argc, char *argv[])
{
  NAT n, s;

  printf("Divisibility by 7 test\n\n");
  if (argc == 2) {
    n = atol(argv[1]);
  } else {
    printf("n = ");
    scanf("%lu", &n);
  }

  printf("%lu is ", n);
  s = 0;
  while (n > 0)
  {
    s += n & 7;                /* extract digit by masking */
    n >>= 3;                   /* shifting number by radix */
  }

  if ((s % 7) == 0)
    printf("divisible by 7\n");
  else
    printf( "NOT divisible by 7\n");
}
```

```
C> 5_5_1
Divisibility by 7 test

n = 555555
555555 is divisible by 7
```

Rather than treating a polynomial $f(x)$ like a number with radix x, we will now look at attempting to find solutions to polynomials. I'll begin by presenting the *Lagrange Factor Theorem*.

Lagrange Factor Theorem (5.4). For all polynomials $f(x)$ with integer coefficients and for all m, if a is a root of the congruence

$$f(x) \equiv 0 \pmod{m},$$

then there exists a polynomial $g(x)$ with integer coefficients such that

$$f(x) \equiv (x - a) \cdot g(x) \pmod{m},$$

and conversely.

Of course to be able to enjoy the benefits of the Lagrange Factor Theorem, we must first find a root. Lagrange's theorem can tell us the maximum number of roots to expect. It states that for prime p, the congruence

$$f(x) \equiv 0 \;(\text{mod } p),$$

in which

$$f(x) = a_0{\cdot}x^n + a_1{\cdot}x^{n-1} + \ldots + a_n, \qquad a_0 \not\equiv (\text{mod } p)$$

has at most n roots.

Application: Error Detection by Casting Out 9s

In the field of telecommunications, it is necessary to detect errors in data transmission. A means of error detection can be created based on a checksum derived from the addition of data elements being transmitted. A difference between the computed checksum and the expected checksum indicates that a data transmission error occurred.

Error detection can be implemented using addition or multiplication and then casting out 9s. Consider the following examples of casting out 9s:

Addition:

$$
\begin{array}{r}
94 \\
+ \quad 83 \\
\hline
177
\end{array}
$$

    94   sum of the digits = 13, sum of the sum of the digits = 4

+   83   sum of the digits = 11, sum of the sum of the digits = 2

177   sum of the digits = 15, sum of the sum of the digits = 6

Not only is it true that $(13 + 11) \equiv 15 \;(\text{mod } 9)$, but $(4 + 2) \equiv 6 \;(\text{mod } 9)$. We can always iteratively sum the digits until we receive a sum that is less than 10 and thereby use the arithmetic residue system.

Multiplication:

    94   sum of the digits = 13, sum of the sum of the digits = 4

*   83   sum of the digits = 11, sum of the sum of the digits = 2

7802   sum of the digits = 17, sum of the sum of the digits = 8

For multiplication we multiply the sums so that $(13 \cdot 11) \equiv 143 \;(\text{mod } 9)$. Reducing to the arithmetic residue system gives $(4 \cdot 2) \equiv 8 \;(\text{mod } 9)$.

Using this technique, we can check streams of data. In the following program we check the sum of the numbers from 1 to 255. The function SumDigits computes the sum of the digits and each sum is reduced to the arithmetic residual system:

Program 5_5_2.c. Demonstrate check sums using casting out 9's

```
/*
** Program 5_5_2 - Check the sum from 1 to 255 by casting out 9's
*/
#include "numtype.h"

void main(void)
{
  NAT i, s, t;

  printf("Check sum from 1 to 255\n\n");

  s = t = 0;
  for (i = 1; i <= 255; i++)
  {
    t += i;                    /* actual sum */
    s += SumDigits(i);         /* sum of digits */
  }

  printf("actual sum = %6ld\n", t);

  while ((t = SumDigits(t)) > 9);
  printf("sum of the digits of the actual sum = %lu\n", t);

  while ((s = SumDigits(s)) > 9);
  printf("sum of the digits of the addends  = %lu\n", s);
}

/*
** SumDigits - adds up the value of the digits in a decimal number
*/
#include "numtype.h"

NAT SumDigits(NAT i)
{
 NAT s = 0;

  while (i > 0)
  {
    s += i % 10;               /* get digit */
    i /= 10;                   /* shift decimal digits right */
  }
  return(s);
}
```

```
C> 5_5_2
Check sum from 1 to 255

actual sum = 32640
sum of the digits of the actual sum = 6
sum of the digits of the addends  = 6
```

Challenges

1. Modify Program 5_5_2 to test by multiplication. What limitations does this technique have?

2. Modify Program 5_5_2 to compute checksums to base 16. What percentage of random errors would you expect to pass the checksum test?

3. There is an old "magic" trick based on casting out 9s. It goes like this:

   a. Have a friend secretly write down a four-digit number that does not contain any zeroes.

   b. Tell your friend to add the digits but keep this sum secret.

   c. Ask your friend to cross out any one of the digits in the four-digit number.

   d. Have your friend subtract the remaining three-digit number from the sum of the digits in (b) and tell you this result.

   e. Astound your friend by "magically" naming the digit crossed out by mentally adding the digits of the result from d) and subtracting that sum from the next higher multiple of 9.

Explain this trick using congruence arithmetic. Can this trick be performed if the four-digit number contains zeroes?

4. All books today are assigned number called the International Standard Book Number (ISBN). The ISBN is a string of 10 digits, x_1, x_2, ... , x_{10}, where the first four digits are the publisher, the next five are the publisher's number, and the last is a check digit. This last digit is chosen so that $x_1 + 2x_2 + 3x_3 + ... + 10x_{10}$ is divisible by 11. If $x_{10} = 10$ then the last digit is an x. Write a program that will correct the ISBN if there is an error in one its digits.

5.6 Exponential Congruences

Exponential congruences have the form

$$a^x \equiv b \pmod{m},$$

where a and b are integer constants and x is the desired solution. Being able to tell if the congruence is solvable is a little more difficult than in the case of the linear congruence. However, like the linear congruence, for small m the exponential congruence can be solved by simple trial.

Program 5_6_1.c. Solve exponential congruences using the trial method.

```
/*
** Program 5_6_1 - Solve for x in the relation: a^x ≡ b mod m
**                 using the trial method
*/
#include "numtype.h"

void main(int argc, char *argv[])
{
   NAT a, b, m, x;

   if (argc == 4) {
     a = atol(argv[1]);
     b = atol(argv[2]);
     m = atol(argv[3]);
   } else {
     printf("Solve for x in: a^x ≡ b (mod m)\n\n");
     printf("a, b, m = ");
     scanf("%ld,%ld,%ld", &a, &b, &m);
   }

   for ( x = 1, b %= m; x < m; x++)
     if (PowMod(a, x, m) == b)
       printf("x = %lu\n", x);
}

/*
** PowMod - computes r for aⁿ ≡ r (mod m) given a, n, and m
*/
#include "numtype.h"

NAT PowMod(NAT a, NAT n, NAT m)
{
 NAT r;

   r = 1;
   while (n > 0)
   {
     if (n & 1)                    /* test lowest bit */
       r = MulMod(r, a, m);        /* multiply (mod m) */
     a = MulMod(a, a, m);          /* square */
     n >>= 1;                      /* divided by 2 */
   }

   return(r);
}

/*
** MulMod - computes r for a * b ≡ r (mod m) given a, b, and m
*/
#include "numtype.h"
```

```
NAT MulMod(NAT a, NAT b, NAT m)
{
  NAT r;

  if (m == 0) return(a * b);                    /* (mod 0) */

  r = 0;
  while (a > 0)
  {
    if (a & 1)                                  /* test lowest bit */
        if ((r += b) > m) r %= m;               /* add (mod m) */
      a >>= 1;                                  /* divided by 2 */
        if ((b <<= 1) > m) b %= m;              /* times 2 (mod m) */
  }

  return(r);
}
```

```
C> 5_6_1
Solve for x in: a^x ≡ b (mod m)

a, b, m = 15,12,19
x = 3
```

Any trial method will be limited in its applicability because such methods are generally wasteful of computer resources. What we need is a more efficient method, but the price of efficiency is complexity. To solve these equations directly, we will use the *index*, sometimes called the *discrete logarithm*.

Recall from Chapter 4 that if a number m has primitive roots and a is one of them, then $\{ a, a^2, a^3, ..., a^{\phi(m)} \}$ are mutually incongruent and form a reduced residue system (mod m). For any number b, where $gcd(b,m) = 1$, it is possible to find a smallest exponent of a power of a that is congruent to b. That is to say

$$a^h \equiv b \;(\text{mod } m)$$

The exponent h is called the index (or discrete logarithm) of b to the base a and is written as

$$h = \text{ind}_a b$$

The fact that the index is similar to the logarithm can be used to solve a variety of congruences. Let's first look at an example with a linear congruence.

Recall that $\phi(19) = 18$. If we have a simple linear congruence

$$7 \cdot x \equiv 10 \;(\text{mod } 19),$$

then

$$\text{ind}_2 7 + \text{ind}_2 x \equiv \text{ind}_2 10 \;(\text{mod } 18)$$

An index table is used to find values of ind_2 7 and ind_2 10 (created using Program 4_4_1):

1	2	3	4	5	6	7	8	9	10	11	12	13	14	15	16	17	18
a	a	a	a	a	a	a	a	a	a	a	a	a	a	a	a	a	a
2	4	8	16	13	7	14	9	18	17	15	11	3	6	12	5	10	1

$$ind_2 \, x \equiv 17 - 6 = 11 \pmod{18}.$$

Now taking the "anti-index" a^{11} gives the correct answer

$$x = 15.$$

Returning to the main point of this section, exponential congruences can be solved using indices. Consider the following congruence:

$$15^x \equiv 12 \pmod{19}.$$

Taking the index yields the linear congruence.

$$x \cdot ind_2 \, 15 \equiv ind_2 \, 12 \pmod{18}$$

$$x \cdot 11 \equiv 15 \pmod{18}$$

$$x = 3.$$

This is how an exponential congruence can be reduced to a linear congruence. From that discussion, we know that this linear congruence could only be solved because $gcd(11, 18) = 1$.

We can use the concept of the index to help us find a solution for the general case of the exponential congruence. If a primitive root g exists for $\phi(m)$, then we can rewrite the congruence:

$$a^x \equiv b \pmod{m}$$

by taking the index to base g giving

$$x \cdot ind_g \, a \equiv ind_g \, b \pmod{\phi(m)}.$$

This linear congruence has a solution if and only if $gcd(ind_g a, \phi(m))$ | $ind_g b$ and the number of incongruent solutions is $gcd(ind_g a, \phi(m))$. When solutions exist, we can use the method described in Section 5.1 for solving linear congruences from this point forward.

The first order of business for solving exponential congruences is finding a primitive root for the given modulus. After finding a primitive root g, the next step in solving the exponential congruence $a^x \equiv b$ (mod m) is to find the index of a and b to the base g. Owing to the ran-

dom nature of the discrete logarithm, the easiest method is to exponentiate the primitive root until the number we are taking the index
of is reached, the exponent is index. On average this is an $O(p/2)$ algorithm with a negligible storage requirement. Other algorithms that
require more storage can operate in $O(p^{1/2})$ time (Pollard, 1979).

With the primitive root g, the exponential congruence is reduced to
a linear congruence

$$x \cdot \mathrm{ind}_g\, a \equiv \mathrm{ind}_g\, b \pmod{\phi(m)}$$

and is solved for x using Gauss's method. Although Program 5_6_2.c
is a bit long, it incorporates many of the functions that we have developed up to this point.

Program 5_6_2.c. Solve exponential congruences of the form $a^x \equiv b$
mod m.

```
/*
** Program 5_6_2 - Solve for x in the relation: a^x ≡ b mod m
**                 using the index method
*/
#include "numtype.h"

void main(int argc, char *argv[])
{
  INT a, b, m, x;

  if (argc == 4) {
    a = atol(argv[1]);
    b = atol(argv[2]);
    m = atol(argv[3]);
  } else {
    printf("Solve for x in: a^x ≡ b (mod m)\n\n");
    printf("a, b, m = ");
    scanf("%ld,%ld,%ld", &a, &b, &m);
  }

  x = SolveExponentialCongruence(a, b, m);

  if (x >= 0)
    printf("x = %ld\n", x);
  else
    printf("no solution\n");
}

/*
** SolveExponentialCongruence - solve an exponential congruence:
**                              a^x ≡ b mod m using the Totient
**                              function
**
** notes:  g = primitive root
**         t = Euler's totient function value for m-1
*/
#include "numtype.h"
```

```
INT SolveExponentialCongruence(a, b, m)
INT a, b, m;
{
   NAT g, x;

/* find a primitive root for m */

   if ((g = PrimitiveRoot(m)) == 0)
     return(-1);                        /* no primitive roots */

/* take the index of a and b */

   a = IndMod(g, a, m);
   b = IndMod(g, b, m);

   if ((a == 0) || (b == 0))
     return(-1);                        /* cannot find index (mod m) */

/* solve the congruence */

   x = SolveLinearCongruence(a, b, EulerTotient(m));

   return(x);
}

/*
** SolveLinearCongruence - solve a linear congruence: ax ≡ b mod m
**                         using Gauss' method
*/
#include "numtype.h"

INT SolveLinearCongruence(INT a, INT b, INT m)
{
   NAT g;

   if ((a %= m) < 0) { a = -a; b = -b; } /* transfer sign to b */
   if ((b %= m) < 0) b += m;             /* make b > 0 */

   if ((b % GCD(a,m)) == 0)
   {
     if ((g = GCD(a,b)) > 1) { a /= g; b /= g; }
     while (a > 1)
     {
       b += m;
       if ((g = GCD(a,b)) > 1) { a /= g; b /= g; }
     }
     return(b);
   }

   return(-1);                          /* solution not possible */
}

/*
** PrimitiveRoot - finds the smallest primitive root for m
*/
#include "numtype.h"

NAT PrimitiveRoot(m)
```

```
NAT m;
{
  NAT g, i, t, fk, f[32];

/* find a primitive root for m */

  if (m == 2)
    g = 1;
  else
  {
    t = EulerTotient(m);
    fk = Factor(t, f);
    for (g = 2; g < m; g++)
    {
      if (GCD(g, m) == 1)
      {
        for (i = 0; i < fk; i++)
          if (PowMod(g, t / f[i], m) == 1) break;
          if (i == fk) break;          /* primitive root found */
      }
    }
  }


  if (g == m)
    return(0);                         /* none */
  else
    return(g);
}

/*
** EulerTotient - compute values for Euler's totient function using a
**                prime factorization by trial division
*/
#include "numtype.h"

NAT EulerTotient(NAT n)
{
  NAT i, j, et, pk, fk, f[32];

  et = 1;                              /* totient value */
  pk = 0;                              /* power count */

  if (n == 1) return(1);

  fk = Factor(n, f);                   /* factor count, factors */
  if (fk > 32) exit(1);

  for (i = 0; i < fk; i++)
  {
    if (f[i] == f[i+1])
    {
      pk++;
    }
    else
    {
      for (j = 0; j < pk; j++) et *= f[i];
      et *= f[i] - 1;
```

```
            pk = 0;
        }
    }

    return(et);
}

/*
** Factor - finds prime factorization by trial division
**            returned values: number of factors, array w/ factors
*/
#include "numtype.h"

NAT Factor(NAT n, NAT f[])
{
    NAT i, fk, s;

    fk = 0;                              /* factor count */

    while (!(n & 1))                     /* even - factor all 2's */
    {
        f[fk++] = 2;
        n /= 2;
    }

    s = sqrt(n);
    for (i = 3; i <= s; i++, i++)        /* factor odd numbers */
        while ( (n % i) == 0L)
        {
            f[fk++] = i;
            n /= i;
            s = sqrt(n);
        }

    if (n > 1L) f[fk++] = n;

    for (i = fk; i < 32; i++) f[i] = 0;  /* zero out the rest */

    return(fk);
}

/*
** IndMod - find the index (discrete logarithm): ind(g) b mod m
**
** input: b = number to find index for
**        g = primitive root for m
**        m = modulus w/ primitive root g
*/
#include "numtype.h"

NAT IndMod(NAT g, NAT b, NAT m)
{
    NAT i, x;

    i = 1;
    x = g;

    if (GCD(b, m) != 1) return(0);
```

```
  while ((x != b) && (i < m))
  {
    x = MulMod(x, g, m);
    i++;
  }

  if (i == m)
    return(0);                        /* index not found */
  else
    return(i);                        /* got it */
}

/*
** PowMod - computes r for a^n ≡ r (mod m) given a, n, and m
*/
#include "numtype.h"

NAT PowMod(NAT a, NAT n, NAT m)
{
  NAT r;

  r = 1;
  while (n > 0)
  {
    if (n & 1)                        /* test lowest bit */
      r = MulMod(r, a, m);            /* multiply (mod m) */
    a = MulMod(a, a, m);              /* square */
    n >>= 1;                          /* divided by 2 */
  }

  return(r);
}

/*
** MulMod - computes r for a * b ≡ r (mod m) given a, b, and m
*/
#include "numtype.h"

NAT MulMod(NAT a, NAT b, NAT m)
{
  NAT r;

  if (m == 0) return(a * b);          /* (mod 0) */

  r = 0;
  while (a > 0)
  {
    if (a & 1)                        /* test lowest bit */
      if ((r += b) > m) r %= m;       /* add (mod m) */
    a >>= 1;                          /* divided by 2 */
    if ((b <<= 1) > m) b %= m;        /* times 2 (mod m) */
  }

  return(r);
}

/*
** GCD - find the greatest common divisor using Euclid's algorithm
*/
```

```
#include "numtype.h"

NAT GCD(NAT a, NAT b)
{
  NAT r;

  while (b > 0)
  {
    r = a % b;
    a = b;
    b = r;
  }

  return(a);
}
```

```
C> 5_6_2
Solve for x in: a^x ≡ b (mod m)

a, b, m = 15,12,19
x = 3
```

For small moduli there is little computational advantage to Program 5_6_2 over Program 5_6_1, but for larger numbers the new program is much quicker. To see this, run both programs against the following congruence:

$$15x \equiv 12 \ (\text{mod } 199831)$$

The output looks identical but Program 5_6_2 is about six times faster than 5_6_1, and the advantage increases as the size of the modulus increases.

```
C> 5_6_2
Solve for x in: a^x ≡ b (mod m)

a, b, m = 15,12,199831
x = 28703
```

Primitive roots and the discrete logarithm problem have direct in public-key cryptography (Odlyzko, 1985; Schneier, 1996). For this reason, algorithms for solving exponential congruences have received a considerable amount of attention (McCurley, 1990).

Challenge

1. Modify Program 5_6_2 to solve congruences of the form:

$$a^x \cdot c \equiv b \ (\text{mod } m)$$

6

Continued Fractions

6.0 Introduction

Number theory has many tools, but probably none is more fascinating or elegant than continued fractions. It was once said that God invented the integers and man invented everything else. It might be added,that God probably invented continued fractions as well.

Continued fractions were developed (or discovered) in part as a response to a need to approximate irrational numbers. Since that time, they have distinguished themselves as important tools for solving problems in probability theory, analysis, and especially number theory. Continued fractions give us something more than solutions to specific problems. They offer another way to represent real numbers, and this affords us an insight obscured by traditional decimal representations.

Because continued fractions are developed using Euclid's algorithm, it is tempting to believe that the great geometrician used them, but this is probably not true. The first known examples come from Rafael Bombelli (ca. 1526–1573), the Italian algebraist best known for having a "wild thought" and developing the concept of complex numbers (Boyer and Merzbach, 1991). In his *Algebra*, published in 1572, he gave a reasonable approximation for $\sqrt{13}$ as

$$3 + \cfrac{4}{6 + \cfrac{4}{6}}.$$

The formal study of continued fractions continued throughout the 17th century with contributions from C. Schwenter (birth-death) and Christiaan Huygens (1629–1695). In 1695 John Wallis (1616–1703) gave continued fractions a thorough analysis and bestowed on them the name they now bear. In their more general form, continued fractions look like

$$b_0 + \cfrac{a_1}{b_1 + \cfrac{a_2}{b_2 + \cfrac{a_3}{b_3 + \cfrac{\cdots}{\quad + \cfrac{a_n}{b_n}}}}},$$

where a_i and b_i can be integers, real or complex numbers. It is often useful to employ the space-saving notation for continued fractions which has the exact same meaning as above:

$$b_0 + \cfrac{a_1}{b_1 +} \; \cfrac{a_2}{b_2 +} \; \cfrac{a_3}{b_3 +} \cdots \cfrac{a_n}{b_n},$$

The numbers a_i and b_i are called the ith partial numerator and denominator of the continued fraction. Finite (or terminating) continued fractions have a finite n and therefore a finite number of terms. If there are an infinite number of terms (i.e., $n \to \infty$) then the continued fractions are called infinite (or nonterminating).

If we consider only those continued fractions $a_i = 1$ for all i:, and b_i are all positive integers (with the possible exception of b_n), then we have the special case of simple or regular *continued fractions:*

$$b_0 + \cfrac{1}{b_1 +} \; \cfrac{1}{b_2 +} \; \cfrac{1}{b_3 +} \cdots \cfrac{1}{b_n}.$$

When looking at simple continued fractions, we can use even more compact notation:

$$[\, b_0, b_1, b_2, b_3, \, ..., b_n \,].$$

The b_i are called the partial quotients. There should be no confusion between this notation and the greatest integer function, which takes only one term.

An important fact about continued fractions is that they can be used to represent *any* real number; a fact that should be stated as a formal theorem.

Theorem 6.1. For every real number a, there is a corresponding unique continued fraction with value equal to a. This fraction is finite if a is rational and infinite if a is irrational.

One final note: Although fractions where a_i and b_i are complex numbers are outside our scope, there are many good books on the subject including Wall (1948), Jones and Thron (1980), and Khinchin, Eagle, and Khinchin (1997). For the rest of this chapter, unless otherwise noted, I will be considering only simple continued fractions with partial quotients having integral values.

6.1 Finite Continued Fractions

Any positive rational number can be written as a finite continued fraction. Recall Euclid's algorithm from Chapter 2 for finding the GCD. The methodology is identical for finding the continued fraction representation of a rational number. In the example from Chapter 2, where we were seeking the GCD for 13,020 and 5797:

$$13{,}020 = 5797 \cdot 2 + 1426$$

$$5797 = 1426 \cdot 4 + 93$$

$$1426 = 93 \cdot 15 + 31$$

$$93 = 31 \cdot 3 + 0.$$

Our goal now is not to find the greatest common divisor but to find the continued fraction representation. Suppose we want to find the continued fraction for the rational number 13,020/5797 Paraphrasing Euclid's algorithm we get:

$$\frac{13{,}020}{5797} = 2 + \frac{1{,}426}{5797}.$$

To achieve the continued fraction format we write:

$$= 2 + \cfrac{1}{\cfrac{5797}{1426}}$$

$$= 2 + \cfrac{1}{4 + \cfrac{93}{1426}}$$

$$= 2 + \cfrac{1}{4 + \cfrac{1}{15 + \cfrac{31}{93}}}$$

$$= 2 + \cfrac{1}{4 + \cfrac{1}{15 + \cfrac{1}{3}}}$$

ending with an integer value greater than 1. In compact notation we have $13{,}020 / 5797 = [2, 4, 15, 3]$. It's easy to see that $[2, 4, 15, 3]$ has the same value as $[2, 4, 15, 2, 1]$; to ensure uniqueness we require that the last value is $b_n > 1$ when $n > 0$.

The number of quotients in a specific continued fraction is highly variable. However, we see from Table 6.1, showing continued fractions for various rational numbers a/b, that there is a distinct pattern. For a given b, the length increases to a certain point and then decreases. Challenge 2 will help you identify exactly what that point is.

Using a variation on the GCD function from Chapter 2, the program that follows computes a continued fraction for a rational number a/b, where a and b are integers.

Program 6_1_1.c. Finds a continued fraction representation of a rational number of the form a/b

```
/*
** Program 6_1_1 - Find a continued fraction representation of a
**                 rational number
*/
#include "numtype.h"
```

Table 6.1. Values for Various Rational Numbers $a / a < 1$ and gcd (a, b)

a/b	Decimal	Continued fraction	Length
1 / 2	0.500000	[0, 2]	2
1 / 3	0.333333	[0, 3]	2
2 / 3	0.666667	[0, 1, 2]	3
1 / 4	0.250000	[0, 4]	2
3 / 4	0.750000	[0, 1, 3]	3
1 / 5	0.200000	[0, 5]	2
2 / 5	0.400000	[0, 2, 2]	3
3 / 5	0.600000	[0, 1, 1, 2]	4
4 / 5	0.800000	[0, 1, 4]	3
1 / 6	0.166667	[0, 6]	2
5 / 6	0.833333	[0, 1, 5]	3
1 / 7	0.142857	[0, 7]	2
2 / 7	0.285714	[0, 3, 2]	3
3 / 7	0.428571	[0, 2, 3]	3
4 / 7	0.571429	[0, 1, 1, 3]	4
5 / 7	0.714286	[0, 1, 2, 2]	4
6 / 7	0.857143	[0, 1, 6]	3
1 / 8	0.125000	[0, 8]	2
3 / 8	0.375000	[0, 2, 1, 2]	4
5 / 8	0.625000	[0, 1, 1, 1, 2]	5
7 / 8	0.875000	[0, 1, 7]	3
1 / 9	0.111111	[0, 9]	2
2 / 9	0.222222	[0, 4, 2]	3
4 / 9	0.444444	[0, 2, 4]	3
5 / 9	0.555556	[0, 1, 1, 4]	4
7 / 9	0.777778	[0, 1, 3, 2]	4
8 / 9	0.888889	[0, 1, 8]	3
1 / 10	0.100000	[0, 10]	2
3 / 10	0.300000	[0, 3, 3]	3
7 / 10	0.700000	[0, 1, 2, 3]	4
9 / 10	0.900000	[0, 1, 9]	3

```
void main(int argc, char *argv[])
{
    INT a, b, q, r;

    if (argc == 3) {
        a = atol(argv[1]);
```

```
    b = atol(argv[2]);
} else {
    printf("Find the CF representation of a / b\n\n");
    printf("a / b = ");
    scanf("%ld / %ld", &a, &b);
}

printf("\n%ld / %ld = [ ", a, b);

if (b < 0) { a = -a; b = -b; }        /* sign transfer */

while (b > 0)
{
    if (a < 0)                        /* negative number */
    {
        q = a / b - 1;                /* greatest integer */
        r = a % b + b;                /* remainder */
    }
    else
    {
        q = a / b;                    /* greatest integer */
        r = a % b;                    /* remainder */
    }
    a = b;
    b = r;
    printf("%ld%s", q, r ? ", " : " ");
}

    printf("]\n");
}
```

In the case in which the rational number is negative, the first partial quotient b_0 is the greatest negative integer and the remainder is forced positive. This permits the all remaining partial quotients to be positive.

Challenges

1. Modify Program 6_1_1 to create output like that shown in Table 6.1. Increase the range of a and b. Examine the output and identify at least four distinct patterns.

2. Modify Program 6_1_1 to find the rational numbers with strictly increasingly greater numbers of partial quotients. What can you say about these numbers?

6.2 Convergents

It is often productive to look at the behavior of a continued fraction as terms are added. We call

$$[\, b_0, b_1, b_2, b_3, \, ..., b_n \,]$$

for $0 \le n \le N$ the nth convergent to $[\, b_0, b_1, b_2, b_3, \, ..., b_N \,]$. We can look at approximations to a continued fractions by computing its convergents. The nth convergent (also called approximant) f_n is the value obtained by evaluating the continued fraction to the nth convergent:

$$f_0 = [\, b_0 \,]$$

$$f_1 = [\, b_0, b_1 \,]$$

$$...$$

$$f_n = [\, b_0, b_1, b_2, b_3, \, ..., b_n \,].$$

Consider the example from the preceding section: $13{,}020 \, / \, 5797 = [2, 4, 15, 3]$. Evaluating at the convergents to six decimal places we have:

$$f_0 = [2] = 2.000000$$

$$f_1 = [2, 4] = 2.250000$$

$$f_2 = [2, 4, 15] = 2.245902$$

$$f_3 = [2, 4, 15, 3] = 2.245989.$$

with the last being equal to $13{,}020/5{,}797$.

Obviously, each additional convergent adds a level of refinement to the approximation. What is probably not apparent is that as the convergents are added, the approximation alternates between values less than and greater than the rational number.

Theorem 6.2. If f_k is the kth convergent to the continued fraction $[\, b_0, b_1, b_2, b_3, \, ..., b_n \,]$, then

$$f_0 < f_2 < f_4 < ... < f_5 < f_3 < f_1$$

The direct consequence of Theorem 6.2 is that all continued fractions, whether finite or infinite, converge. Put another way,

$$\lim_{n \to \infty} [\, b_0, b_1, b_2, b_3, \, ..., b_n \,] \text{ exists and is finite}$$

Backward Recursive Algorithm for Evaluating Continued Fractions

Although it would appear that the only way to compute a decimal representation of a continued fraction is from the bottom up, this is not the case. Continued fractions can be evaluated by a backward recursive algorithm and a forward recursive algorithm. The backward recursive algorithm is the simplest. Given the finite continued fraction $[b_0, b_1, b_2, b_3, ..., b_n]$, set

$$g_{n+1} = 0$$

and compute

$$g_k = \frac{1}{b_k + g_{k+1}} \; , k = n, n - 1, ..., 1,$$

and finally

$$f_n = b_0 + g_1.$$

Only n division and addition operations are needed for this efficient algorithm. However the intermediate convergents $f_0, f_1, ..., f_{n-1}$ are not obtainable.

A few programming notes: I use the predefined macro BUFSIZ for the character string that holds the input. This is usually defined in STDIO.H as 512. Also, I use the string function strtoul() to parse the values from the input string. Since continued fractions never have a zero-value partial quotient, input ends when a zero is read or any nondigit character is encountered

Program 6_2_1.c. Finds the decimal number represented by a continued fraction using backward recursion

```
/*
** Program 6_2_1 - Find the decimal number represented by
**                 a continued fraction using backward recursion
*/
#include "numtype.h"

void main(int argc, char *argv[])
{
   char cf[BUFSIZ], *cp;
   INT b[BUFSIZ/2], i, k;
   double g, f;

   if (argc > 1) {
      memset(cf, 0, sizeof(cf));
```

```
    for (i = 1; i < argc; i++) strcat(cf, argv[i]);
  } else {
    printf("Find the real number represented by a CF =
      [b0,b1,...,bn]\n\n");
    printf("cf = ");
    if (gets(cf) == NULL) exit(1);
  }

  if ((cp = strchr(cf, '[')) == NULL)   /* find starting [ */
  {
    printf("error: input not in CF format\n");
    exit(2);
  }
  else
    cp++;                               /* next char after [ */

/* read the partial quotients from the input string */

  for (k = 0; k < BUFSIZ/2; k++)
  {
    b[k] = strtoul( cp, &cp, 10);
    if (k > 0 && b[k] == 0) break; else cp++;
  }

/* perform the backward recursion algorithm */

  g = 0.0;
  for (i = k-1; i >= 1; i--)            /* fraction part */
    g = 1.0 / (b[i] + g);

  f = b[0] + g;                         /* add integer part to fraction */

  printf("cf = %.12f\n", f);
}
```

Using the continued fraction [1, 2, 2, 3, 1, 5] as input, we find its real value:

```
C> 6_2_1
Find the real number represented by a CF [b0,b1,...,bn]

cf = [ 1, 2, 2, 3, 1, 5 ]
cf = 1.4094488189
```

Forward Recursive Algorithm for Evaluating Continued Fractions

It is usually more convenient to evaluate continued fractions using the forward recursive algorithm, although it is a more slightly complex procedure. To evaluate the continued fraction from "head-to-tail," we compute the following:

$$p_{-1} = 1 \qquad\qquad q_{-1} = 0$$

$$p_0 = b_0 \qquad\qquad q_0 = 1$$

$$p_1 = b_1 \cdot p_0 + p_{-1} \qquad\qquad q_1 = b_1 \cdot q_0 + q_{-1}$$

$$p_2 = b_2 \cdot p_1 + p_0 \qquad\qquad q_2 = b_2 \cdot q_1 + q_0$$

...

$$p_n = b_n \cdot p_{n-1} + p_{n-2} \qquad\qquad q_n = b_n \cdot q_{n-1} + q_{n-2}$$

or in the (sometimes) more convenient matrix notation

$$\begin{bmatrix} p_n & p_{n-1} \\ q_n & q_{n-1} \end{bmatrix} = \begin{bmatrix} b_0 & 1 \\ 1 & 0 \end{bmatrix} \begin{bmatrix} b_1 & 1 \\ 1 & 0 \end{bmatrix} \cdots \begin{bmatrix} b_n & 1 \\ 1 & 0 \end{bmatrix}.$$

At any time, the convergent f_i can be computed by $f_i = p_i / q_i$.

p_n and q_n themselves have some interesting properties. For example, if $k \geq 1$ then

$$p_{k-1} \cdot p_k - p_k \cdot q_{k-1} = (-1)^k.$$

Also, the pairs (p_k, q_k), (p_{k-1}, p_k), and (q_{k-1}, q_k) are all relatively prime (Shockley, 1967).

In the next program I use the variables pm2 to indicate p_{n-2}, p_{n1} for p_{n-1}, with similar notation for q. In this way the values can be saved for use in the recursive relationship.

Program 6_2_2.c. Finds the decimal number representation by a continued fraction using forward recursion

```
/*
** Program 6_2_2 - Find the decimal number represented by
**                 a continued fraction using forward recursion
*/
#include "numtype.h"

void main(int argc, char *argv[])
{
  char cf[BUFSIZ], *cp;
  INT b[BUFSIZ/2], i, k, p, pm1, pm2, q, qm1, qm2;
  double f;

  if (argc > 1) {
    memset(cf, 0, sizeof(cf));
    for (i = 1; i < argc; i++) strcat(cf, argv[i]);
  } else {
    printf("Find the real number represented by a CF =
      [b0,b1,...,bn]\n\n");
    printf("cf = ");
    if (gets(cf) == NULL) exit(1);
  }
```

```
if ((cp = strchr(cf, '[')) == NULL)   /* find starting [ */
{
  printf("error: input not in CF format\n");
  exit(2);
}
else
  cp++;                          /* next char after [ */

/* read the partial quotients from the input string */

for (k = 0; k < BUFSIZ/2; k++)
{
  b[k] = strtoul( cp, &cp, 10);
  if (k > 0 && b[k] == 0) break; else cp++;
}

/* perform the forward recursion algorithm */

pm2 = 0;                    /* p[-2] (penultimate) */
pm1 = 1;                    /* p[-1] (previous) */
qm2 = 1;                    /* q[-2] (penultimate) */
qm1 = 0;                    /* q[-1] (previous) */
for (i = 0; i < k; i++)
{
  p = b[i] * pm1 + pm2;
  q = b[i] * qm1 + qm2;
  pm2 = pm1;
  pm1 = p;
  qm2 = qm1;
  qm1 = q;
}

f = (double) p / (double) q;

printf("%lu / %lu = %.12f\n", p, q, f);
}
```

Using the same input as in the previous example, we find [1, 2, 2, 3, 1, 5] = 1.409449. Although the real number obtained from Program 6_2_2 is the same as Program 6_2_1, we now know the rational number that the continued fraction represents: 179/127. The last iteration of the forward recursive algorithm yields the rational number's numerator p and denominator q.

```
C> 6_2_2
Find the real number represented by a CF [b0,b1,...,bn]

cf = [ 1, 2, 2, 3, 1, 5 ]
179 / 127 = 1.409448818898
```

Challenges

1. Modify program 6_2_2 to show the convergents each time they are computed. Note how the convergents alternate between being less than and greater than the rational number.

2. Show that the following formula leads to the more general case for continued fractions in which partial numerators are not all 1:

$$\begin{bmatrix} p_n & p_{n-1} \\ q_n & q_{n-1} \end{bmatrix} = \begin{bmatrix} b_0 & 1 \\ 1 & 0 \end{bmatrix} \begin{bmatrix} b_1 & 1 \\ a_1 & 0 \end{bmatrix} \cdots \begin{bmatrix} b_n & 1 \\ a_n & 0 \end{bmatrix}$$

6.3 Infinite Continued Fractions

It is possible to define continued fractions that are infinite sequences of numbers, and this leads to even more interesting results. In fact, any irrational number can be written as an infinite continued fraction and that representation is unique (Shockley, 1967). Furthermore, if c is an infinite continued fraction, then c is irrational.

But what is the basic procedure for obtaining a continued fraction expansion from an irrational number? Given the irrational number ξ or ξ_0, we define

$$b_0 = [\, \xi_0 \,],$$

meaning the least integer. The fractional part of the irrational number $\xi_0 = (\xi_0 - b_0)$ so the next partial quotient is

$$\xi_1 = 1 \, / \, (\xi_0 - b_0)$$

Next let

$$b_1 = [\, \xi_1 \,]$$

and

$$\xi_2 = 1 \, / \, (\xi_1 - b_1)$$

and so on. By induction we arrive at the following definition (Niven and Zuckerman, 1980)

$$b_i = [\, \xi_i \,], \qquad \xi_{i+1} = 1 \, / \, (\xi_i - b_i)$$

where b_i are all integers and ξ_i are all irrational.

An infinite continued fraction is periodic if after some point $b_i = b_{i+km}$ for all $k = 1, 2, \ldots$. The smallest integer m satisfying the relationship is the period. In the compact notation, the periodicity is indicated by writing down the part that is periodic only once and placing a bar above it. For example,

$$[\, b_0, b_1, b_2, b_3, \overline{b_4, b_5, b_6}, \ldots \,].$$

In this case the period $m = 3$.

All quadratic irrationals are infinite periodic continued fractions and have the form

$$\sqrt{d} = [\, b_0, \overline{b_1, b_2, b_3, \dots, b_n, 2b_0}, \dots],$$

where $b_0 = [\sqrt{d}\,]$. Why this is so can be best be seen by example. Let us write $\sqrt{3}$ as a continued fraction:

$$\sqrt{3} = 1 + (\sqrt{3} - 1) = 1 + \cfrac{1}{\cfrac{1}{\sqrt{3} - 1}}$$

$$\frac{1}{\sqrt{3} - 1} = \frac{1 + \sqrt{3}}{2} = \frac{2 + (\sqrt{3} - 1)}{2} = 1 + \cfrac{1}{\cfrac{2}{\sqrt{3} - 1}}$$

$$\frac{2}{\sqrt{3} - 1} = \sqrt{3} + 1 = 2 + (\sqrt{3} - 1) = 2 + \cfrac{1}{\cfrac{1}{\sqrt{3} - 1}}.$$

Because the same denominator appears in the last two equations, we have found the periodic continued fraction [1, 1, 2, 1, 2, 1, 2, ...]. Therefore, $\sqrt{3} = [1, \overline{1, 2}]$. A proof of the general case for irrational numbers can be found in Shockley (1967),

Table 6.2 shows the period length of the continued fractions of the square root of numbers 2 to 25. Like other functions in number theory, the period length exhibits what appears to be random behavior. It has long been pondered what controls the period of the continued fraction expansions; however, few results beyond basic observations are known.

Computing the continued fraction expansion of an irrational quadratic is not a difficult procedure. Theorem 6.3 describes the algorithm.

Theorem 6.3. The complete quotients of the irrational number $\sqrt{d}$ $= [\, b_0, b_1, b_2, b_3, \dots, b_n, \dots]$ are given by the following forward recursion algorithm. Let

$$x_n = \frac{p_n + \sqrt{d}}{q_n}$$

where p_n and q_n are integers such that

$$p_0 = 0, \quad q_0 = 1, \quad p_{n+1} = a_n \cdot q_n - p_n, \quad q_{n+1} = (d - p_{n+1})/q_n, \quad b_n = [\, x_n\,]$$

Table 6.2. Continued Fraction Expansions for $\sqrt{d}$, $2 \leq d \leq 25$

d	Period	Continued fraction expansion for $\sqrt{d}$
2	1	[1, 2, ...]
3	2	[1, 1, 2, ...]
4	0	[2]
5	1	[2, 4, ...]
6	2	[2, 2, 4, ...]
7	4	[2, 1, 1, 1, 4, ...]
8	2	[2, 1, 4, ...]
9	0	[3]
10	1	[3, 6, ...]
11	2	[3, 3, 6, ...]
12	2	[3, 2, 6, ...]
13	5	[3, 1, 1, 1, 1, 6, ...]
14	4	[3, 1, 2, 1, 6, ...]
15	2	[3, 1, 6, ...]
16	0	[4]
17	1	[4, 8, ...]
18	2	[4, 4, 8, ...]
19	6	[4, 2, 1, 3, 1, 2, 8, ...]
20	2	[4, 2, 8, ...]
21	6	[4, 1, 1, 2, 1, 1, 8, ...]
22	6	[4, 1, 2, 4, 2, 1, 8, ...]
23	4	[4, 1, 3, 1, 8, ...]
24	2	[4, 1, 8, ...]
25	0	[5]

Proof for this algorithm can be found in Hardy and Wright (1979) or Niven and Zuckerman (1980).

The following program implements the recursive algorithm for quadratic irrational numbers of the form $\sqrt{d}$. It ends after computing one period of the continued fraction. I use the iSquareRoot() function to compute the integer square root.

Program 6_3_1.c. Finds a continued fraction representation of a quadratic irrational number

```
/*
** Program 6_3_1 - Find a continued fraction representation of
**                 an quadratic irrational number
*/
#include "numtype.h"
```

```c
void main(int argc, char *argv[])
{
  INT b, d, i, k, p, q, r, s;

  printf("CF representation of ûd\n\n");
  if (argc == 2) {
    d = atol(argv[1]);
  } else {
    printf("d = ");
    scanf("%ld", &d);
  }

  b = iSquareRoot(d, &s, &r);
  k = 0;

  printf("\n√%ld = [ %ld", d, b);

  if (r > 0)
  {
    p = 0;
    q = 1;
    do
    {
      p = b * q - p;
      q = (d - p * p) / q;
      k++;
      b = (p + s) / q;
      printf(", %ld", b);
    } while (q != 1);
    printf(", ... ");
  }

  printf("] (period = %lu)\n", k);
}

/*
** iSquareRoot - finds the integer square root of a number
**               by solving: a = q" + r
*/
#include "numtype.h"

NAT iSquareRoot(NAT a, NAT *q, NAT *r)
{
  *q = a;

  if (a > 0)
    while (*q > (*r = a / *q))      /* integer root */
      *q = (*q + *r) >> 1;          /* divided by 2 */

  *r = a - *q * *q;                 /* remainder */

  return(*q);
}
```

Challenge

1. Write a program that will generate a table of continued fraction expansions for $\sqrt{d}$ (similar to Table 6.2) for $1 < \sqrt{d} < 1{,}000$. Find the 10 numbers that have the longest period and try to draw some conclusions about them.

6.4 Continued Fraction Expansions for Special Numbers

Continued fractions representations for well-known constants have long been investigated. When they can be expressed in closed form, they have the ability to provide approximations to irrational numbers within any desired degree of precision. Also the rate of convergence is high so it doesn't take too many iterations to compute the number to any desired degree of precision.

The *golden ratio*, called the precious jewel of geometry by Kepler, is the talisman of the aesthetic beauty of mathematics. The golden ratio, defined as

$$\phi = (1 + \sqrt{5}) / 2,$$

is arrived at as one of two solutions to geometric ratio problem

$$\frac{\phi + 1}{\phi} = \frac{\phi}{1}, \quad \text{i.e.,} \quad \phi^2 - \phi - 1 = 0.$$

It is rationally approximated by successive numbers from the Fibonacci sequence (see Chapter 10). Of continued fractions,

$$\phi = [\, 1, 1, 1, 1, 1, 1, 1, 1, \dots \,]$$

has the distinct honor of being the slowest to converge of all continued fractions, taking 26 iterations to converge to 10 decimal places (see Table 6.3).

An *algebraic number* is a number that satisfies an algebraic equation with integer coefficients. Clearly ϕ is an algebraic number because it satisfies $\phi^2 - \phi - 1 = 0$. Numbers that do not satisfy algebraic equations are called *transcendental numbers*. While the proof that a given number is transcendental is beyond the scope of this book, there are numerous transcendental constants known. Hermite was the first to prove in 1873 that e, the natural logarithm base, is a

Table 6.3. Approximation of ϕ to its Continued Fraction Expansion. ϕ is Rounded in the 10th Decimal Place and is Taken to Be 1.61803 39887

n	p_n	q_n	f_n	$\phi - f_n$
0	1	1	1.0000000000	0.6180339887
1	2	1	2.0000000000	−0.3819660113
2	3	2	1.5000000000	0.1180339887
3	5	3	1.6666666667	−0.0486326779
4	8	5	1.6000000000	0.0180339887
5	13	8	1.6250000000	−0.0069660113
6	21	13	1.6153846154	0.0026493734
7	34	21	1.6190476190	−0.0010136303
8	55	34	1.6176470588	0.0003869299
9	89	55	1.6181818182	−0.0001478294
10	144	89	1.6179775281	0.0000564607
11	233	144	1.6180555556	−0.0000215668
12	377	233	1.6180257511	0.0000082377
13	610	377	1.6180371353	−0.0000031465
14	987	610	1.6180327869	0.0000012019
15	1597	987	1.6180344478	−0.0000004591
16	2584	1597	1.6180338134	0.0000001753
17	4181	2584	1.6180340557	−0.0000000670
18	6765	4181	1.6180339632	0.0000000256
19	10946	6765	1.6180339985	−0.0000000098
20	17711	10946	1.6180339850	0.0000000037
21	28657	17711	1.6180339902	−0.0000000014
22	46368	28657	1.6180339882	0.0000000005
23	75025	46368	1.6180339890	−0.0000000002
24	121393	75025	1.6180339887	0.0000000001
25	196418	121393	1.6180339888	−0.0000000000
26	317811	196418	1.6180339887	0.0000000000

transcendental number. Cantor later proved that almost all real numbers are transcendental (Hardy and Wright, 1979).

With respect to continued fractions, the natural logarithm base e can be represented by

$$e = [\, 2, 1, 2, 1, 1, 4, 1, 1, 6, 1, 1, 8, ...\,]\,.$$

Although ϕ is a periodic continued fraction where the b_n are all 1, e has a recognizable pattern in b_n. Knowing this regularity permits us to compute e to any degree of precision. The continued fraction converges to e to 10 decimal places in 15 steps (see Table 6.4).

Table 6.4. Approximation of e to Its Continued Fraction Expansion. e is Rounded in the 10th Decimal Place and Is Taken to Be 2.7182818285.

n	p_n	q_n	f_n	$e - f_n$
0	2	1	2.0000000000	0.7182818285
1	3	1	3.0000000000	−0.2817181715
2	8	3	2.6666666667	0.0516151618
3	11	4	2.7500000000	−0.0317181715
4	19	7	2.7142857143	0.0039961142
5	87	32	2.7187500000	−0.0004681715
6	106	39	2.7179487179	0.0003331105
7	193	71	2.7183098592	−0.0000280307
8	1,264	465	2.7182795699	0.0000022586
9	1,457	536	2.7182835821	−0.0000017536
10	2,721	1,001	2.7182817183	0.0000001102
11	23,225	8,544	2.7182818352	−0.0000000067
12	25,946	9,545	2.7182818229	0.0000000055
13	49,171	18,089	2.7182818287	−0.0000000003
14	517,656	190,435	2.7182818284	0.0000000001
15	566,827	208,524	2.7182818285	0.0000000000

Unlike ϕ and e, the transcendental π umber ϕ cannot be expressed in a general formula for b_n. This fact leads us to ask "why" and I regret that no satisfactory explanation exists. It does however suggest that there are something like degrees of "transcendentalism," which can be discerned through the lens of continued fractions. Under this method of examination we might say that π then is more transcendental than e.

Even though it cannot be expressed in closed form, the representation for π has been studied for many, many years. As a continued fraction, its rate of convergence is quite fast. In fact,

$$\pi = [\ 3, 7, 15, 1, 292, 1, 1, 1, 2, 1, 3, 1, 14, \ldots]$$

can be computed to 10 decimal places of accuracy in just seven iterations of the forward recursion formula (see Table 6.5).

Table 6.5. Approximation of π to Its Continued Fraction Expansion. π Is Rounded in the 10th Decimal Place and Is Taken to Be 3.1415926536.

n	p_n	q_n	f_n	$e - f_n$
0	3	1	3.0000000000	0.1415926536
1	22	7	3.1428571429	−0.0012644893
2	333	106	3.1415094340	0.0000832196
3	355	113	3.1415929204	−0.0000002668
4	103,993	33,102	3.1415926530	0.0000000006
5	104348	33,215	3.1415926539	−0.0000000003
6	208,341	66,317	3.1415926535	0.0000000001
7	312,689	99,532	3.1415926536	−0.0000000000

Challenge

1. For each of the special numbers presented in this section, compute how many quotients are necessary to achieve five and ten decimal places of accuracy.

6.5 Continued Fraction Expansions for Functions

There are many formulae that have been developed over the years for continued fraction expansions of functions. Easily derived representations for rational numbers include:

$$1 / a = [\, 0, a \,]$$

$$(a - 1) / a = [\, 0, 1, a - 1 \,]$$

$$a / (2a + 1) = [\, 0, 2, a \,].$$

We already know that quadratic irrational numbers can be expressed in closed form, where $a \in Z$. Quadratic irrational numbers having a particular form can be stated even more explicitly:

$$\sqrt{a^2 + 1} = [\, a, 2a, 2a, 2a, \ldots]$$

$$\sqrt{a^2 + 2} = [\, a, a, 2a, a, 2a, a, 2a, \ldots].$$

Many analytic function are known to have continued fraction representations. If we permit ourselves to move away from simple continued fractions and to consider the more general case in which continued fractions are defined for $z \in C$, then a much greater number of functions can be expressed in a closed form.

Euler was able to show that if a convergent series has the form

$$a_1 + a_1 \cdot a_2 + a_1 \cdot a_2 \cdot a_3 + a_1 \cdot a_2 \cdot a_3 \cdot a_4 + \ldots,$$

then it can be expressed in a continued fraction:

$$\cfrac{a_1}{1 - \cfrac{a_2}{(1 + a_2) - \cfrac{a_3}{(1 + a_3) - \cfrac{a_4}{(1 + a_4) - \ldots}}}}$$

This identity gives rise to many continued fraction formulae for functions. For example, the general binomial can be expressed as

$$(1 + z)^n = 1 + \cfrac{(-n)\,z}{1 +} \ \cfrac{1(1+n)\,z}{2 +} \ \cfrac{1(1-n)\,z}{3 +} \ \cfrac{2(2+n)\,z}{4 +} \ \cfrac{2(2-n)\,z}{5 +} \ \ldots$$

where $n \in C$ and $n \neq 0, \pm 1, \pm 2, \ldots$.

The trigonometric functions tangent and arctangent are represented by the following continued fraction expansions:

$$\tan z = \cfrac{z}{1 -} \ \cfrac{z^2}{3 -} \ \cfrac{z^2}{5 -} \ \cfrac{z^2}{7 -} \ \cfrac{z^2}{9 -} \ \ldots$$

$$\arctan z = \cfrac{z}{1 +} \ \cfrac{1^2 z^2}{3 +} \ \cfrac{2^2 z^2}{5 +} \ \cfrac{3^2 z^2}{7 +} \ \cfrac{4^2 z^2}{9 +} \ \ldots$$

Here an interesting case arises. Since $\arctan 1 = \pi / 4$, we can further develop a continued fraction representation o π in which the elements can be expressed in closed form:

$$\pi = \cfrac{4}{1 +} \ \cfrac{1}{3 +} \ \cfrac{2^2}{5 +} \ \cfrac{3^2}{7 +} \ \ldots$$

Knowing a regular formula exists for π compels one to conjecture if transcendental numbers exist that cannot be expressed with a general formula as ordered pairs (a_n, b_n).

The truth is that it is possible to construct a number more transcendental than π; all we need is to make a continued fraction that cannot be expressed in closed form under any circumstances. A simple example I shall call h, is defined by

$$h = 2 + \cfrac{3}{5 +} \cfrac{7}{11 +} \cfrac{13}{17 +} \cfrac{19}{23 +} \cdots,$$

where each integral a_n and b_n are successive prime numbers. Using just integers, and the continued fraction representation as a defining characteristic, we must conclude that h is "more transcendental" than π. In fact, all such numbers defined in a similar manner would have a degree of "transcendentalism" higher than π. Incidentally, $h \approx 2.53602$ $70816\ 89339$.

Forms involving the natural logarithm and e are shown below:

$$e^z = \cfrac{1}{1 -} \cfrac{z}{1 +} \cfrac{z}{2 -} \cfrac{z}{3 +} \cfrac{z}{2 -} \cfrac{z}{5 +} \cfrac{z}{2} - \cdots$$

$$\log(1 + z) = \cfrac{z}{1 +} \cfrac{1^2 z}{2 +} \cfrac{1^2 z}{3 +} \cfrac{2^2 z}{4 +} \cfrac{2^2 z}{5 +} \cfrac{3^2 z}{6 +} \cfrac{3^2 z}{7 +} \cdots$$

Of use in statistics is the error function and its complement. The *error function* erfc z is defined by:

$$\operatorname{erf} z = \frac{2}{\sqrt{\pi}} \int_0^z e^{-t^2 \text{ or } -l^2}\, dt$$

$$= \frac{2 e^{-z^2}}{\sqrt{\pi}} \left(\cfrac{1}{1 -} \cfrac{z^2}{\frac{3}{2} +} \cfrac{1 \cdot z^2}{\frac{5}{2} -} \cfrac{\frac{3}{2} \cdot z^2}{\frac{7}{2} +} \cfrac{2 \cdot z^2}{\frac{9}{2} -} \cfrac{\frac{5}{2} \cdot z^2}{\frac{11}{2} +} \cfrac{3 \cdot z^2}{\frac{13}{2} -} \cfrac{\frac{7}{2} \cdot z^2}{\frac{15}{2} +} \cdots \right)$$

The *complementary error* function erfc z is defined by

$$\operatorname{erfc} z = 1 - \operatorname{erf} z = \frac{2}{\sqrt{\pi}} \int_z^\infty e^{-t^2 \text{ or } -l^2}\, dt$$

$$= \frac{e^{-z^2}}{\sqrt{n}} \left(\cfrac{1}{z +} \cfrac{\frac{1}{2}}{z +} \cfrac{1}{z +} \cfrac{\frac{3}{2}}{z +} \cfrac{2}{z +} \cfrac{\frac{5}{2}}{z +} \cfrac{3}{z +} \cfrac{\frac{7}{2}}{z +} \cdots \right), \text{ for } \operatorname{Re}(z) > 0$$

7

Seeking Prime Numbers

*"There is divinity in odd numbers,
either nativity, chance, or death."*

SHAKESPEARE, *THE MERRY WIVES OF WINDSOR*

7.0 Introduction

It has long been known that some integers have the property that they are evenly divisible by integers less than themselves (and other than 1) whereas other integers are not. This idea is so central to the study of integers that it is called the *fundamental theorem of arithmetic*, which I shall now state.

Fundamental Theorem of Arithmetic (7.1). For each integer $n >$ 1, there exist primes $p_1 \leq p_2 \leq p_3 \leq \ldots \leq p_r$, such that $n = p_1 \cdot p_2 \cdot p_3 \cdot \ldots \cdot p_r$, and this factorization is unique.

An integer that has divisors is called composite. If it does not have a divisor other than 1 and itself, then it is a prime. In this chapter, we will hunt for primes using a variety of techniques. The general principle will be to identify all the composites we can, and those that remain are (one hopes) primes.

It's a curious fact that it is possible to test for a given number for primality without actually factoring it. Although these tests provide answers to the immediate question whether a number is prime or not, the tests say nothing about the factors the number may have. A very frustrating situation indeed.

Which applications should pique our interest in prime numbers and factorization? The most important these days is cryptography. Because it is generally believed that the larger a number is, the longer it takes to factor, cryptographic algorithms based on very large primes are thought to encrypt data securely. Several of these methods will be explored in this chapter.

Are there infinitely many primes? Most people intuitively believe that there are. This question was settled long ago by Euclid, whose analysis is a model of elegance in mathematical thinking.

Theorem 7.2. There exist an infinite number of prime numbers.

To prove this, assume that there are only a finite number of primes and index them: $p_1, p_2, ..., p_n$. Let $M = p_1 \cdot p_2 \cdot ... \cdot p_{n+1}$. Clearly, if we were to divide M by any of the primes, the remainder would be 1. Since none of the primes, from our finite set factor M, we must conclude that either there exists at least one more prime number or that M itself is prime. In either case, we must conclude that our premise, that there are finitely many primes, is false. Therefore, there must be infinitely many primes.

Now that we know that every number has a unique prime factorization and that there are infinitely many primes, we are ready to look at their distribution, density, and methods for finding them.

7.1 The Distribution of the Primes

Although there are an infinite number of primes, we are often interested in the number of primes less than a certain number. This function, usually indicated by $\pi(x)$, is called the prime counting function. It has some interesting properties that we shall investigate.

As proved by Euclid in Theorem 7.2 discussed in the preceding section,

$$\lim_{n \to \infty} \pi(x) = +\infty.$$

Historically, the value of $\pi(x)$ was obtained identifying primes using the sieve of Eratosthenes or the trial division method. However, these methods have the unfortunate problem of being slow, computationally. Obviously approximations to $\pi(x)$ would be useful.

What was conjectured by Gauss and later shown after significant effort by Atle Selberg and Paul Erdös now called the prime number theorem.

Prime Number Theorem (7.3).

$$\lim_{n \to \infty} \frac{\pi(x)}{x / \log(x)} = 1.$$

From this we can conclude that $\pi(x) \approx x / \log(x)$.

As it turns out, a better approximation of $\pi(x)$ can be obtained using the so-called logarithmic integral of x, denoted $Li(x)$:

$$Li(x) = \int_0^x du / \log(u) \cong 1.05 + \log(\log(x)) + \log(x) + \frac{\log(x)^2}{2 \cdot 2!} + \frac{\log(x)^3}{3 \cdot 3!} + \cdots,$$

where $Li(x)$ is evaluated only for the greatest integer part. Compare the actual values of $\pi(x)$ versus the functional estimates:

Table 7.1. Actual Values of $\pi(x)$ Versus Functional Estimates

x	$\pi(x)$	$Li(x)$	$Li(x) - \pi(x)$	Relative error
10	4	6	2	0.5
100	25	30	5	0.20
1,000	168	178	10	0.060
10,000	1,229	1,246	17	0.014
100,000	9,592	9,630	38	0.0040
1,000,000	78,498	78,628	130	0.0017
10,000,000	664,579	664,918	339	0.00051
100,000,000	5,761,455	5,762,209	754	0.00013
1,000,000,000	50,847,534	50,849,235	1,701	0.000033

A fair amount of effort has been expended to refine approximations to $\pi(x)$. Other techniques and refinements have been developed by Meissel (1870, 1985), Lagarias, Miller, and Odlyzko (1985), and Deleglise and Rivat (1996).

The Density of the Primes

Using $Li(x)$, we can estimate the density of the primes over any interval fairly easily. The number of primes in the interval $[a,b] = Li(b) - Li(a)$. A histogram is a useful graphical tool for showing the distribution, or in this case a count, of a single variable over a given range.

Program 7_1_1.c. Creates a histogram showing the distribution of the primes

```
/*
** Program 7_1_1 - Create a histogram showing the distribution of
**                 primes
*/
#include "numtype.h"

void main(void)
{
   NAT   a, b, h, l[20], m;
   INT   i, j;
   float f;
   char  *y = " number of primes ";

   printf("Histogram prime distribution\n\n");
```

```
   printf("interval size = ");
   scanf("%lu", &h);

/* compute the number of primes in each interval using Li() */

   b = 0;
   for (i = 0; i < 20; i++)
   {
     a = b;
     b += h;
     l[i] = Li(b) - Li(a);
   }

/* draw a histogram showing the distribution */

   printf("\33[2J");              /* erase display */

   f = (float) Li(h) / 20.0;   /* marker interval */

   for (j = 20; j >= 0; j--)
   {
     m = f * j;

     if (j > 0)
       printf("%c%6lu: ", *y++, m);
     else
       printf("     ");

     for (i = 0; i < 20; i++)
       if (j > 0)
         printf(" %c", l[i] >= m ? '*' : ' ');
       else
         printf("%3lu", i+1);

     printf("\n");
   }
   printf("\t\t\tinterval size = %-6lu", h);
}

/*
** Li - evaluate the logarithmic integral (prime counting function)
*/
#include "numtype.h"

NAT Li(NAT x)
{
   NAT i, j;
   double a, l, lx;

   if (x < 2) return(0);

   lx = log(x);
   a = log(lx) + 1.05;

   for ( i = 1; i <= 100; i++)
   {
     l = lx / (double) i;
     for ( j = 2; j <= i; j++)
```

```
    l *= lx / (double) j;
   a = a + l;
 }
 i = floor(a);

 return(i);
}
```

In our example, each interval is 2000 numbers. This means that we will estimate the number of primes using the $Li()$ function in the intervals: $0 - 2000$, $2000 - 4000$, $4000 - 6000$, etc. Output is a histogram showing the number of primes expected in each interval:

```
C> 7_1_1
Histogram prime distribution

interval size = 2000

    315:  *
    299:  *
  n 283:  *
  u 267:  *
  m 252:  *
  b 236:  *  *
  e 220:  *  *  *  *  *
  r 204:  *  *  *  *  *  *  *  *  *
    189:  *  *  *  *  *  *  *  *  *  *  *  *  *  *  *  *  *  *  *  *
  o 173:  *  *  *  *  *  *  *  *  *  *  *  *  *  *  *  *  *  *  *  *
  f 157:  *  *  *  *  *  *  *  *  *  *  *  *  *  *  *  *  *  *  *  *
    141:  *  *  *  *  *  *  *  *  *  *  *  *  *  *  *  *  *  *  *  *
  p 126:  *  *  *  *  *  *  *  *  *  *  *  *  *  *  *  *  *  *  *  *
  r 110:  *  *  *  *  *  *  *  *  *  *  *  *  *  *  *  *  *  *  *  *
  i  94:  *  *  *  *  *  *  *  *  *  *  *  *  *  *  *  *  *  *  *  *
  m  78:  *  *  *  *  *  *  *  *  *  *  *  *  *  *  *  *  *  *  *  *
  e  63:  *  *  *  *  *  *  *  *  *  *  *  *  *  *  *  *  *  *  *  *
  s  47:  *  *  *  *  *  *  *  *  *  *  *  *  *  *  *  *  *  *  *  *
     31:  *  *  *  *  *  *  *  *  *  *  *  *  *  *  *  *  *  *  *  *
     15:  *  *  *  *  *  *  *  *  *  *  *  *  *  *  *  *  *  *  *  *
          1  2  3  4  5  6  7  8  9 10 11 12 13 14 15 16 17 18 19 20
                   interval size = 2000
```

What can be seen on this histogram is what we inferred from the prime number theorem; the frequency of prime numbers diminishes as they get larger. More specifically, based on the density of the primes, the probability that a number in the vicinity of x is prime is $1/\log x$.

We can turn this concept around and ask how large would we expect x to be when the average gap between primes is 100. This means we are looking for x where

$$\frac{1}{\log x} = 0.01.$$

This yields $x \cong 2.69 \cdot 10^{43}$, where the average gap between primes is 100.

On the other hand, suppose we want to manufacture a sequence of 99 composites. This is done fairly easily in the sequence

$$\{ x : 100! + 2, \ 100! + 3, \ 100! + 4, \ ..., \ 100! + 100 \ \}$$

The first number is divisible by 2, the second by 3, and so on. However, these numbers are much larger than the previous estimate, $x \cong 9.33 \cdot 10^{157}$. But do we need such numerical extravagance?

In fact, the first gap that contains *exactly* 99 composites occurs between the prime numbers 396,733 and 396,833, where the average gap length is about 13. Also, there is a larger gap of 111 composites between two primes: 370,261 and 370,373. This should remind us to be careful not be misled about the relative randomness of prime numbers, given the monotonic character of the $Li()$ function.

The Sum of the Reciprocals of the Primes

There are many famous infinite series, you may remember seeing this one before:

$$\sum_{n=1}^{\infty} 1/2^n = 1/2 + 1/4 + 1/8 + ... = 1.$$

In this case, the sum converges to the value 1 and is said to be *convergent*. Other series are *divergent*; they do not converge to a finite sum:

$$\sum_{n=1}^{\infty} 1/n = 1 + 1/2 + 1/3 + 1/4 + ... = \infty.$$

What about primes? Does the sum of the reciprocals of the primes converge? Given $M > 1$ and $p_1, p_2, p_3, \ldots, p_n$, are all the primes in the set $\{ 1, 2, 3, ..., M\}$? If we let $M \to \infty$; we know that $n \to \infty$.

$$\sum_{n=1}^{\infty} 1/p_n = 1/2 + 1/3 + 1/5 + 1/7 + 1/11 + ... = \infty.$$

Although not immediately apparent, the answer is no. This conclusion is a consequence of a fact relating to infinite series:

$$\sum_{n=1}^{M} 1/n < \frac{1}{(1 - 1/p_1)(1 - 1/p_2) \ ... \ (1 - 1/p_n)}.$$

We also know that

$$\prod_{n=1}^{M} (1 - 1/p_n) = 0$$

Because either $\prod (1 - a_n)$ and $\Sigma\, a_n$ are both convergent or both divergent $(1 > a_n \geq 0)$, we conclude that $\Sigma\, 1/p_n$ is divergent (Andrews, 1971).

Challenges

1. Having a means to estimate the number of primes gives us a way to estimate how long it would take to compute them. Using the function that evaluates $Li(x)$, estimate how long it would take to factor 30- 50-, and 100-digit prime numbers. Assume that primes can be found at a rate of 108 per second.

2. Using the `Factor()` function presented in Chapter 3, make a table of strictly increasing prime- pair gaps.

7.2 Elementary Sieving Methods

It is an unfortunate fact that in computer science, the algorithms easiest to program are too often inefficient. The traditional factoring methods are easy to implement, but generally are not useful for numbers greater than the standard precision of the computer. The basic method for finding primes in this section is sieving. Sieving is generally used for finding blocks of primes and identifying composites. This goal is different from trial division, discussed in the next section, which is to find prime factors for a specific number.

In the sections that follow, I present a number of variations of similar methodologies. This is done to acquaint you with various programming devices that can be used to both speed algorithms or lower memory requirements. Combining some of these variations would be instructive programming projects to undertake.

The Sieve of Eratosthenes

The sieve of Eratosthenes is the first known systematic means for identifying prime numbers. Although it is not practical for huge numbers, the theory is sound and certainly worth knowing about. To sieve for primes, we begin by writing down all numbers up to the maximum number we wish to test, starting at 2. As we count up, any number that has not been crossed out previously is a prime. Counting up in increments of that number we cross out all of the subsequent multiples.

As we begin the procedure, we will find 2 at once, and then proceed to cross out 4, 6, 8, ... Next we will find 3 and then cross out 6, 9, 12, ... The number 4 has already been crossed out, so it cannot be prime.

The next number not crossed out is 5, so we cross out 10, 15, 20, ... Once we have crossed out all of the multiples of primes less than the square root of the maximum, we have found all the primes.

Here is a list of the numbers from 2 to 25. Since $5 = \sqrt{25}$, we only need to test 2, 3, and 5. Any number that is *not* crossed out after testing 2, 3, and 5 must be prime:

Table 7.2. Sieve of Eratosthenes for $n \le 25$

2	3	4	5	6	7	8	9	10
prime		X		X		X		X
	prime			X			X	
			prime					X
PRIME	PRIME		PRIME		PRIME			

11	12	13	14	15	16	17	18	19	20
	X		X		X		X		X
	X			X			X		
				X					X
PRIME		PRIME				PRIME		PRIME	

21	22	23	24	25
	X		X	
X			X	
				X
PRIME		PRIME		

This technique is relatively easy to program. All that is needed as a means of knowing if a number has been crossed out or not. For this we use an array that will keep track of which numbers are prime and which are not.

Program 7_2_1.c. Find primes using the sieve of Eratosthenes for $n \le$ 10,000

```
/*
** Program 7_2_1 - Find primes < n by Eratosthenes Sieve
**                 (max n = 10,000)
*/
#include "numtype.h"

#define MAXN 10000

void main(int argc, char *argv[])
{
  NAT i, k, n;
  char p[MAXN+1];

  if (argc == 2) {
    n = atol(argv[1]);
  } else {
    printf("Find primes (Eratosthenes sieve)\n\n");
```

```
      printf("n = ");
      scanf("%lu", &n);
  }
  if (n > MAXN) n = MAXN;

  memset(p, 0, sizeof(p));

  for (i = 2; i*i <= n; i++)              /* first mark them out */
      if (!p[i])
          for ( k = i+i; k < n; k += i)
              p[k] = 1;                   /* mark composite */

  for (i = 2; i < n; i++)                 /* now print the primes */
      if ( !p[i])
          printf("%lu\n", i);
}
```

Enter the number below which primes are to be found. The program outputs a listing of primes less than or equal to that number:

```
C> 7_2_1
Find primes (Eratosthenes sieve)

n = 25
2
3
5
7
11
13
17
19
23
```

We can improve the execution speed of the sieve of Eratosthenes a couple of different ways. First, we can consider only odd numbers since only 2 is the only even prime. Testing only odd numbers is an easy trick since it affords an immediate 50% savings on the numbers to be tested.

Second, we do not need to waste a whole byte (i.e., char) to store a true or false value – a single bit will do. Thus, 1 byte holds information about 8 odd numbers (or 16 total numbers). To find the correct byte and bit, we use bit shifting functions. Since each byte holds 16 numbers, to find the byte that contained the bit corresponding to our number, we right shift 4 bit places. Shifting right 4 places is the same as dividing by 16 without the normal overhead associated with general division.

Next, to find the appropriate bit to flag, we mask the remainder (mod 16) and shift right by 1 bit. The remainder gives all numbers, and since we are only interested half of them (i.e., the odd numbers), we can divide by 2 by shifting right 1 bit. If it sounds complicated, it is; but the memory saving is 16-fold! This illustrates a useful programming technique that has many applications beyond sieving.

An example may help make this methodology clear. Consider the following byte array. Each byte holds information for 8 odd numbers. The range for each byte is

```
byte #    0       1       2       3       4       5       6        7     ...
range  [0-15]  [16-31] [32-47] [48-63] [64-79] [80-95] [96-111] [112-127] ...
```

If our number happens to be 107, we compute the byte offset by

$$107 / 16 = 107 >> 4 = 6,$$

or byte number 6. Next, to compute the bit,

$$107 \% 16 = 107 \& 16 = 11 \text{ remainder.}$$

Because we are not interested in even numbers, we correct the remainder

$$11 / 2 = 11 >> 1 = 5,$$

or bit number 5. If we look "microscopically" at byte number 6, we see the following correspondence between the bits and the odd numbers they flag:

```
bit #     0  1   2   3   4   5   6   7
number  [ 97 99 101 103 105 107 109 111 ].
```

The following program uses this methodology to find all primes less than a given number. The maximum number for the program is 1,000,000, although this could certainly be increased. (Note: If you are using an older version of Microsoft C/C++, you will need to compile with the large memory model.)

Program 7_2_2.c. Find primes using the sieve of Eratosthenes for $n \le$ 1,000,000

```c
/*
** Program 7_2_2 - Find primes < n by Eratosthenes Sieve
**                 (max n = 1,000,000)
*/
#include "numtype.h"

#define MAXN 1000000

void main(int argc, char *argv[])
{
  NAT i, k, n, q;
  unsigned int b, c;
```

```
  unsigned char *p;
  unsigned char m[8] = { 0x01, 0x02, 0x04, 0x08, 0x10, 0x20, 0x40,
                         0x80 };

  if (argc == 2) {
    n = atol(argv[1]);
  } else {
    printf("Find primes (Eratosthenes sieve)\n\n");
    printf("n = ");
    scanf("%lu", &n);
  }
  if (n > MAXN) n = MAXN;
/* get memory for this task */

  p = malloc((unsigned int) 62500);
  memset(p, 0, (unsigned int) 62500);

/* loop on each possible prime */

  printf("2\n");                     /* first prime is skipped */

  q = sqrt(n);
  for (i = 3; i <= q; i++, i++)
  {
    c = i >> 4;                      /* byte offset */
    b = (i & 0x0f) >> 1;             /* bit offset */

    if (!(*(p + c) & m[b]))          /* if not composite then prime */
    {
      for (k = i+i+i; k <= n; k += i+i)
      {
        c = k >> 4;                  /* byte offset */
        b = (k & 0x0f) >> 1;         /* bit offset */
        *(p + c) |= m[b];            /* set composite indicator bit
                                        to true */

      }
    }
  }

/* now print the primes */

  for (i = 3; i < n; i++, i++)
  {
    c = i >> 4;                      /* byte offset */
    b = (i & 0x0f) >> 1;             /* bit offset */
    if ( !(*(p + c) & m[b]))
      printf("%lu\n", i);
  }

  free(p);                           /* release memory */
}
```

The output from this program and others in this section is omitted since it is simply a list of primes and is identical to the output from Program 7_2_1.

When looking at elementary prime-generating algorithms, the problem with the sieve of Eratosthenes is more related to storage than to run-time. As a general rule, it is more efficient than trial division methods.

The Segmented Sieve

To reduce the space demands of the traditional sieve of Eratosthenes, Bays and Hudson (1977) developed a method called the segmented sieve. The principle is fairly simple, although the code looks a bit complex. First the primes less than $\sqrt{n}$ are found using the sieve of Eratosthenes. Then the remaining interval over which primes are sought $[\sqrt{n} + 1, n]$ is broken into equal-size intervals holding composite flags for d numbers. Each interval is then sieved separately. The total storage requirement is $(\sqrt{n} + d)$ bits.

Program 7_2_3.c. Find primes using the segmented sieve for $n \le$ 1,000,000

```
/*
** Program 7_2_3 - Find primes < n by segmented sieve
**                 (max n = 1,000,000)
*/
#include "numtype.h"

#define MAXN 1000000
#define SQRTN 1000
#define INTSIZ 100

void main(int argc, char *argv[])
{
  NAT b, e, i, k, n, o, r;
  char p[SQRTN];                 /* primes < sqrt(MAXN) */
  char pi[INTSIZ];               /* primes on interval */

  if (argc == 2) {
    n = atol(argv[1]);
  } else {
    printf("Find primes (segmented sieve)\n\n");
    printf("n = ");
    scanf("%lu", &n);
  }
  if (n > MAXN) n = MAXN;

/* find all the primes less than sqrt(n) */

  memset(p, 0, sizeof(p));

  for (i = 2; i*i <= n; i++)     /* first mark them out */
    if (!p[i])
      for (k = i+i; k*k <= n; k += i)
        p[k] = 1;                /* mark composite */

  for (i = 2; i*i <= n; i++)     /* print the first primes */
```

```
    if ( !p[i])
      printf("%lu\n", i);

/* next find the primes on each interval */

  r = i;                              /* number to check */
  b = i;                              /* interval beginning */
  if ((e = i + INTSIZ) > n) e = n;    /* interval end */

  while (r < n)
  {
    memset(pi, 0, sizeof(pi));

    for ( ; r < e; r++)
      if (!pi[r-b])
        for (i = 2; i*i <= n; i++)     /* check against sieved primes */
          if (!p[i] && !(r % i))       /* if prime and multiple */
          {
            for ( k = r; k < e; k += i)
              pi[k-b] = 1;             /* mark composite */
            break;                     /* next r */
          }

    for (i = b; i < e; i++)            /* print primes on interval */
      if (!pi[i-b])
        printf("%lu\n", i);

    b = e;
    if ((e += INTSIZ) > n) e = n;
  }
}
```

```
C> 7_2_3
Find primes (segmented sieve)

n = 25
2
3
5
7
11
13
17
19
23
```

In the preceding program, I used a character array to hold the composite flag —a dreadful waste of space. This situation could be dramatically improved using the technique for storing single-bit values described in the previous section.

Challenges

1. Modify 7_2_1 so that the sieve of Eratosthenes uses a file for the prime/composite flags instead of an array.

2. Modify Program 7_2_3 so that it will use bit shifting and even-number skipping instead of the wasteful character array.

7.3 Trial Division with Composite Skipping

Although the sieving methods are faster computationally than trial division, they are not generally used to factor numbers. Trial division, on the other hand, is usually used to find prime factors for a given number rather than blocks of primes. It is a prime-seeking strategy from antiquity. Trial division means attempting division with all the numbers less than the square root of the number being tested for primality. If there are no proper divisors, then the number is a prime. In these programs you will notice that division is actually the modulus operation because we are interested only in the remainder, not the quotient.

Obviously, the simplest trial division technique is to trial divide all numbers less than the square root of n until a divisor is found. If a divisor is not found, the number is prime.

Program 7_3_1.c. Identifies primes by trial division

```
/*
** Program 7_3_1 - Identify primes by trial division
*/
#include "numtype.h"

void main(int argc, char *argv[])
{
  NAT i, n;

  if (argc == 2) {
    n = atol(argv[1]);
  } else {
    printf("Identify a prime (trial division)\n\n");
    printf("n = ");
    scanf("%lu", &n);
  }

  for (i = 2; i*i <= n; i++)
    if ((n % i) == 0) break;

  if (i*i > n)
    printf("%lu is prime\n", n);
  else
    printf("%lu is composite\n", n);
}
```

```
C> 7_3_1
Identify a prime (trial division)

n = 19
19 is prime
```

Patterned Skipping

Methods that employ trial division can be improved by patterned skipping, a fact that has long been known. As we observed when we implemented the sieve of Eratosthenes, once we have found a prime, all the multiples of that prime are marked as composite. For finding blocks of primes, if we could skip testing common composites a significant savings can be achieved. For example, if we test only odd numbers greater than 2, we obtain a 50% savings. If we skip numbers divisible by 3 as well, our computational savings jumps to 67%. If we add 5 to our list, then we do not need to test 73% of the numbers.

Skipping even numbers is easy; we only need to increment by 2. However, to skip testing numbers divisible by 3, it is necessary to increment alternately by 2 and 4. Since no number is tested that is divisible by 2 or 3, there's no need to test these divisors. Similarly, the divisors are also alternated using an increment of either 2 or 4. Thus, the numbers 2 and 3 and their multiples are never tested as composites and never tested as divisors.

Program 7_3_2.c. Find primes less than n using trial division and patterned skipping

```
/*
** Program 7_3_2 - Find primes < n by trial division and
**                 patterned skipping
*/
#include "numtype.h"

void main(int argc, char *argv[])
{
  NAT i, j, n, si, sj;

  if (argc == 2) {
    n = atol(argv[1]);
  } else {
    printf("Find primes (trial division)\n\n");
    printf("n = ");
    scanf("%lu", &n);
  }

  printf("2\n3\n");               /* skip first two primes */

  si = 4;                         /* skip for i */

  for (i = 5; i <= n; i+= si)
  {
    sj = 4;                       /* skip for j */

    for (j = 5; j*j <= i; j += sj)
    {
      if ((i % j) == 0)
        break;
```

```
        else
            sj = (sj == 2) ? 4 : 2;
    }

    if (j*j > i) printf("%lu\n", i);

        si = (si == 2) ? 4 : 2;
    }
}
```

```
C> 7_3_2
Find primes (trial division)

n = 25
2
3
5
7
11
13
17
19
23
```

The problem with skipping multiples of primes is that the more one adds to the list of numbers to be skipped, the more complex the skipping pattern becomes. Also, the incremental savings in testing become smaller and smaller with each prime added to the list. The skipping patterns incorporating the first four primes are shown in Table 7.3.

Table 7.3. Period Lengths for Various Composite Skipping Patterns

Multiples to skip	Total savings	Period length	Skipping pattern	Number sequence generated (note initial value)
2	50%	2	+2	3, 5, 7, 9, 11, …
2, 3	67%	6	+2, +4	5, 7, 11, 13, 17, …
2, 3, 5	73%	30	+4, +2, +4, +2, +4, +6, +2, +6	7, 11, 13, 17, 19, 23, 29, 31, 37, …
2, 3, 5, 7	77%	210	+2, +4, +2, +4, +6, +2, +6, +4, +2, +4, +6, +6, +2, +6, +4, +2, +6, +4, +6, +8, +4, +2, +4, +2, +4, +8, +6, +4, +6, +2, +4, +6, +2, +6, +6, +4, +2, +4, +6, +2, +6, +4, +2, +4, +2, +10, +2, +10	11, 13, 17, 19, 23, 29, 31, 37, 41, 43, 47, 53, 59, 61, 67, 71, 73, 79, 83, 89, 97, 101, 103, 107, 109, 113, 121, 127, 131, 137, 139, 143, 149, 151, 157, 163, 167, 169, 173, 179, 181, 187, 191, 193, 197, 199, 209, 211, 221, …

You will no doubt see in Table 7.3 that each skipping pattern has a particular period. Once you have incremented completely through the pattern, the pattern begins again. In fact, each skipping pattern has a period equal to the product of the multiples being skipped.

The possibility of finding primes (or factoring) using trial division is somewhat limited. This derives from the fact that the trial division running time is proportional to (and in fact bounded above by) $O(n \cdot \sqrt{n})$. Any algorithm employing trial division will face this polynomial growth of running time with increasing number size.

One obvious strength of finding primes using trial division is that no memory of previous results is necessary. Although skipping, in a way, incorporates exclusion of composites like sieving, we can save even more if we are willing to expend some memory.

Challenges

1. Modify Program 7_3_1 so that only odd numbers are tested.

2. Modify Program 7_3_2 so that the multiples of 5 are skipped from being tested and skipped as divisors.

Trial Division with History and History Files

If you keep in memory the primes that you found, they can be used to test other numbers by trial division. For example, since we know that 2 and 3 are primes, we can use that fact to discern all the primes less than 10. We could then use the primes less than 10 to find all the primes less than 100. The primes less than 100 can then be used to find primes less than 10,000, and so on.

In the following program, we use an array to hold the primes we have already found. We begin by seeding the array with a single prime: 2. From this humble beginning, we will find all the primes less than the number input.

Program 7_3_3.c. Find primes less than n using a history array

```
/*
** Program 7_3_3 - Find primes < n using history in an array
*/
#include "numtype.h"

#define MAXN 10000
#define NPRIME 1229

void main(int argc, char *argv[])
{
  NAT i, j, k, n, p[NPRIME];

  if (argc == 2) {
    n = atol(argv[1]);
  } else {
```

```
    printf("Find primes (history array)\n\n");
    printf("n = ");
    scanf("%lu", &n);
}
if (n > MAXN) n = MAXN;

p[0] = 2;                       /* first prime */
k = 1;                          /* count number of primes found */

for (i = 3; i <= n; i++)
{
    for (j = 0; p[j]*p[j] < i; j++)
        if ((i % p[j]) == 0) break;
    if (p[j]*p[j] > i)
        p[k++] = i;
    if (k == NPRIME) break;
}

for (j = 0; j < k; j++)     /* print the primes */
    printf("%lu\n", p[j]);
}
```

As noted above, the primes found are held in memory, which presents its own problems. The array that holds the primes must be declared large enough to hold the number of primes less than *n* (see Challenge 4). If you cannot declare an array large enough, you will need to write the primes to a file as you find them. You can then read them back when you need to.

There are other reasons you may want to save the primes you have found in a file. Since you already went to the trouble of finding these primes, you may want to use them to help you in other applications, not just to find other primes. Of course the easiest way to obtain a listing of primes is to use one of the preceding programs and redirect its output to a file. For example, using the program shown above (Program 7_3_3), you can create an ASCII text history file simply by typing the following:

```
C> 7_3_3 1000 > primes.dat
```

Because the program accepts command line arguments instead of prompting, you will not need to edit the output file primes.dat to remove any extraneous text. Reading this text file can be done using C run-time library functions: fopen(), fscanf(), and fclose().

Generally speaking, a heavy penalty is always paid for reading and writing a file repeatedly because of the computer operating system's overhead. This is especially true when converting an ASCII text string into a number. Within the standard precision of the computer, it is always more efficient to work in memory versus performing input and output (I/O) tasks.

To improve throughput when using a history file, you can use binary I/O and at least save the overhead of converting ASCII to binary numbers. Of course you still have to do the I/O using *C* run-time library functions: fopen(), fread(), fwrite(), and fclose(). In the example that follows, a binary history file, "primes.bin," is created while primes are being found. When a new prime is discovered, it is appended to the file:

Program 7_3_4.c. Find primes less than *n* using a binary history file

```c
/*
** Program 7_3_4 - Find primes < n using a binary history file
*/
#include "numtype.h"

void main(int argc, char *argv[])
{
  NAT i, j, n;
  FILE *fp;

  if (argc == 2) {
    n = atol(argv[1]);
  } else {
    printf("Find primes (binary history)\n\n");
    printf("n = ");
    scanf("%lu", &n);
  }

/* create a binary prime history file */

  if ((fp = fopen("primes.bin", "w+b")) == NULL)
  {
    printf("\nError opening output file\n");
    exit(1);
  }

  j = 2;                               /* first prime */
  fwrite(&j, sizeof(NAT), 1, fp);

/* test for primality using the history file and append new primes
*/

  for (i = 3; i <= n; i++, i++)
  {
    fseek(fp, 0L, SEEK_SET);          /* goto BOF */
    do
    {
      if (fread(&j, sizeof(NAT), 1, fp) != 1) break;
      if ((i % j) == 0) break;
    } while (j*j < i);

    if (j*j > i)
    {
      fseek(fp, 0L, SEEK_END);        /* goto EOF */
      fwrite(&i, sizeof(NAT), 1, fp); /* i is prime */
    }
```

```
    }

/* print the contents of the history file and close it */

    fseek(fp, 0L, SEEK_SET);              /* goto beginning of file */

    while (fread(&j, sizeof(NAT), 1, fp) == 1)
      printf( "%lu\n", j);

    fclose(fp);
}
```

Running Program 7_3_4 with $n = 25$ gives the following output to the screen:

```
C> 7_3_4
Find primes (binary history)

n = 25
2
3
5
7
11
13
17
19
23
```

while at the same time creating the file "prime.bin". Because there are 9 primes less than 25, Program 7_3_4 will create a file with these data:

2
3
5
7
11
13
17
19
23

Because we do not know how many primes are in the file, we need to test for the end-of-file condition. Note the while loop conditional where the file contents are printed near the end of Program 7_3_4:

```
    while (fread(&j, sizeof(NAT), 1, fp) == 1)
```

If we cannot read a prime, the file is exhausted. It is true that other conditions may cause an error, but in the context of this program they are far too contrived to bear mentioning.

Challenges

1. Write a program to compute the skip pattern period and the savings when skipping the first five, six, and seven prime numbers.

2. Modify Program 7_3_3 to allocate memory dynamically based on an estimate of the number of primes less than n.

Application: Factoring by Trial Division

Program 3_0_1 used a trial division algorithm for factoring a number. This basic function is enhanced in this application using the patterned skipping algorithm employed in Program 7_3_2. Following the example given in Chapter 3, suppose we wish to factor the number 2,000,000,003. The factors of 2,000,000,003 —(17, 211, 233, and 2393) can be discovered with 232 trial divisions when every possible divisor is tested. Skipping even numbers greater than 2 gives the factors in 117 trial divisions. If we further skip testing multiples of 3, then it takes only 77 attempted division operations.

Program 7_3_5.c. Factoring by trial division and patterned skipping

```
/*
** Program 7_3_5 - Factor by trial division and patterned skipping
*/
#include "numtype.h"

main(int argc, char *argv[])
{
  NAT/ i, n, fk, f[32];

  if (argc == 2) {
    n = atol(argv[1]);
  } else {
    printf("Factor by trial division\n\n");
    printf("n = ");
    scanf("%lu", &n);
  }

  fk = Factor(n, f);

  for (i = 0; i < fk; i++)
    printf("%lu\n", f[i]);
}

/*
** Factor - finds prime factorization by trial division
**          returned values: number of factors, array w/ factors
*/
#include "numtype.h"
```

```
NAT Factor(NAT n, NAT f[])
{
  NAT i, fk, q, si;

  fk = 0;                               /* factor count */

  while (!(n & 1))                      /* factor all 2's */
  {
    f[fk++] = 2;
    n /= 2;
  }

  while (!(n % 3))                      /* factor all 3's */
  {
    f[fk++] = 3;
    n /= 3;
  }

  si = 4;                               /* skip for i */

  q = sqrt(n);
  for (i = 5; i <= q; i += si)
  {
    while ( (n % i) == 0L)
    {
      f[fk++] = i;
      n /= i;
      q = sqrt(n);
    }
    si = (si == 2) ? 4 : 2;
  }

  if (n > 1L) f[fk++] = n;

  for (i = fk; i < 32; i++) f[i] = 0;   /* zero out the rest */

  return(fk);
}
```

```
C> 7_3_5
Factor by trial division

n = 2000000003
17
211
233
2393
```

Application: Public-Key Data Encryption

A tremendous amount of technical literature exists on the subject of public-key encryption (Rivest, Shamir, and Adleman, 1978, 1979; Schneier, 1996) . The first of these is the well-known Rivest-Shamir Adelman (RSA) cryptosystem (named after the developers). It is based on the *assumption* that it is extremely difficult to factor large numbers; the larger the number, the greater the difficulty. As we have seen so far, there is good empirical evidence to support this conjecture.

The RSA cryptosystem employs a fairly simple algorithm. The first step is to find a pair of relatively large (100 to 200 decimal digits, possibly larger) prime numbers. Establishing the primality of large numbers without factoring them is another task we will discuss. Assume for the moment that we have a pair of primes, p and q, roughly equal in size. Next find r such that

$$r = p \cdot q.$$

Also needed is an encrypting exponent e, which is chosen at random so that $gcd(e, \phi(r)) = 1$. $\phi(r)$ is Euler's totient function since p and q are prime $\phi(r) = (p - 1) \cdot (q - 1)$. What you, the person who wishes to receive secret messages, publish is r and e. Anybody who wishes to send you a encoded message can do so.

But what about an eavesdropper? Can he decrypt your message from knowledge of r and e. The answer is qualified no – unless factoring very large numbers is easier than everybody in the world suspects. This is what is meant by a trap-door or one-way function. It is easy to go through in the forward direction but impossible (or nearly so) to go backward.

To recover the original message, a decrypting exponent is needed, called d. The decryption exponent must satisfy the following congruence:

$$d \cdot e \equiv 1 \ (\text{mod} \ \phi(r)).$$

ComputeInverse () d can be computed two different ways: (1) by using the function developed in Chapter 5, or (2) by evaluating $d \equiv e^{\phi(\phi(r))-1}$ (mod $\phi(r)$). Of course if e is relatively prime to r, then d will be as well. Also, because of the symmetrical nature of the algorithm, you can encrypt, with d and decrypt with e. At this point, the knowledge of p and q can be destroyed.

Given a message M, compute the ciphertext C by

$$M^e \equiv C \ (\text{mod} \ r)$$

and decrypt ciphertext through

$$C^d \equiv M \ (\text{mod} \ r).$$

Look at this simple example. Let $p = 23$ and $q = 43$, then

$$r = p \cdot q = 989$$

and

$$\phi(r) = (p - 1) \cdot (q - 1) = 924.$$

Now we choose e at random such that $gcd(e, \phi(r)) = 1$; 47 will do nicely. Next we compute the decryption exponent d (using Program 3_3_4) from the congruence

$$47 \cdot d \equiv 1 \; (\text{mod } 924),$$

which yields

$$d = 59.$$

Having derived r, d, and e, we enter 007 mode and start encrypting and decrypting secret messages. Given a message $M = 112101097099101$, break it into blocks smaller than r, say three digits per block:

$$M_1 = 112$$

$$M_2 = 101$$

$$M_3 = 097$$

$$M_4 = 099$$

$$M_5 = 101$$

Apply the encryption exponent e and encryption modulus r to each block to get the encrypted text

$$C_1 = 112^{47} \; (\text{mod } 989) \equiv 433$$

$$C_2 = 101^{47} \; (\text{mod } 989) \equiv 683$$

$$C_3 = 097^{47} \; (\text{mod } 989) \equiv 102$$

$$C_4 = 099^{47} \; (\text{mod } 989) \equiv 504$$

$$C_5 = 101^{47} \; (\text{mod } 989) \equiv 683$$

So the ciphertext message is $C = 433683102504683$. To decrypt the message, we perform the complementary operation using the super-secret decryption exponent $d = 59$:

$$M_1 = 433^{59} \; (\text{mod } 989) \equiv 112$$

$$M_2 = 683^{59} \; (\text{mod } 989) \equiv 101$$

$$M_3 = 102^{59} \; (\text{mod } 989) \equiv 097$$

$M_4 = 504^{59} \pmod{989} \equiv 099$

$M_5 = 683^{59} \pmod{989} \equiv 101$

If you use the PowMod and MulMod functions (developed in Chapter 3), the public-key encryption algorithm presented here is easy to implement, although for a serious implementation you would use large integers. Also, you would want to consider a larger block size to go with your larger r.

It would not be a good practice to have one program to both encode and decode messages because of the need to keep the value of d secret. Here I present them as two separate programs. To encrypt data use Program 7_3_6:

Program 7_3_6.c. Perform public key encryption

```
/*
** Program 7_3_6 - An RSA encryption program for text files
*/
#include "numtype.h"

void main(int argc, char *argv[])
{
  NAT c, m, e = 47, r = 989;    /* encryption exponent, modulus */
  char M[128];
  FILE *fpi, *fpo;
  int i;

/* check input, open files */

  printf("RSA encryption\n");
  if (argc != 3) {
    printf("Usage: 7_3_6 input_file output_file\n");
    exit(1);
  }
  if ((fpi = fopen(argv[1], "r")) == NULL) exit(2);
  if ((fpo = fopen(argv[2], "w+")) == NULL) exit(3);

/* process the plaintext file */

  while ( fgets(M, sizeof(M), fpi) != NULL )
    for (i = 0; i < strlen(M); i++)
      fprintf(fpo, "%03lu", PowMod((NAT) M[i], e, r));

  fclose(fpi);
  fclose(fpo);
}

/*
** PowMod - computes r for a^n ≡ r (mod m) given a, n, and m
*/
#include "numtype.h"

NAT PowMod(NAT a, NAT n, NAT m)
{
```

```
  NAT r;

  r = 1;
  while (n > 0)
  {
    if (n & 1)                    /* test lowest bit */
      r = MulMod(r, a, m);        /* multiply (mod m) */
    a = MulMod(a, a, m);          /* square */
    n >>= 1;                      /* divided by 2 */
  }

  return(r);
}


/*
** MulMod - computes r for a * b ≡ r (mod m) given a, b, and m
*/
#include "numtype.h"

NAT MulMod(NAT a, NAT b, NAT m)
{
  NAT r;

  if (m == 0) return(a * b);      /* (mod 0) */

  r = 0;
  while (a > 0)
  {
    if (a & 1)                    /* test lowest bit */
      if ((r += b) > m) r %= m;   /* add (mod m) */
    a >>= 1;                      /* divided by 2 */
    if ((b <<= 1) > m) b %= m;    /* times 2 (mod m) */
  }

  return(r);
}
```

Program 7_3_6 takes two command-line arguments. The first is the input file and the second is the output file. In this example, the input file —that is, the plaintext—, is called "PLAIN.TXT" and the output file or ciphertext is called "CIPHER.TXT".

```
C> TYPE PLAIN.TXT
Pumpkin Pie

C> 7_3_6 PLAIN.TXT CIPHER.TXT

C> TYPE CIPHER.TXT
48035306043347728801384448028868385
```

Decrypt data using Program 7_3_7:

Program 7_3_7.c. Perform public key decryption of data encrypted with Program 7_3_7

```
/*
** Program 7_3_7 - An RSA decryption program for text files
*/
#include "numtype.h"

void main(int argc, char *argv[])
{
  NAT c, d = 59, r = 989;    /* decryption exponent, modulus */
  char C[384];
  FILE *fpi, *fpo;
  int i;
/* check input, open files */
  printf("RSA decryption\n");
  if (argc != 3) {
    printf("Usage: 7_3_7 input_file output_file\n");
    exit(1);
  }
  if ((fpi = fopen(argv[1], "r")) == NULL) exit(2);
  if ((fpo = fopen(argv[2], "w+")) == NULL) exit(3);
/* process the ciphertext file */
  while ( fgets(C, sizeof(C), fpi) != NULL )
    for (i = 0; i < strlen(C); i+=3)
    {
      sscanf(&C[i], "%03lu", &c);
      fprintf(fpo, "%c", (char) PowMod(c, d, r));
    }

  fclose(fpi);
  fclose(fpo);
}

/*
** PowMod - computes r for a^n ≡ r (mod m) given a, n, and m
*/
#include "numtype.h"

NAT PowMod(a, n, m)
NAT a, n, m;
{
  NAT r;

  r = 1;
  while (n > 0)
  {
    if (n & 1)                   /* test lowest bit */
      r = MulMod(r, a, m);       /* multiply (mod m) */
    a = MulMod(a, a, m);         /* square */
    n >>= 1;                     /* divided by 2 */
  }

  return(r);
}

/*
```

```
** MulMod - computes r for a * b ≡ r (mod m) given a, b, and m
*/
#include "numtype.h"

NAT MulMod(a, b, m)
NAT a, b, m;
{
  NAT r;

  if (m == 0) return(a * b);        /* (mod 0) */

  r = 0;
  while (a > 0)
  {
    if (a & 1)                      /* test lowest bit */
      if ((r += b) > m) r %= m;     /* add (mod m) */
    a >>= 1;                        /* divided by 2 */
    if ((b <<= 1) > m) b %= m;      /* times 2 (mod m) */
  }

  return(r);
}
```

```
C> TYPE CIPHER.TXT
4803530604334772880138444802886883885

C> 7_3_7 CIPHER.TXT PLAIN.TXT

C> TYPE PLAIN.TXT
Pumpkin Pie
```

7.4 A Famous Formula for Primes

It takes no effort whatsoever to produce an infinite sequence of numbers that are composite and nothing but composite. However, it is (seemingly) impossible produce an infinite sequence of primes and nothing but primes, although many have tried. A sequence is a function that, when we put in positive integers (zero sometimes being included), out will pop numbers having a desired attribute.

An interesting "phenomenon" was observed in 1963 by Stanislaw Ulam, a brilliant Los Alamos mathematician, while reportedly doodling. As recounted by Paul Hoffman, in writing down integers in a square spiral that radiates outward, many of the prime numbers are observed to fall on diagonals. What does this square spiral look like?

We can recreate his now famous doodles, called the Ulam spiral, using the next program. This program is a text-graphic program that uses American National Standards Institute (ANSI) standard escape sequences to erase the screen and locate printed text on the screen. If you are not familiar with them, they are a simple standardized system for erasing the screen, controlling cursor movements, and setting text

attributes. To use them on a Personal Computer (PC), you will have to
have the device driver ANSI.SYS (or an equivalent) installed. You may
wish to look over the format of ANSI standard escape sequences
explained in Appendix C while reviewing the next program.

With that said, here is the program that generates the Ulam spiral
of integers:

Program 7_4_1.c. Plot a Ulam spiral using text

```
/*
** Program 7_4_1 - Plot a Ulam spiral text graph of prime numbers using
**                 ANSI escape sequences to locate the text and set colors
*/
#include "numtype.h"

main()
{
  NAT        i, n, m, p, q;
  short int  k, l, x, y, dx, dy;

              /* initialize graphics, compute center of the graph */

  x = 10;                         /* x center */
  y = 13;                         /* y center */
  dx = 0;                         /* x increment */
  dy = 1;                         /* y location */
  l = 0;                          /* initial spiral level */
  m = 400;                        /* total primes */
  printf("\33[2J");               /* clear screen */

/* create spiral text graphics function */

  for (n = 0; n < m; n++)
  {
    p = n + 2;                    /* primes */
    k = 7;                        /* 7 = reverse video */
    if (p & 1)
    {
      q = sqrt(p);
      for (i = 3; i <= q; i++, i++)
        if ((p % i) == 0) { k = 0; break; }
    }
    else
      if (p > 2) k = 0;           /* 0 = normal */
/* compute pixel location for the spiral */

  if (dx == 0 && dy == 0)
    dx = dy = ++l;

  if (dx > 0)
  {
    if (l & 1) x++; else x--;
    dx--;
  }
  else
  {
```

```
    if (1 & 1) y++; else y--;
    dy--;
}

/*
** print the text: white text on black background is composite (normal)
**                 black text on white background is prime (reverse video)
*/

    printf("\33[%dm\33[%d;%dH%4ld", k, y, 4*x -3, p);
}
while(!kbhit()) ;
}
```

The spiral starts with the number 2 in its very center and winds its way outward in a counterclockwise manner. Because of screen space limitations, only the first 400 numbers in the sequence are displayed. Primes are shown in reverse video (black text on a white background), and composites are printed with normal video attributes (white text on a black background) (Figure 7.1):

Figure 7.1. Ulam spiral for the first 400 numbers.

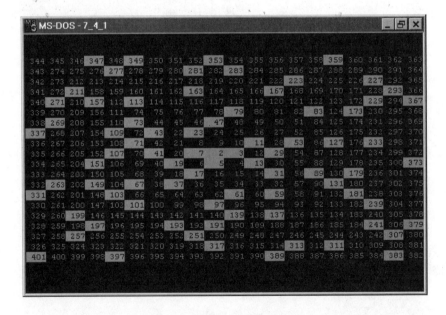

There are some significant strings of prime numbers that can be observed along diagonals. The two longest runs displayed are five primes long: 3, 5, 11, 17, 37 and 17, 19, 41, 71, 109. This occurrence might make one think that longer runs might be possible.

To see more numbers, we'll resort to using a graphics program similar to the one listed above. However, instead of printing text, we will now plot a single pixel. Using only a single pixel will allow us to display 40,000 numbers. Analogously, we will plot a white pixel when we have a prime and a black pixel when we have a composite.

To perform the required graphics, I will use the basic Microsoft C/C++ graphics library. In this book I used the so-called medium-resolution graphics mode (320 × 200 pixels) because the pixels are relative large and therefore more visible. You may get a different look based on the graphics capability of your video adapter.

Program 7_4_2 is a relatively simple graphics program for plotting an Ulam spiral using several of the Microsoft C/C++ run-time library graphics routines. The functions used are: _setvideomode, _setcolor, and _setpixel. Please refer to Appendix B for a description of these graphics functions and for information on their use. Also, you may want to consult a runtime library reference manual for more detailed information regarding the various graphics commands.

Program 7_4_2.c. Plot an Ulam spiral using pixels

```
/*
** Program 7_4_2 - Plot a Ulam spiral pixel graph of prime numbers
**                 using Microsoft C graphics functions
*/
#include "numtype.h"

void main(void)
{
  struct videoconfig vc;
  NAT        i, n, m, p, q, np;
  short int  k, l, x, y, dx, dy;

/* initialize graphics, compute center of the graph */

  _setvideomode( _MRESNOCOLOR );
  _getvideoconfig( &vc);

  x = vc.numxpixels / 2;          /* x center */
  y = vc.numypixels / 2;          /* y center */

  dx = 0;                         /* x increment */
  dy = 1;                         /* y location */
  l = 0;                          /* initial spiral level */
  m = 40000;                      /* total primes */
  np = 0;

/* create spiral graphic function */
  for (n = 0; n < m; n++)
  {
    p = n + 2;                    /* primes */
    /* p = n * (n - 79) + 1601; /* modified Euler's polynomial */
```

```
k = 7;        /* 7 = white */
if (p & 1)
{
  q = sqrt(p);
  for (i = 3; i <= q; i++, i++)
    if ((p % i) == 0) { k = 0; break; }
    if (i > q) np++;
}
else
  if (p > 2) k = 0;    /* 0 = black */

/* compute pixel location for the spiral */

  if (dx == 0 && dy == 0)
    dx = dy = ++1;

  if (dx > 0)
  {
    if (1 & 1) x++; else x--;
    dx--;
  }
  else
  {
    if (1 & 1) y++; else y--;
    dy--;
  }

/* set the pixel: black is composite, white is prime */

  _setcolor(k);
  _setpixel(x,y);

  if (kbhit()) break;
}

while(!kbhit());
_setvideomode( _DEFAULTMODE);

printf("%lu / %lu\n", np, m);
}
```

In this display, white means prime and black means composite. Of the first 40,000 numbers (n = 2, 3, ...), 4,203 (11%) are prime (Figure 7.2).

Is there any way to explicitly generate primes? Consider the fact that 25% of the first 100 numbers are prime. If we look only at the first 100 odd numbers, we would see that 45% of these are prime. This observation can be written using a slightly modified $\pi(n)$ notation as:

$$\overset{99}{\underset{n=0}{\pi}} (n) < \overset{99}{\underset{n=0}{\pi}} (2n+1)$$

In our new notation, we are counting the number of primes that are generated in the sequences { n } and { $2n + 1$ }, for n = 0, 1, 2, ..., 99.

Figure 7.2. Ulan spiral for the first 40,000 numbers.

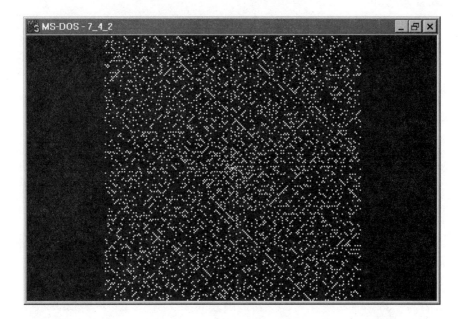

As another example, a sequence of strictly composite numbers could be produced from

$$\{ a_n \} = \{ 2n + 4 \},$$

where $n = 0, 1, 2, 3, ...$ Here we have a sequence of even numbers greater than 2: $a_n = 4, 6, 8, ...$ Of course, in this case

$$\prod_{n=0}^{99} (2n+4) = 0.$$

For quite a long time, it was believed that such a sequence generator for primes could be found. One of the better attempts to create a sequence of prime numbers was put forward by Euler in his prime-generating quadratic polynomial:

$$\{ a_n \} = \{ n^2 + n + 41 \}.$$

Euler's polynomial will produce an uninterrupted sequence of 80 primes for $n = -40, -39, ..., 39$. Modifying the quadratic's range so that $n = 0, 1, 2, ...,$

$$\{ a_n \} = \{ n^2 - 79 \cdot n + 1601 \}.$$

The first 100 values of this function show how well Euler's polynomial produces primes:

0	1601	prime	33	83	prime	66	743	prime
1	1523	prime	34	71	prime	67	797	prime
2	1447	prime	35	61	prime	68	853	prime
3	1373	prime	36	53	prime	69	911	prime
4	1301	prime	37	47	prime	70	971	prime
5	1231	prime	38	43	prime	71	1033	prime
6	1163	prime	39	41	prime	72	1097	prime
7	1097	prime	40	41	prime	73	1163	prime
8	1033	prime	41	43	prime	74	1231	prime
9	971	prime	42	47	prime	75	1301	prime
10	911	prime	43	53	prime	76	1373	prime
11	853	prime	44	61	prime	77	1447	prime
12	797	prime	45	71	prime	78	1523	prime
13	743	prime	46	83	prime	79	1601	prime
14	691	prime	47	97	prime	80	1681	composite
15	641	prime	48	113	prime	81	1763	composite
16	593	prime	49	131	prime	82	1847	prime
17	547	prime	50	151	prime	83	1933	prime
18	503	prime	51	173	prime	84	2021	composite
19	461	prime	52	197	prime	85	2111	prime
20	421	prime	53	223	prime	86	2203	prime
21	383	prime	54	251	prime	87	2297	prime
22	347	prime	55	281	prime	88	2393	prime
23	313	prime	56	313	prime	89	2491	composite
24	281	prime	57	347	prime	90	2591	prime
25	251	prime	58	383	prime	91	2693	prime
26	223	prime	59	421	prime	92	2797	prime
27	197	prime	60	461	prime	93	2903	prime
28	173	prime	61	503	prime	94	3011	prime
29	151	prime	62	547	prime	95	3121	prime
30	131	prime	63	593	prime	96	3233	composite
31	113	prime	64	641	prime	97	3347	prime
32	97	prime	65	691	prime	98	3463	prime
						99	3581	prime

As you can see, the polynomial generates nothing but primes for the first 80 entries ($n = 0, ..., 79$), and for $n < 100$, 95% are primes. The formula fails for the first time at $n = 80$: that is, $1{,}681 = 41^2$.

Recalling our original polynomial before we modified the range, when $n = 41$, we can say that Euler's formula fails catastrophically. Clearly any "prime-generating" polynomial will always generate a composite when n is equal to the constant term. More formally, we can see why it is not possible for a single polynomial to generate only primes. Suppose we had a prime-generating polynomial of the form

$$G(x) = a_0 + a_1 \cdot x + a_2 \cdot x^2 + ... + a_n \cdot x^n.$$

If we let $G(b) = p$, what we need to show is that p divides $G(b + m \cdot p)$.

$$G(b + m \cdot p) = a_0 + a_1 \cdot (b + m \cdot p) + a_2 \cdot (b + m \cdot p)^2 + \ldots + a_n \cdot (b + m \cdot p)^n.$$

Expanding the terms in parentheses and isolating the first term from each expansion, we get

$$G(b + m \cdot p) = [\, a_0 + a_1 \cdot b + a_2 \cdot b^2 + \ldots + a_n \cdot b^n \,]$$

$$+ [\text{ many terms that all contain } p \text{ as a factor }],$$

but the first bracketed term is nothing more than $G(b) = p$. Since p divides the first bracketed term and p divides the second bracketed term, we can conclude that p is a factor of $G(b + m \cdot p)$, thus contradicting our initial assumption that all $G(x)$ are prime.

Even though a single polynomial that generates only primes may not exist, we cannot resist asking if there is a polynomial that will generate nothing but primes over a given interval. If so, then one can imagine, if not create, an extended family of piecewise continuous quadratics that would generate nothing but primes.

Returning to Euler's (modified) polynomial, compare the density of primes shown above to the density of the primes found using Euler's formula. Program 7_4_2 can be modified by uncommenting the line that computes the value of p for the modified Euler's function. For the first 40,000 numbers generated, 14,065 (35%) are prime (Figure 7.3).

Figure 7.3. Ulam spiral on the modified Euler's prime generating function.

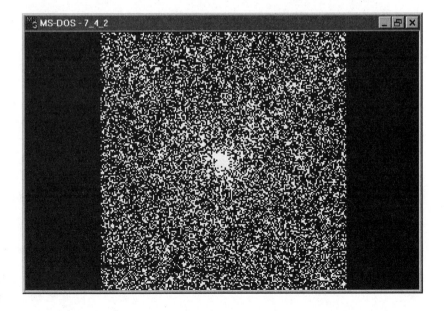

Evaluating expressions having the form { $an^2 + bn + c$ }, for $0 < a \le 100$, $0 < b \le 100$, and $0 < c \le 100$ reveals how efficiently each of the quadratics generate primes. Since this represents distinct 1,000,000 polynomials, only those that generated more than 50% primes for $0 \le n \le 1000$ are listed:

a	b	c	Primes	Composites	Percent
2	40	1	592	408	59.2
2	44	43	591	409	59.1
2	48	89	591	409	59.1
1	1	41	581	419	58.1
1	3	43	581	419	58.1
1	5	47	580	420	58.0
1	7	53	579	421	57.9
1	13	83	578	422	57.8
1	9	61	578	422	57.8
1	11	71	578	422	57.8
1	15	97	577	423	57.7
4	84	43	562	438	56.2
8	88	43	551	449	55.1
8	80	1	539	461	53.9
8	96	89	538	462	53.8
3	51	17	538	462	53.8
3	57	71	538	462	53.8
1	51	31	533	467	53.3
1	53	83	532	468	53.2
1	97	59	530	470	53.0
2	40	19	519	481	51.9
2	44	61	519	481	51.9
1	61	47	518	482	51.8
4	10	47	515	485	51.5
4	18	61	515	485	51.5
4	2	41	515	485	51.5
4	26	83	515	485	51.5
6	30	67	509	491	50.9
6	18	43	509	491	50.9
6	6	31	509	491	50.9
12	0	59	508	492	50.8
12	24	71	507	493	50.7
4	6	43	506	494	50.6
4	30	97	505	495	50.5
4	14	53	505	495	50.5
4	22	71	505	495	50.5

Challenges

1. All odd primes must be of the form $4n+1$ or $4n-1$. Show that the sequence produced by $4n-1$ for $n = 0, 1, 2, \ldots$ contains an infinite number of primes.

2. For the function _setvideomode Program 7_4_2 uses the parameter _MRESNOCOLOR, a video mode from the "Stone Age." If you want to increase the number of numbers plotted and also take better advantage of your video card's capabilities, modify the program to use _MAXRESMODE instead.

3. Try different formulas for generating primes — watch out for overflow conditions.

4. Modify Program 7_4_2 to plot pixels in different colors or gray scale, depending on the number of divisors.

8

Advanced Factoring

*A problem left to itself dries up or goes rotten. But fertilize
a problem with a solution –you'll hatch out dozens.*

N. F. SIMPSON, *A RESOUNDING TINKLE*

8.0 Introduction

If you think of the elementary factoring methods as being a method
where factors are sought based on linear multiples of numbers, the
more advanced methods are related to powers of numbers and, in particular, quadratics. It is important to note at this point that factoring
is not the same as finding primes, although clearly the two are related. As we have seen in the preceding chapter, small numbers (read
long int's) can be completely tested fairly easily by trial division. Also,
as noted in Chapter 3, having a complete factorization is especially
useful in evaluating some number-theoretic functions as well.
However, for larger numbers, and in particular multiple-precision
integers, different strategies are called for.

The two factoring algorithms in widest use today (for large numbers) are based on Pomerance's quadratic sieve factoring method
(abbreviated QS) and Lenstra's elliptic curve factoring method (abbreviated EC). These represent two distinctly different types of methods,
the former has been called "combinations of congruences," whereas
the latter has been called "groups of smooth order." Other algorithms
worth noting are the continued fraction method and the (general)
number field sieve. Like the QS and EC, each of these last two techniques mentioned have shown success with composite numbers having a distinctive format to their factorization.

8.1 Fermat's Factoring Method

The next factorization method was developed by Pierre de Fermat (1601–1665), the French mathematician and lawyer. His method was never actually published but was discovered in a letter to the Franciscan friar Mersenne (1588–1648), after whom Mersenne numbers are named (Ore, 1948). Fermat's factorization method rests on the fact that if n is any odd number, then it can be written as

$$n = x^2 - y^2,$$

the difference of two squares. The obvious factorization is

$$n = (x + y) \cdot (x - y).$$

There is always a trivial solution to this equation whenever $y = x - 1$. This follows from the fact that the difference between any two successive squares is an odd number. For this reason, *any* odd number, including the primes, can be obtained given a particular x. However, when n is a composite odd number a non trivial solution can be found for some $y < x - 1$.

To find an appropriate x and y, we rewrite the preceding expression as

$$y^2 = x^2 - n,$$

where x is an integer greater than or equal to $\sqrt{n}$. Unless n is a perfect square, x^2 will always be greater than n and therefore $y^2 > 0$.

Fermat's procedure begins with the initial value $x = [\sqrt{n}]$. If $x^2 = n$, then we are done. From this point on, we compute increment x and compute $y^2 = x^2 - n$ until y^2 is the square of an integer. Once that square has been discovered, a factor has been found. When the number is a square, its last two digits must be one of only 22 values (i.e., y^2 mod 100): 00, 01, 04, 09, 16, 21, 24, 25, 29, 36, 41, 44, 49, 56, 61, 64, 69, 76, 81, 84, 89, 96. Fermat knew this fact and used it to speed computations by hand. It is worth noting that a number could have any of these last two digits and still not be a square. Even with this limitation, eliminating nonsquares reduces the field of candidates by 78%.

Looking at only the last two digits, Fermat determined if the number $x^2 - n$ was *possibly* a perfect square. If it was a possible square, then he would continue the computation. Consider the following example, where $n = 2,027,651,281$, a number that Fermat himself applied his method to (Table 8.1).

In Table 8.1, the last column indicates whether the last two digits rule out y^2 completely or whether it needs to be tested as a square. The

Table 8.1. Factoring 2,027,651,281 Using Fermat's Method

x	$y^2 = x^2 - n$	Last two digits	Is y^2 a perfect square?
45,030	49,619	19	Not square
45,031	139,680	80	Not square
45,032	229,743	43	Not square
45,033	319,808	08	Not square
45,034	409,875	75	Not square
45,035	499,944	44	Possible square
45,036	590,015	15	Not square
45,037	680,088	88	Not square
45,038	770,163	63	Not square
45,039	860,240	40	Not square
45,040	950,319	19	Not square
45,041	1,040,400	00	Square

last entry, 1,040,400, is the square of the integer 1020. Using $x =$ 45,041 and $y = 1020$ yields the factorization

$$n = (45{,}041 + 1{,}020) * (45{,}041 - 1{,}020) = 46{,}061 * 44{,}021.$$

The program that follows implements Fermat's method. To make it more efficient, it takes advantage of the fact that the difference between two successive squares, x and $x + 1$, is $2 \cdot x + 1$. Obviously for factoring, this procedure works best when n is the product of two numbers that are relatively equal in magnitude. If y is small, then the solution will be arrived at relatively soon.

Program 8_1_1.c. Find factors of composites using Fermat's method

```
/*
** Program 8_1_1.c — Find factors of a number using Fermat's Method,
**                    where n = (x + y) * (x - y)
*/
#include "numtype.h"

void main(int argc, char *argv[])
{
  NAT n, x, y, xx, yy;

  if (argc == 2) {
    n = atol(argv[1]);
  } else {
    printf("Find x, y, such that n = x^2 - y^2\n\n");
    printf("n = ");
    scanf("%ld", &n);
  }
```

```
x = sqrt(n);
yy = x * x - n;
if (yy == 0)
{
  printf("%lu = %lu²\n", n, x);
  exit(0);
}

while (x < n)
{
  yy += x + x + 1;                /* y squared */
  x++;                           /* increment x */
  y = sqrt(yy);
  if (yy == y*y)
  {
      if (y == x - 1) break;  /* trivial solution */
      printf("%lu = (%lu + %lu) * (%lu - %lu) = %lu * %lu\n",
          n, x, y, x, y, (x + y), (x - y));
      exit(0);
  }
}
printf("no factors found\n");
}
```

The program output shows the factors found:

```
C> 8_1_1
Find x, y, such that n = x^2 - y^2

n = 2027651281
2027651281 = (45041 + 1020) * (45041 - 1020) = 46061 * 44021
```

Although the factors discovered may be prime, they are not necessarily prime. This technique only finds two factors. For example, if we choose $n = 2695$ we would get

$$2695 = (52 + 3) * (52 - 3) = 55 * 49,$$

where neither 55 nor 49 are prime. Of course one could iterate on each factor found until the prime factorization has been found.

Challenges

1. What are the possible numbers for the last one-, two-, and three decimal digits of a perfect square? Write a program that finds all of these.

2. It is easy to verify that for all x, $x^2 \equiv 0$ or 1 (mod 4). Of course this does not mean that every number that equals 0 or 1 mod 4 is a square. Use your conclusion to speed up Program 8_1_1.

3. Modify Program 8_1_1 to find prime factors by recursively invoking Fermat's method.

8.2 A Simple $x^2 \equiv y^2$ (mod n) Factoring Method

Suppose we are trying to factor the odd composite number n. We can modify Fermat's method to look for finding x and y in the congruence $x^2 \equiv y^2$ (mod n). The method described here is similar to the one proposed by Kraitchik in 1926. Unlike Kraitchik, we have the advantage of being able to use a computer to test values of x.

To implement this method, we square each x in the interval $\sqrt{n} < x < n$. Next, x^2 (mod n) is tested to see if the remainder is a perfect square, that is, y^2. If we find a suitable y, then the $gcd(x - y, n)$ is a divisor of n. However, with this method it is possible that this divisor may be equal to 1 or n.

If n is not a prime or a power of a prime, then at least 1/2 of the xs and ys found satisfying $x^2 \equiv y^2$ (mod n) will be a proper divisor of n. To ensure that we have obtained a proper divisor, we test for $1 < gcd(x - y, n) < n$.

Program 8_2_1.c. Find a factor of n using the $x^2 \equiv y^2$ (mod n) method

```
/*
** Program 8_2_1 - Find a factor of n using the x^2 ≡ y^2 (mod n)
**                 method
*/
#include "numtype.h"

void main(int argc, char *argv[])
{
  NAT n, g, r, x, y;

  if (argc == 2) {
    n = atol(argv[1]);
  } else {
    printf("Factor n using x^2 ≡ y^2 (mod n)\n\n");
    printf("n = ");
    scanf("%lu", &n);
  }

  for (x = iSquareRoot(n, &y, &r) + 1; x < n; x++)
  {
    iSquareRoot((x * x % n), &y, &r);
    if (r == 0)
      if ((g = GCD(x - y, n)) > 1)
        {
          printf("%lu is a factor of %lu\n", g, n);
          exit(0);
        }
  }
  printf("no factor found\n");
}

/*
** iSquareRoot - finds the integer square root of a number
**               by solving: a = q^2 + r
```

```
*/
#include "numtype.h"

NAT iSquareRoot(NAT a, NAT *q, NAT *r)
{
   *q = a;

   if (a > 0)
     while (*q > (*r = a / *q))      /* integer root */
        *q = (*q + *r) >> 1;         /* divided by 2 */

   *r = a - *q * *q;                 /* remainder */

   return(*q);
}

/*
** GCD - find the greatest common divisor using Euclid's algorithm
*/
#include "numtype.h"

NAT GCD(NAT a, NAT b)
{
   NAT r;

   while (b > 0)
   {
     r = a % b;
     a = b;
     b = r;
   }

   return(a);
}
```

The output shown below demonstrates that although the technique finds proper divisors, there is no guarantee that they are prime divisors:

```
C> 8_2_1
Factor n using x^2 ≡ y^2 (mod n)

n = 1000
20 is a factor

C> 8_2_1
Factor n using x^2 ≡ y^2 (mod n)

n = 2773
47 is a factor
```

Program 8_2_1 uses the function first introduced with Program 6_3_1 called iSquareRoot() — the integer square root. By way of definition, iSquareRoot(n) = [$\sqrt{n}$], the greatest integer less than or equal to the square root of n. Although we could employ the C/C++ function `sqrt()`, isquareRoot() has the added feature of returning the remainder— the difference between the number and its integer root squared. Obviously a perfect square has 0 remainder. The algorithm for isquareRoot() is based on a modified Newton's method where we are given a and seeking x, such that $x_{n+1} = x_n$ in the following iteration:

$$x_{n+1} = (x_n + a / x_n) / 2.$$

When $x_{n+1} = x_n$, then we have found x, the square root of a. Newton's method and the isquareRoot() function are discussed in more detail as a Diophantine equation in Chapter 9.

A variation of this method is called Dixon's method (Dixon, 1981), where we try to factor n using random integers x in

$$f(x_i) \equiv x_i^2 \pmod{n}.$$

If $f(x_i)$ is not easily factored, then another x_i is tried. This method leads directly to the quadratic sieve method discussed in the next section.

8.3 The Quadratic Sieve Factoring Method

The QS and similar algorithms share a common strategy based in congruence arithmetic. Based generally on Fermat's method, the idea is to multiply congruences of the form $a \equiv b \pmod{n}$, where n is the number to be factored and $a \neq b$. The product obtained is used to produce a congruence $x^2 \equiv y^2 \pmod{n}$. This congruence implies that $n \mid (x^2 - y^2) = (x + y)(x - y)$. The greatest common divisors of $(x + y, n)$ and $(x - y, n)$, which can be found by Euclid's algorithm, are nontrivial factors n. It is worth noting that these factors are not necessarily prime, although the process can be repeated until only primes are found.

The $x^2 \equiv y^2 \pmod{n}$ factoring method described in the previous section is in some ways analogous to the quadratic sieve (QS). The obvious problem in the previous method is that we increment the xs by 1s in the hope of stumbling across a suitable y. I suppose that the same can be said of Fermat's method as well. The QS is more directed, more complicated, and still has a possibility of not finding a factor.

The QS was first suggested in 1981 as a modification of Schroeppel's linear sieve factoring method (Pomerance, 1981). The QS algorithm and some of its variations is described in Pomerance (1984). Varieties of QS have been used extensively since its introduction (Gerver, 1983; Davis, Holdbridge, and Simmons, 1985). The multiple polynomial version has proved to be a significant enhancement to the QS (Silverman, 1987).

The goal of the QS method is to enhance the technique of finding x and y. The method described here is similar to the one published by

Davis and Holdridge (1983). To begin to try solving the congruence $x^2 \equiv y^2 \pmod{n}$, we use the polynomial

$$Q(x) = (\, x + [\sqrt{n}\,]\,)^2 - n.$$

We can at once observe two facts. The quadratic $Q(x)$ has integer coefficients. Also, if z is an integer, then

$$(\, z + [\sqrt{n}\,]\,)^2 \equiv Q(z) \pmod{n}.$$

If we could find a set of integers $z_1, z_2, ..., z_k$ such that

$$Q(z_1) \cdot Q(z_2) \cdot \,...\, \cdot Q(z_k) = y^2$$

and if we let

$$(\, z_1 + [\sqrt{n}\,]\,) \cdot (\, z_2 + [\sqrt{n}\,]\,) \cdot \,...\, \cdot (\, z_k + [\sqrt{n}\,]\,) = x,$$

then from our first congruence we get $x^2 \equiv y^2 \pmod{n}$. As before, $gcd(x - y, n)$ is a nontrivial factor of n.

How shall we find our integers $z_1, z_2, ..., z_k$ that will permit us to try to factor n? Rather than the brute force of trial and error for finding $Q(x)$ and ultimately y, we will use a factor base to help us. A factor base is nothing more than known primes, $p_1, p_2, p_3, ...$

Some of the primes (in the set of all primes less than a certain number) will not divide $Q(z)$. They are easily identified using the Legendre symbol (first encountered in Chapter 3): $(n \,|\, p) = -1$. On the other hand, those for which $(n \,|\, p) = 1$ have a chance of yielding a proper divisor.

We begin with a factor base of k primes, $p_1, p_2, ..., p_k$. The factor base is derived from the set of primes greater than 2, where $(n \,|\, p) = 1$. We then search for integers $a_1, a_2, ..., a_{k+1}$ and $b_1, b_2, ..., b_{k+1}$ such that for each i, $a_i^2 \equiv b_i \pmod{n}$. Also, we want b_i to be completely factored over the factor base

$$b_i = \prod_{j=1}^{k} p_j^{\,c_{ij}} .$$

For each i, let $e_i = (\, e_{i1}, e_{i2}, ..., e_{ik} \,)$ be a vector over the integers (mod 2) such that for each j, $e_{ij} \equiv c_{ij} \pmod{2}$. If all of the e_{ij} are even in a given row, then b_i is a square, and we do not need to consider that row further. If not, then we must seek linear combinations of the rows to produce a square.

To find linear combinations of rows, it is necessary to perform a type of Gaussian elimination (mod 2) on the exponent array. Also, we will need a "history" array to keep track of the row reductions we perform. Consider the following 6x7 exponent array and its companion 6×6 history array:

e — exponent array

	p_1	p_2	p_3	p_4	p_5	p_6	p_7
Q_1	1	0	1	0	0	0	0
Q_2	0	1	0	1	1	0	0
Q_3	0	0	1	0	0	1	0
Q_4	1	0	1	1	1	0	0
Q_5	1	0	1	0	0	0	1
Q_6	0	0	0	1	1	0	1

e — history array

	Q_1	Q_2	Q_3	Q_4	Q_5	Q_6
Q_1	1	0	0	0	0	0
Q_2	0	1	0	0	0	0
Q_3	0	0	1	0	0	0
Q_4	0	0	0	1	0	0
Q_5	0	0	0	0	1	0
Q_6	0	0	0	0	0	1

In the exponent array, one row represents the square-free factorization of Q over the factor base. For example,

$$Q_1 = p_1 \cdot p_3.$$

We perform row operations using the XOR operator ($\wedge$) since we are really only toggling a single bit. To reduce a row, first find the pivot element — the first nonzero element on the row. Next scan up and down the columns looking for another nonzero element. If one is found, XOR the row with the pivot with the other. In this example, e_{11} is a pivot, so we XOR row Q_1 with Q_4 in both the exponent and history arrays:

	p_1	p_2	p_3	p_4	p_5	p_6	p_7
Q_1	1	0	1	0	0	0	0
Q_2	0	1	0	1	1	0	0
Q_3	0	0	1	0	0	1	0
Q_4	0	0	0	1	1	0	0
Q_5	1	0	1	0	0	0	1
Q_6	0	0	0	1	1	0	1

	Q_1	Q_2	Q_3	Q_4	Q_5	Q_6
Q_1	1	0	0	0	0	0
Q_2	0	1	0	0	0	0
Q_3	0	0	1	0	0	0
Q_4	1	0	0	1	0	0
Q_5	0	0	0	0	1	0
Q_6	0	0	0	0	0	1

Next we XOR row Q_1 with Q_5 to yield:

	p_1	p_2	p_3	p_4	p_5	p_6	p_7
Q_1	1	0	1	0	0	0	0
Q_2	0	1	0	1	1	0	0
Q_3	0	0	1	0	0	1	0
Q_4	0	0	0	1	1	0	0
Q_5	0	0	0	0	0	0	1
Q_6	0	0	0	1	1	0	1

	Q_1	Q_2	Q_3	Q_4	Q_5	Q_6
Q_1	1	0	0	0	0	0
Q_2	0	1	0	0	0	0
Q_3	0	0	1	0	0	0
Q_4	1	0	0	1	0	0
Q_5	1	0	0	0	1	0
Q_6	0	0	0	0	0	1

Scanning row Q_2 for a pivot, we find one at e_{22}. However, no other Qs contain this factor (p_2), so we move on to row Q_3, which can be XORed with row Q_1:

	p_1	p_2	p_3	p_4	p_5	p_6	p_7
Q_1	1	0	0	0	0	1	0
Q_2	0	1	0	1	1	0	0
Q_3	0	0	1	0	0	1	0
Q_4	0	0	0	1	1	0	0
Q_5	0	0	0	0	0	0	1
Q_6	0	0	0	1	1	0	1

	Q_1	Q_2	Q_3	Q_4	Q_5	Q_6
Q_1	1	0	1	0	0	0
Q_2	0	1	0	0	0	0
Q_3	0	0	1	0	0	0
Q_4	1	0	0	1	0	0
Q_5	1	0	0	0	1	0
Q_6	0	0	0	0	0	1

Continuing in a like manner Q_4 can be XORed with row Q_2; Q_4 can be XORed with row Q_6; and finally Q_5 can be XORed with row Q_6 giving the final result:

	p_1	p_2	p_3	p_4	p_5	p_6	p_7
Q_1	1	0	0	0	0	1	0
Q_2	0	1	0	0	0	0	0
Q_3	0	0	1	0	0	1	0
Q_4	0	0	0	1	1	0	0
Q_5	0	0	0	0	0	0	1
Q_6	0	0	0	0	0	0	0

	Q_1	Q_2	Q_3	Q_4	Q_5	Q_6
Q_1	1	0	1	0	0	0
Q_2	1	1	0	1	0	0
Q_3	0	0	1	0	0	0
Q_4	1	0	0	1	0	0
Q_5	1	0	0	0	1	0
Q_6	0	0	0	1	1	1

Scanning the exponent array, we find that row Q_6 exponents are all zero. From the history array, we arrive at the desired congruence

$$(Q_4 \cdot Q_5 \cdot Q_6)^2 \equiv (b_4 \cdot b5 \cdot b_6)^2 \pmod{n}.$$

All that remains is to compute $g = gcd((Q_4 \cdot Q_5 \cdot Q_6 - b_4 \cdot b_5 \cdot b_6)$, n). If $1 < g < n$, then a proper divisor of n has been found. If not, then another linearly dependent row will need to be found. If no other linearly dependent rows exist in the exponent array, then the factor base will need to be enlarged and the process begun anew.

The program then follows in a basic implementation of the quadratic sieve. It is quite long because it has many of the congruence arithmetic functions developed in earlier chapters. I suggest that you use the source code on the disk accompanying this book rather than try to type it in.

Program 8_3_1.c. Factoring using the quadratic sieve

```
/*
** Program 8_3_1 - Find a factor of n using the quadratic sieve
**                   method
*/
#include "numtype.h"

#define MAXFB   (NAT) 100

void main(int argc, char *argv[])
{
   INT a, b, c, m, n, r, s, p[MAXFB], q[MAXFB], x[2];
   int i, j, k, l, pk, vk, sum;
   char e[MAXFB][MAXFB], h[MAXFB][MAXFB], v[MAXFB][MAXFB];

   printf("Factor n (quadratic sieve)\n\n");
   if (argc == 2) {
     n = atol(argv[1]);
   } else {
     printf("n = ");
     scanf("%lu", &n);
   }

/*
** part 0:initialize the integer square root of n and
**         coefficients of Q(x) = (x + [ √n̄])² - n
*/

   iSquareRoot(n, &s, &r);

   if (r == 0)              /* what luck, a perfect square! */
   {
     printf("%ld = %ld²\n", n, s);
     exit(0);
   }

   s++;
   a = 1;                   /* coefficients of the quadratic Q() */
   b = 2 * s;
   c = s * s - n;

/*
** part 1: build factor base for n
*/

   pk = BuildFactorBase(n, p, MAXFB);

/*
** part 2: solve for each x in the congruence (x + [√n̄])² ≡ n (mod p)
**          and factor the solution of the factor base
*/

   memset(v, 0, sizeof(v));
   vk = 0;                  /* count rows in exponent matrix */

   for (i = 0; i < pk && vk < MAXFB; i++)
   {
```

```
    SolveQuadraticCongruence(a, b, c, p[i], &x[0], &x[1]);

    for (j = 0; j < 2 && vk < MAXFB; j++)
    {
      q[vk] = (a * x[j] + b) * x[j] + c;
      if (!SearchArray(q[vk], q, vk))
        if (SieveOverFactorBase(q[vk], p, v[vk], pk))
          vk++;
    }
  }

/*
** part 3: find linear combinations within the exponent matrix
*/

  for (i = 0; i < vk; i++)            /* rows (Q's) */
    for (j = 0; j < pk; j++)          /* columns (factors) */
    {
      e[i][j] = v[i][j] & 1;          /* exponent (mod 2) */
      if (i == j)
        h[i][j] = 1;                  /* history */
      else
        h[i][j] = 0;
    }

  for (i = 0; i < vk; i++)            /* rows (Q's) */
    for (j = 0; j < pk; j++)          /* columns (factors) */
      if (e[i][j])                    /* pivot found */
      {
        for (k = 0; k < vk; k++)        /* check remaining rows */
          if ((k != i) && (e[k][j]))
          {
            for (l = 0; l < pk; l++)
              e[k][l] ^= e[i][l];
            for (l = 0; l < vk; l++)
              h[k][l] ^= h[i][l];
          }
        break;
      }

/*
** part 4: scan the reduced exponent array for linear dependencies
**         and find GCD's
*/

  for (i = 0; i < vk; i++)
  {
    for (j = 0, sum = 0; j < pk; j++)
      sum += e[i][j];
    if (sum == 0)                     /* found one */
    {
      for (k = 0, a = 1; k < vk; k++)
        if (h[i][k])
        {
          a = MulMod(a, q[k], n);
          for (l = 0, b = 1; l < pk; l++)
            if (v[k][l])
              b = MulMod(b, p[l], n);
```

```
        }
      printf("%5ld² ≡ %5ld² (mod %ld) ==> ", a, b, n);
      printf("gcd = %lu\n", GCD(ABS(a - b), n));
    }
  }
}

/*
** BuildFactorBase - builds the factor base with small odd primes
*/
#include "numtype.h"

NAT BuildFactorBase(INT n, INT p[], NAT maxfb)
{
  INT i, k;

  k = 0;

  for (i = 3; k < maxfb; i++, i++)
    if ( IsPrime(i) )
      if ( LegendreSymbol(n, i) == 1 ) p[k++] = i;

  return(k);
}

/*
** SearchArray - searchs an integer to see if a number is in it
*/
#include "numtype.h"

int SearchArray(NAT k, NAT a[], NAT n)
{
  NAT i;

  for (i = 0; i < n; i++)
    if (k == a[i]) return(1);

  return(0);
}

/*
** SieveOverFactorBase - finds the factorization over a factor base
*/
#include "numtype.h"

int SieveOverFactorBase(INT q, INT p[], char v[], NAT maxfb)
{
  int i;

  for (i = 0; i < maxfb; i++)
  {
    v[i] = 0;
    while ((q % p[i]) == 0L)
    {
      v[i]++;
      q /= p[i];
      if (q == 1)
        return(1);             /* done */
```

```
        }
     }
     return(0);                    /* unsuccessful over factor base */
}

/*
** SolveQuadraticCongruence — find x such that a*x² + b*x + c ? 0
**                            (mod p) by the RESSOL algorithm and
**                            Gauss' method
*/
#include "numtype.h"

INT SolveQuadraticCongruence(INT a, INT b, INT c, INT p, INT *x1,
INT *x2)
{
   INT d, u, z;

/* first find the discriminant and the quadratic residue */

   d = (b * b - 4 * a * c) % p;
   if (d < 0) d += p;                /* make d > 0 */
   u = QuadraticResidue(d, p);

/* find the z, inverse of a */

   z = PowMod(a, p - 2, p);

/* now find the solutions */

   if ((b + u) & 1)
   {
     *x1 = (z * ( - b + u + p) / 2) % p;
     *x2 = (z * ( - b - u + p) / 2) % p;
   }
   else
   {
     *x1 = (z * ( - b + u) / 2) % p;
     *x2 = (z * ( - b - u) / 2) % p;
   }
   if (*x1 < 0) *x1 += p;
   if (*x2 < 0) *x2 += p;

   return(0);
}

/*
** QuadraticResidue - find x such that x² ≡ a (mod p) using the
                      RESSOL algorithm (Shanks, 1972)
**
*/
#include "numtype.h"

NAT QuadraticResidue(NAT a, NAT p)
{
   NAT b, c, i, l, k, n, q, r, s, x, z;

   if ( LegendreSymbol(a, p) != 1)
   {
```

```
    printf("%ld is not a quadratic residue of %ld\n", a, p);
    exit(1);
  }

/* find initial value for s, k, such that p = 2^s * (2*k + 1) - 1
*/

  s = 0;
  k = p - 1;
  while ( !(k & 1) ) { s++; k /= 2; }   /* reduce to 2*k + 1 */
  k = (k - 1) / 2;                        /* solve for k */

/* find z such that z is not a quadratic residue of p */

  z = 2;
  while ( LegendreSymbol(z, p) == 1) z++;

/* find initial values for r, n, and c */

  r = PowMod(a, k + 1, p);
  n = PowMod(a, 2*k + 1, p);
  c = PowMod(z, 2*k + 1, p);

/* derive the solution */

  while (n > 1)
  {
    l = s;
    b = n;
    for ( i = 0; i < l; i++)
      if ( b == 1)
      {
        b = c;
        s = i;
      }
      else
      {
        b = (b * b) % p;
      }

    c = (b * b) % p;
    r = (b * r) % p;
    n = (c * n) % p;
  }

  return(r);
}

/*
** LegendreSymbol — evaluates the Legendre Symbol using modulus
**                  and Euler's criterion for quadratic congruences
**
** returns:  0 if p divides a
**           1 if a is a quadratic residue modulo p
**          -1 otherwise
*/
#include "numtype.h"
int LegendreSymbol(NAT a, NAT p)
```

```
{
  if ( (a % p) == 0) return(0);
  if ( PowMod(a, (p - 1)/2, p) == 1) return(1);
  return(-1);
}

/*
** iSquareRoot - finds the integer square root of a number
**               by solving: a = q² + r
*/
#include "numtype.h"

NAT iSquareRoot(NAT a, NAT *q, NAT *r)
{
  *q = a;

  if (a > 0)
    while (*q > (*r = a / *q))        /* integer root */
      *q = (*q + *r) >> 1;           /* divided by 2 */

  *r = a - *q * *q;                   /* remainder */

  return(*q);
}

/*
** PowMod - computes r for aⁿ ≡ r (mod m) given a, n, and m
*/
#include "numtype.h"

NAT PowMod(NAT a, NAT n, NAT m)
{
  NAT r;

  r = 1;
  while (n > 0)
  {
    if (n & 1)                        /* test lowest bit */
      r = MulMod(r, a, m);           /* multiply (mod m) */
    a = MulMod(a, a, m);             /* square */
    n >>= 1;                         /* divided by 2 */
  }

  return(r);
}

/*
** MulMod - computes r for a * b ≡ r (mod m) given a, b, and m
*/
#include "numtype.h"

NAT MulMod(NAT a, NAT b, NAT m)
{
  NAT r;

  if (m == 0) return(a * b);          /* (mod 0) */

  r = 0;
```

```
while (a > 0)
{
  if (a & 1)
    if ((r += b) > m) r %= m;        /* test lowest bit */
                                     /* add (mod m) */
  a >>= 1;                           /* divided by 2 */
  if ((b <<= 1) > m) b %= m;         /* times 2 (mod m) */
}

return(r);
}

/*
** IsPrime - determines if a number is prime by trial division
*/
#include "numtype.h"

int IsPrime(NAT n)
{
  NAT i;

  if (n == 1)  return(0);
  if (n == 2)  return(1);
  if (!(n & 1)) return(0);           /* n is even */

  for (i = 3; i*i <= n; i++, i++)    /* n is odd */
    if ((n % i) == 0) return(0);

  return(1);
}

/*
** GCD - find the greatest common divisor using Euclid's algorithm
*/
#include "numtype.h"

NAT GCD(NAT a, NAT b)
{
  NAT r;

  while (b > 0)
  {
    r = a % b;
    a = b;
    b = r;
  }

  return(a);
}
```

The size of the factor base used in Program 8_3_1 is controlled by the parameter MAXFB which can be changed in the define statement. In the example below a factor base of 100 primes was used to find the factor 31 for 50003.

```
C> 8_3_1
Factor n (quadratic sieve)

n = 50003

31125² ≡ 39870² (mod 50003) ==> gcd = 1
32940² ≡ 38861² (mod 50003) ==> gcd = 31
```

As noted above, many variations have been tried on the basic algorithm. One of the more common is to add the numbers −1 and 2 to the factor base (Davis and Holdridge, 1983). Another is to use a small multiplier λ so that instead of sieving n, you would sieve $\lambda \cdot n$. This technique produces an altogether different factor base that offers another chance to find a factor.

Challenges

1. Modify Program 8_3_1 so that a multiplier is used whenever no factor is found.

2. Modify Program 8_3_1 by adding −1 and 2 to the factor base.

8.4 Pollard's $p-1$ Factoring Method

Pollard's $p-1$ factoring method is a slightly different approach to factoring than the preceding QS and related methods. If a large integer n has a prime factor p with the property that $p-1$ has no large prime factors, then this method can be very successful. The algorithm proceeds as follows.

Let n be an odd composite that is to be factored and let p be a prime factor of n. Define l as the least common multiple of the first k integers

$$l = lcm(1, 2, 3, ..., k).$$

Recall from Chapter 3 Fermat's little theorem, which states that if p is prime and n is a positive integer, then $n^{p-1} \equiv 1 \pmod{p}$. So if $(p-1) \mid l$, then from Fermat's little theorem we can construct the following congruence:

$$2^l \equiv 1 \pmod{p}.$$

Now suppose that

$$2^l \not\equiv 1 \pmod{n}.$$

Then

$$1 < gcd(2^l - 1, n) < n.$$

How does this work? The first congruence tells us that the p divides 2^l $- 1$ and the second says n does not. As noted in Chapter 2, computing the gcd does not involve knowing p. It is possible for the gcd to yield a trivial divisor of n. However, Pomerance (1990) asserted that it is highly likely that the gcd will be p, so a prime factor will have been found.

Program 8_4_1.c. Pollard's $p - 1$ factoring method

```
/*
** Program 8_4_1 - Factor using the p-1 method
*/
#include "numtype.h"

void main(int argc, char *argv[])
{
  NAT g, k, l, n, p;

  if (argc == 2) {
    n = atol(argv[1]);
  } else {
    printf("Factor n (p-1 method)\n\n");
    printf("n = ");
    scanf("%lu", &n);
  }

/* compute various values for m(k) */

  for (k = 2, l = 2; k <= 22; k++)
  {
    l = LCM(k, l);
    p = PowMod(2L, l, n);
    if (p < 2 ) p += n;
    p--;                        /* p - 1 */
    g = GCD(p, n);
    if (g > 1 && p < n)         /* factor found */
    {
      printf("%lu is a factor of %lu\n", g, n);
      exit(0);
    }
  }

  printf("no factor found\n");
}

/*
** PowMod - computes r for a^n ≡ r (mod m) given a, n, and m
*/
#include "numtype.h"

NAT PowMod(NAT a, NAT n, NAT m)
```

```
{
  NAT r;

  r = 1;
  while (n > 0)
  {
    if (n & 1)                    /* test lowest bit */
      r = MulMod(r, a, m);        /* multiply (mod m) */
    a = MulMod(a, a, m);          /* square */
    n >>= 1;                      /* divided by 2 */
  }

  return(r);
}

/*
** MulMod - computes r for a * b ≡ r (mod m) given a, b, and m
*/
#include "numtype.h"

NAT MulMod(NAT a, NAT b, NAT m)
{
  NAT r;

  if (m == 0) return(a * b);      /* (mod 0) */

  r = 0;
  while (a > 0)
  {
    if (a & 1)                    /* test lowest bit */
      if ((r += b) > m) r %= m;   /* add (mod m) */
    a >>= 1;                      /* divided by 2 */
    if ((b <<= 1) > m) b %= m;    /* times 2 (mod m) */
  }

  return(r);
}

/*
** LCM - finds the least common multiple for two integers
**       LCM = a * b / GCD(a, b)
*/
#include "numtype.h"

NAT LCM(NAT a, NAT b)
{
  b /= GCD(a,b);

  if ( (0xffffffff / b) < a )
    return(0);                    /* overflow */
  else
    a *= b;                       /* ok to multiply */

  return(a);
}

/*
** GCD - find the greatest common divisor using Euclid's algorithm
```

```
*/
#include "numtype.h"

NAT GCD(NAT a, NAT b)
{
  NAT r;

  while (b > 0)
  {
    r = a % b;
    a = b;
    b = r;
  }

  return(a);
}
```

```
C> 8_4_1
n = 858058123
13 is a factor of 858058123
```

But why should the algorithm be blind to some primes and not to other (significantly larger) primes? The reason is that p must be "special" and hold for a fairly small k. For this application $k \leq 22$ because $lcm(1, 2, 3, ..., 22) = 232{,}792{,}560$; a very small, k indeed. Any larger k would exceed the fixed precision of fair 4-byte integers in the current implementation. ps that are less special require ks closer to $p-1$ to discover since $(p - 1)$ needs to be a proper divisor of $lcm(1, 2, 3, ..., k)$.

Look again at our example. The number in the example 858,058,123 can be factored because

$$858{,}058{,}123 = 13^5 \cdot 2{,}311$$

and for $p - 1$,

$$13 - 1 = 12 = 2^2 \cdot 3$$

$$2311 - 1 = 2310 = 2 \cdot 3 \cdot 5 \cdot 7 \cdot 11.$$

All of these prime factors are less than $k = 22$, so a prime factor can be found.

The $p-1$ method, as implemented in Program 8_4_1 (and in general), cannot factor all numbers. For example, it fails to find a factor for 3901 since

$$3901 = 47 \cdot 83$$

and

$$47 - 1 = 46 = 2 \cdot 23$$

$$83 - 1 = 82 = 2 \cdot 41.$$

Because the factorization of $p - 1$ of both prime factors does not factor into numbers less than $k = 22$, a factorization of n cannot be found. Thus, the algorithm is "blind" to some integers. Not surprisingly, looking at a distribution of number that can be factored one finds that the blind gaps increase in size with increasing n. To factor these, we must use ever-increasing k.

As seen in the preceding examples and as noted by Pomerance (1990), if just one number $(p - 1)$ can be factored into small primes less than k, then a factor of n can be found. However, for an arbitrary integer this means that this algorithm has a large element of luck associated with it. Sadly, when it comes to factoring algorithms that rely on luck, one more often tends to encounter the bad variety.

One final note about Pollard's method. The main weakness in Pollard's $p - 1$ method, that it cannot factor n where $p - 1$ is divisible by a large prime, that it is generally viewed as a strength by cryptographers. In public-key encryption (see below), p and q are selected such that $p - 1$ and $q - 1$ have large prime power divisors and therefore are more likely to resist factorization by Pollard's method.

Challenge

1. Modify the $p-1$ algorithm to use congruence arithmetic to find factors so that larger ps can be examined.

8.5 The Elliptic Curve Factoring Method

The $p-1$ factoring method was a predecessor of the elliptic curve method (EC), the method that we now turn our attention to. It was originally proposed by Lenstra (1987) and has enjoyed significant success in factoring large integers. In a way, EC is only a variation on the $p-1$ method. The difference is that the multiplicative group from the $p-1$ method is replaced by the group of points on random elliptic curves. If a single elliptic curve is used, then the properties of the algorithm are exactly the same as the $p-1$ method (Lenstra, 1987).

Before beginning with the algorithm, we need one new definition and some background discussion. Recalling the definition of a field (Chapter 2) we now define the characteristic of a field. Let F be an arbitrary field and 1 be an identity element. For any positive integer n the number k defined by

$$k = \sum_{i=1}^{n} 1$$

is also contained in F. If $k \neq 0$ for all positive integers in F, then F is said to have *characteristic* 0. However, if there exists a least positive integer p, such that $k = 0$, then F is said to have *characteristic p*. The fields of rational numbers Q, the real numbers R, and complex numbers C all have *characteristic O*. For a prime p, the finite field Z_p, having p elements, is an example of a field with characteristic p.

Let K be a field with characteristics different from 2 or 3 and let

$$f(x) = x^3 + a \cdot x + b \text{ (where } a, b \in K).$$

f is a cubic polynomial over K with no repeated roots if and only if the discriminant

$$4 \cdot a^3 + 27 \cdot b^2 \neq 0.$$

An *elliptic curve* is the set of points $(x, y) \in K$ satisfying

$$y = x^3 + a \cdot x + b,$$

including a single element O called the point at infinity. Elliptic curves make interesting graphs. To visualize this family of curves, look at three elliptic curves graphed over the real numbers R (Figure 8.1), where $b = -1, 0,$ and 1.

Figure 8.1. Elliptic curves: $y^2 = x^3 - x - 1$, $y^2 = x^3 - x$, $y^2 = x^3 - x + 1$.

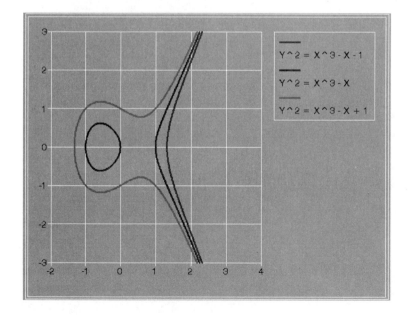

The solutions in integers to a given elliptic curve can be seen where the curves intersect the grid line intersections (i.e., lattice points).

The rational points on a elliptic curve form a commutative (or abelian) group by defining an operation denoted by + using the *tangent-and-chord* method. To see how this works, consider P and Q, two points on an elliptic curve E over a field K, the characteristic of which is not equal to 2 or 3. Let $P = (x, y)$, $Q = (x', y')$, and $-P = (x, -y)$. If $P \neq Q$, then the (nontangent) line passing through P and Q intersects the elliptic curve at a single point of projection R, and the sum of P and Q is

$$-R = P + Q.$$

In terms of (x, y) and (x', y'), $P + Q = (x'', y'')$ is defined as

$$x'' = \left(\frac{y' - y}{x' - x}\right)^2 - x - x'$$

$$y'' = \left(\frac{y' - y}{x' - x}\right) \cdot (-x - x'') - y.$$

If it happens that the line is tangent to either P or Q, then R equals the point of tangency. If $P = Q$, then R is a point different from P, where the line tangent to P intersects the curve. $P + Q = (x'', y'')$ is defined as

$$x'' = \left(\frac{3x^2 + a}{2y}\right)^2 - 2x$$

$$y'' = \left(\frac{3x^2 + a}{2y}\right) \cdot (x - x'') - y.$$

And lastly, if $Q = -P$, then Q is the point symmetric with respect to the x-axis. Q is said to be the inverse of P and $P + Q = O$, the point at infinity. For elliptic curve $y^2 = x^3 - 2 \cdot x + 5$, let $P = (-2, 1)$ and $Q = (1/4, 2\,1/8)$, then from the foregoing formula we compute $P + Q = (2, -3)$ and $2 \cdot P = (29, -156)$. Figure 8.2 shows a graphic interpretation of $P + Q$ and how R serves as a point of projection.

For additional discussion of the nature of groups constructed using elliptic curves, see Ellis (1992) and Atkin and Morain (1993b). With this background, we are now ready to develop the elliptic curve factoring method.

Lenstra's EC method operates over the rational numbers, Q. Given a rational number $b\,/\,a$ where a is relatively prime to an integer n, we can find an integer x such that

$$x \equiv b\,/\,a \pmod{n}$$

or, put it another way,

$$a \cdot x \equiv b \pmod{n}.$$

Figure 8.2. The sum of two points on the elliptic curve: $y^2 = x^3 - 2 \cdot x + 5$.

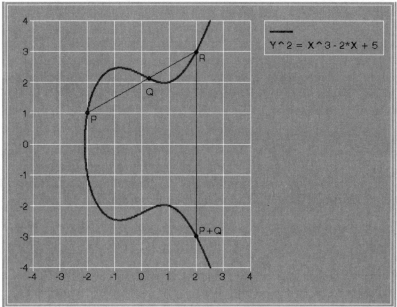

For this application, we will prefer that x be from the arithmetic residue system (i.e., least positive remainders).

Suppose we are trying find a factor d for n. Because our elliptic curves work in a field with characteristics other than 2 or 3, we first must test to see if n is divisible by either of these. If so, then $d = 2$ or 3 and we stop. If not, then choose a "random" elliptic curve; guidelines for selecting an appropriate elliptic curve can be found in Atkin and Morain (1993a). For this example, I'll choose the coefficients $a = -2$, $b = 5$. At this point, we may actually get lucky. If the greatest common divisor g for the discriminant $4a^3 + 27b^2$ and n, that is,

$$g = gcd(4a^3 + 27b^2, n)$$

is greater than 1, then $d = g$ is a factor of n.

Needless to say, this will not be the case very often, so, like the $p - 1$ method, we select a k and compute $lcm(1, 2, ..., k)$. Next we compute

$$kP = P + P + P + ... + P,$$

all the while working modulo n. We use an algorithm similar to MulMod, halving and doubling to achieve the sum. Each time an intermediate sum is computed, there exists a possibility that the integer denominator will have a factor in common with n. When this event happens, the inverse cannot be computed and a factor can be found by either

$$d = gcd(x', n)$$

or

$$d = gcd(2y, n),$$

depending on which set of equations we are using.

In the program that follows, the risk of overflow is great. It should be considered illustrative of the algorithm rather than a factoring program. In fact, many enhancements could be made to the program to extend its range as well as to optimize calculations.

Program 8_5_1.c. Factoring using elliptic curves

```
/*
** Program 8_5_1 - Factor using the elliptic curves having
**                 the form y^2 = x^3 + ax + b
*/
#include "numtype.h"

void main(int argc, char *argv[])
{
  NAT k, l;
  INT a, b, g, n, p;
  LPOINT P, Q;

  if (argc == 2) {
    n = atol(argv[1]);
  } else {
    printf("Factor n (elliptic curve)\n\n");
    printf("n = ");
    scanf("%ld", &n);
  }

/* first check 2 and 3 as divisors */

  if ((n % 2) == 0)
  {
    printf("2 is a factor of %ld\n", n);
    exit(0);
  }
  else if ((n % 3) == 0)
  {
    printf("3 is a factor of %ld\n", n);
    exit(0);
  }

/* select random a and b for an elliptic curve */
```

```c
  a = -2;
  b = 5;
  if ((g = GCD(4*a*a*a + 27*b*b, n)) > 1)
  {
    printf("%ld is a factor of %ld\n", g, n);
    exit(0);
  }

/* find lcm for some arbitrary k */

  for (k = 2, l = 2; k <= 22; k++)
    l = LCM(k, l);

/* choose random starting point on the elliptic curve */

  P.x = 0; P.y = 5;
  P.x = Q.x = -2; P.y = Q.y = 1;

  while (l > 0)
  {
    if (l & 1)                    /* test lowest bit */
    {
      p = ECAddPointsMod(&P, &Q, &P, a, b, n); /* add (mod n) */
      if (p) break;
    }
    l /= 2;                                     /* divided by 2 */
    ECAddPointsMod(&Q, &Q, &Q, a, b, n);        /* times 2 (mod n) */
  }

/* check to see if we found a non-trivial divisor */

  if (((g = GCD(p, n)) != n) && (g > 1))
    printf("%ld is a factor of %ld\n", g, n);
  else
    printf("no factor found\n");
}

/*
** EDAddPointsMod - Add two points on an elliptic curve (mod n)
*/
#include "numtype.h"

INT ECAddPointsMod( LPOINT *P, LPOINT *Q, LPOINT *R, INT a, INT b,
INT n)
{
  INT x, y, d, dd;

  if (P->x != Q->x)
  {
    d = Q->x - P->x;
    dd = d * d;

    x = (Q->y - P->y) * (Q->y - P->y) - (P->x - Q->x) * dd;
    x = SolveLinearCongruence(dd, x, n);
    if (x == -1) return(d);

    y = (Q->y - P->y) * (P->x - x) - P->y * d;
    y = SolveLinearCongruence(d, y, n);
```

```
    if (y == -1) return(d);
  }
  else
  {
    d = 2 * P->y;
    dd = d * d;

    x = (3 * P->x * P->x + a) * (3 * P->x * P->x + a) - 2 * P->x
* dd;
    x = SolveLinearCongruence(dd, x, n);
    if (x == -1) return(d);

    y = (3 * P->x * P->x + a) * (P->x - x) - P->y * d;
    y = SolveLinearCongruence(d, y, n);
    if (y == -1) return(d);
  }

/* return the sum */

  R->x = x;
  R->y = y;

  return(0);
}

/*
** SolveLinearCongruence - solve a linear congruence: ax ≡ b mod m
**                         using Gauss' method
*/
#include "numtype.h"

INT SolveLinearCongruence(INT a, INT b, INT m)
{
  NAT g;

  if ((a %= m) < 0) { a = -a; b = -b; }    /* transfer sign to b */
  if ((b %= m) < 0) b += m;                /* make b > 0 */

  if ((b % GCD(a,m)) == 0)
  {
    if ((g = GCD(a,b)) > 1) { a /= g; b /= g; }
    while (a > 1)
    {
      b += m;
      if ((g = GCD(a,b)) > 1) { a /= g; b /= g; }
    }
    return(b);
  }

  return(-1);                    /* solution not possible */
}

/*
** LCM - finds the least common multiple for two integers
**       LCM = a * b / GCD(a, b)
*/
#include "numtype.h"

NAT LCM(NAT a, NAT b)
{
```

```
  b /= GCD(a,b);

  if ( (0xffffffff / b) < a )
    return(0);                    /* overflow */
  else
    a *= b;                       /* ok to multiply */

  return(a);
}

/*
** GCD - find the greatest common divisor using Euclid's algorithm
*/
#include "numtype.h"

NAT GCD(NAT a, NAT b)
{
  NAT r;

  while (b > 0)
  {
    r = a % b;
    a = b;
    b = r;
  }

  return(a);
}
```

```
C> 8_5_1
Factor n (elliptic curve)

n = 26411
11 is a factor of 26411
```

Of all the factoring methods discussed, it would be nice to know which is best. Alas, this question cannot be answered definitively. Of the QS and EC, each is suitable for particular situations. Like Fermat's method, the QS seems to perform best if the number is the product of two primes roughly equal in size. Code breakers favor this algorithm because these are the types of numbers usually sought for public-key encryption systems. For applications that are not predisposed to composites in a certain form, the EC is usually selected.

Challenge

1. Modify the EC algorithm to use congruence arithmetic to find factors so that larger ps can be examined.

8.6 Primality Testing

Over the centuries, methods have been devised to test the primality of a particular number. Obviously this is not the same as finding a number's factors but for certain applications, particularly those relating to public key encryption, knowledge of primality is sufficient. In this section several tests are presented that can be divided into two types: deterministic and probabilistic.

Deterministic tests provide the definitive answer to the question, is n prime. They have the unfortunate limitation that they require a factorization of a number related to n. Obtaining this factorization can be nearly as difficult as factoring n itself.

Probabilistic tests are easier and faster give us hope that the number tested is prime, but cannot guarantee it absolutely. With probabilistic tests we can test much larger numbers trading the risk of certainty. No free lunches here either.

The Lucas-Lehmer (Deterministic) Test

An old and well-known theorem related to Fermat's little theorem regarding orders of integers and primitive roots gives us a useful primality test.

Lucas-Lehmer Primality Test. If n is a positive integer and if an integer x exists such that

$$x^{n-1} \equiv 1 \ (\text{mod } n)$$

and

$$x^{(n-1)/q} \not\equiv 1 \ (\text{mod } n)$$

for all prime divisors q of $n - 1$, then n is prime.

This theorem can be simplified for the purpose of our primality test. The corollary is numerically simpler and better suited for implementation.

Corollary: If n is an odd positive integer and if an integer x exists such that

$$x^{(n-1)/2} \equiv -1 \ (\text{mod } n)$$

and

$$x^{(n-1)/q} \not\equiv 1 \ (\text{mod } n)$$

for all odd prime divisors q of $n - 1$, then n is prime. Since $x^{n-1} = (x^{(n-1)/2})^2 \equiv (-1)^2 \ (\text{mod } n) \equiv 1 \ (\text{mod } n)$. The remaining condition is satisfied by the theorem.

The corollary (and theorem) can actually work in two different ways; we can test a number n if we can factor $n - 1$ or we can create a simple $n - 1$ and test n for primality. Consider the case for testing primality: let $n = 443$ and $x = 5$. The first condition is satisfied:

$$5^{(443-1)/2} \quad = 5^{221} \quad \equiv 442 \ (\text{mod } 443).$$

From the odd divisors of 442, 13 and 17, we get the following results:

$$5^{(443-1)/13} \quad = 5^{34} \quad \equiv 56 \ (\text{mod } 443)$$

$$5^{(443-1)/17} \quad = 5^{26} \quad \equiv 267 \ (\text{mod } 443);$$

therefore 443 is prime.

Suppose we want to construct an n and test its primality. Considering numbers of the form $n = 2 \cdot 3' + 1$ with $i = 5$ as an example. If we let $n = 487$ ($2 \cdot 3^5 + 1$) and let $x = 5$, the primality test fails:

$$5^{(487-1)/2} \quad = 5^{243} \quad \equiv 486 \ (\text{mod } 487)$$

$$5^{(487-1)/3} \quad = 5^{162} \quad \equiv 1 \ (\text{mod } 487),$$

even though 487 is prime. On the other hand, if $x = 11$ then

$$11^{(487-1)/2} \quad \equiv 11^{243} \quad \equiv 486 \ (\text{mod } 487)$$

$$11^{(487-1)/3} \quad \equiv 11^{162} \quad \equiv 232 \ (\text{mod } 487),$$

then 487 passes the test. Keep in mind that we only need to find one x for which n passes the test to prove that n is prime.

These examples highlight the two serious limitations to this test. In the first example, we needed to know beforehand the factorization of $n - 1$. From the previous discussions, we know that finding this factorization is often difficult and time-consuming. In the second example we construct $n - 1$ from factors 2 and 3. In this case, the problem is with x, The test will fail for some xs and pass for others. Even so, the fact remains that if n is prime, then a common x will exist.

Notwithstanding the limitations, the method is straightforward to implement. To work in the arithmetic residue system, note that -1 (mod n) = $(n-1)$ mod n. The inputs are x and n. The program can be easily modified to construct and test numbers without factorization (see exercises).

Program 8_6_1. Test primality using the Lucas-Lehmer test

```
/*
** Program 8_6_1 - Test primality using the Lucas-Lehmer test
*/
#include "numtype.h"
```

```
void main(int argc, char *argv[])
{
  NAT f[32], i, k, n, r, x;

  if (argc == 3) {
    n = atol(argv[1]);
    x = atol(argv[2]);
  } else {
    printf("Test primality using the Lucas-Lehmer test\n\n");
    printf("n = ");
    scanf("%lu", &n);
    printf("x = ");
    scanf("%lu", &x);
  }

  k = Factor((n-1), f);
  for (i = 0; i < k; i++)
  {
    if (f[i] != f[i+1])        /* skip duplicates */
    {
      r = PowMod(x, (n-1)/f[i], n);

      printf("%lu^(%lu-1)/%lu = %lu^%lu ≡ %lu (mod %lu)",
             x, n, f[i], x, (n-1)/f[i], r, n);

      if (f[i] == 2)
        if ( r != (n - 1))
          printf("\t...fail\n");
        else
          printf("\t...pass\n");
      else
        if ( r == 1)
          printf("\t...fail\n");
        else
          printf("\t...pass\n");
    }
  }
}

/*
** Factor - finds prime factorization by trial division
**          returned values: number of factors, array w/ factors
*/
#include "numtype.h"

NAT Factor(NAT n, NAT f[])
{
  NAT i, fk, s;

  fk = 0;                     /* factor count */

  while (!(n & 1))            /* even - factor all 2's */
  {
    f[fk++] = 2;
    n /= 2;
  }

  s = sqrt(n);
```

```
  for (i = 3; i <= s; i++, i++)      /* factor odd numbers */
    while ( (n % i) == 0L)
    {
      f[fk++] = i;
      n /= i;
      s = sqrt(n);
    }

  if (n > 1L) f[fk++] = n;

  for (i = fk; i < 32; i++) f[i] = 0;   /* zero out the rest */

  return(fk);
}


/*
** PowMod - computes r for a^n ≡ r (mod m) given a, n, and m
*/
#include "numtype.h"

NAT PowMod(NAT a, NAT n, NAT m)
{
 NAT r;

  r = 1;
  while (n > 0)
  {
    if (n & 1)                    /* test lowest bit */
      r = MulMod(r, a, m);        /* multiply (mod m) */
    a = MulMod(a, a, m);          /* square */
    n >>= 1;                      /* divided by 2 */
  }

  return(r);
}

/*
** MulMod - computes r for a * b ≡ r (mod m) given a, b, and m
*/
#include "numtype.h"

NAT MulMod(NAT a, NAT b, NAT m)
{
  NAT r;

  if (m == 0) return(a * b);      /* (mod 0) */

  r = 0;
  while (a > 0)
  {
    if (a & 1)                    /* test lowest bit */
      if ((r += b) > m) r %= m;   /* add (mod m) */
    a >>= 1;                      /* divided by 2 */
    if ((b <<= 1) > m) b %= m;    /* times 2 (mod m) */
  }
```

```
    return(r);
}
```

```
C> 8_6_1
Test primality using the Lucas-Lehmer test

n = 32413
x = 5
5^(32413-1)/2  = 5^16206 ≡ 32412 (mod 32413)    ...pass
5^(32413-1)/3  = 5^10804 ≡ 3948  (mod 32413)    ...pass
5^(32413-1)/37 = 5^876   ≡ 9127  (mod 32413)    ...pass
5^(32413-1)/73 = 5^444   ≡ 14137 (mod 32413)    ...pass
```

The Miller-Rabin (Probabilistic) Compositeness Test

The Miller-Rabin test is one of those good news-bad news situations. The good news is that the test does not require a factorization of $n - 1$; the bad news is that it does not guarantee that the number tested is prime. However, it can conclude that a number is composite. Given these limitations, why bother with it then, you may ask. What the Miller-Rabin test offers is an extremely high degree of probability of primality, and short of actually finding a factor, this may be the best we can hope for. Because of the rest's nature — of determining with certainty that a number is composite but only probably prime — it is called a compositeness test.

The algorithm begins with Fermat's little theorem, which states that if n is an odd prime and b is a positive integer such that $gcd(b, n) = 1$, then $b^{n-1} \equiv 1 \pmod{n}$. If n is prime then this will certainly be true. Since n is odd, $n-1$ is even and can be expressed as $n - 1 = 2^t \cdot q$. If $b^q \equiv 1 \pmod{n}$, we conclude that n is probably prime — that is to say, we cannot prove n is composite. Otherwise we begin to compute the series of values (mod n) for $e \le t - 1$: b^{2q}, b^{4q}, ..., $b^{2^e q}$. If any $b^{2e q} \equiv -1 \pmod{n}$, then n is probably prime. Otherwise n must be composite.

There are a couple ways to implement the Miller-Rabin test. In one approach, we halve the exponent $n - 1$ repeatedly until it is odd (Schroeder, 1997; Giblin, 1993). This approach is simpler but generally less efficient because we repeatedly exponentiate b. The algorithm presented here is the alternative but more efficient method where b is squared. Multiplication is always computationally faster than exponentiation. This technique requires a little extra work; we first need to find t and q such that $n - 1 = 2^t \cdot q$ (Cohen, 1995; Bach and Shallit, 1996).

The following algorithm and its implementation in Program 8_6_2 represent a single iteration of the Miller-Rabin to base b. Base b can be selected using a random number generator. In the program a random number generator if $b = 0$ is used. However, without randomization the program will always get the same pseudorandom number.

1. set $q = n - 1, t = 0$

2. while q is even

   shift right t bits until q is not (This gives $n - 1 = 2^t \cdot q$.)

3. Choose b such that $1 < b < n$

4. Compute $b^q \equiv r \pmod n$

   if $r = 1$ then goto step 8

5. set $e = 0$

6. while $e < t - 1$

   if $r \equiv n - 1$

      square $r \pmod n$ and increment e

   else

      break

7. if $r \neq n - 1$ then n is composite (end algorithm)

8. n is probably prime (end algorithm)

An additional step to the algorithm has been added after the base b has been randomly chosen but before it is tested. It is possible (although unlikely) that b may contain a common factor with n. If we get this lucky, the algorithm can end immediately without performing the test.

Program 8_6_2.c. Test compositeness with the Miller-Rabin test

```
/*
** Program 8_6_2 - Test compositeness with the Miller-Rabin test
*/
#include "numtype.h"

void main(int argc, char *argv[])
{
  NAT b, n;

  if (argc == 3) {
    n = atol(argv[1]);
    b = atol(argv[2]);
  } else {
    printf("Test compositeness using the Miller-Rabin test\n\n");
    printf("n = ");
    scanf("%lu", &n);
```

```
        printf("b = ");
        scanf("%lu", &b);
    }

    if (MillerRabin(n, b))
        printf("%lu is composite\n", n);
    else
        printf("%lu is probably prime\n", n);
}

/*
** MillerRabin — compositeness test
**                  returns 1 if number is composite
**                          0 if number is prime or strong pseudo
                            prime
*/
#include "numtype.h"

int MillerRabin(NAT n, NAT b)
{
    NAT e, q, r, t;

    if (!(n & 1)) return(1);            /* n is even - composite */

    if (!(b %= n))
        while ((b = rand() + 2) >= n);  /* choose: 1 < a < n */
    if ((GCD(b, n) > 1)) return(1);     /* lucky guess - composite */

    q = n - 1;
    for(t = 0; !(q & 1); t++) q >>= 1; /* while q is even */

    if ((r = PowMod(b, q, n)) != 1)     /* check q, the odd power */
        for (e = 0; e < t - 1; e++)     /* check even powers */
            if (r != n - 1)
                r = MulMod(r, r, n);
            else
                break;

    if (r == 1 || r == n - 1)
        return(0);                      /* probably prime */
    else
        return(1);                      /* composite */
}

/*
** PowMod - computes r for a^n ≡ r (mod m) given a, n, and m
*/
#include "numtype.h"

NAT PowMod(NAT a, NAT n, NAT m)
{
    NAT r;

    r = 1;
    while (n > 0)
    {
        if (n & 1)                      /* test lowest bit */
            r = MulMod(r, a, m);        /* multiply (mod m) */
```

```
      a = MulMod(a, a, m);              /* square */
      n >>= 1;                          /* divided by 2 */
  }

  return(r);
}

/*
** MulMod - computes r for a * b ≡ r (mod m) given a, b, and m
*/
#include "numtype.h"

NAT MulMod(NAT a, NAT b, NAT m)
{
  NAT r;

  if (m == 0) return(a * b);           /* (mod 0) */

  r = 0;
  while (a > 0)
  {
    if (a & 1)                         /* test lowest bit */
      if ((r += b) > m) r %= m;        /* add (mod m) */
    a >>= 1;                           /* divided by 2 */
    if ((b <<= 1) > m) b %= m;         /* times 2 (mod m) */
  }

  return(r);
}

/*
** GCD - find the greatest common divisor using Euclid's algorithm
*/
#include "numtype.h"

NAT GCD(NAT a, NAT b)
{
  NAT r;

  while (b > 0)
  {
    r = a % b;
    a = b;
    b = r;
  }

  return(a);
}
```

The Miller-Rabin test will always correctly identify an odd prime. n as an odd composite will be correctly identified as composite at least 75% of the time. At the very most 25% of the time, the test will indicate that n is "prime" when it is not.

Here is the output from Program 8_6_2:

C> **8_6_2**
Test compositeness using the Miller-Rabin test

```
n = 185
b = 43
185 is probably prime

C> 8_6_2
Test compositeness using the Miller-Rabin test

n = 257
b = 43
257 is probably prime

C> 8_6_2
Test compositeness using the Miller-Rabin test

n = 325
b = 43
325 is composite
```

The number 257 is actually prime, whereas 185 and 325 are obviously composite. In the next section, I will discuss how 185 managed to fool the Miller-Rabin test.

The algorithm can be significantly enhanced by calling the function more than once to obtain different bases (Cohen, 1995). This requires an additional loop that terminates after an arbitrary number of bases have been tried. Clearly if a single execution of this test indicates that n is probably prime, then increasing the number of bases likewise greatly increases the probability that n is prime. If the error probability is bounded above by one quarter then testing 10 different bases would give an error probability of $1/4^{10} = 0.000000954$, although in practice it is much, much lower.

Pseudoprimes, Absolute Pseudoprimes, and Strong Pseudoprimes

No discussion of primality testing would be complete without a discussion of pseudoprimes. Pseudoprimes, as their name suggests, are composite numbers that behave like primes under certain circumstances. They are perennial objects of study because of the problems they create for probabilistic primality tests and, in particular, for the Miller-Rabin test.

Recall Fermat's little theorem, which states that if n is an odd prime and b is a positive integer such that $gcd(b, n) = 1$, then $b^{n-1} \equiv 1 \pmod{n}$. Although this theorem holds fast for any prime n, it unfortunately is true for some composite numbers as well, thus proving that the converse is not true. Composite numbers that exhibit this prime-like property are called pseudoprimes. More specifically, Fermat b-pseudoprimes are composite odd numbers n, such that $b^{n-1} \equiv 1 \pmod{n}$.

Consider, for example, $2^{340} \equiv 1 \pmod{341}$. That is to say, 341 is a pseudoprime in base 2. A pseudoprime in base 2, called a *Poulet number* (Ribenboim, 1991), is any composite odd number n such that $2^{n-1} \equiv 1 \pmod{n}$. There are far fewer pseudoprimes to the base 2 than primes. Of the numbers less than 1 million, there are 245 pseudoprimes to the base 2 but 78,498 actual primes. Therefore, any number that gives the appearance of being a Fermat 2-pseudoprime has a high probability of being truly prime (roughly 99.7%). This fact lends some strength to probabilistic primality tests that rely on Fermat's little theorem.

Crocker (1962) showed how to generate an infinite sequence of pseudoprimes. Other results on pseudoprimes (base b) explore the number of factors and aspects of innumerability (Schnitzel 1958; Lieuwens, 1971; Rotkiewicz, 1972).

Strong pseudoprimes have the characteristic that they pass the Miller-Rabin test to base b. As you might expect, there are fewer strong pseudoprimes than pseudoprimes and all strong pseudoprimes are pseudoprimes. For example, there are 142 strong pseudoprimes to the base 2 less than 1 million; that is, 58% of the 2-pseudoprimes are also strong pseudoprimes. Table 8.2 shows the first 10 pseudoprimes to bases less than or equal to 25. Strong pseudoprimes are marked with an "*".

Table 8.2. The First 10 Pseudoprimes to Bases Less than or Equal to 25

Base	First 10 pseudoprimes (to Bases Less than or Equal to 25)...									
2:	341*	561*	645	1105*	1387	1729*	1905*	2047*	2465*	2701
3:	91	121*	671	703*	949	1105	1541*	1729*	1891*	2465*
4:	15	85	91	341*	435	451	561*	645*	703	1105*
5:	217*	561	781*	1541*	1729*	1891	2821	4123	5461*	5611*
6:	35	185*	217*	301*	481*	1105	1111*	1261*	1333*	1729*
7:	25*	325*	561	703*	817*	1105	1825*	2101*	2353*	2465*
8:	9*	21*	45	63	65*	105*	117	133*	153	231
9:	91*	121*	205	511	671*	697	703*	949*	1105*	1387
10:	9*	33*	91*	99	259	451	481*	561	657*	703
11:	15	133*	259	305*	481*	645*	703	793*	1105	1729*
12:	65*	91*	133*	143	145*	247*	377*	385*	703	1045
13:	21*	85*	105*	231	357	427	561*	1099*	1785*	1891
14:	15*	39	65*	195	481*	561	781*	793*	841*	985*
15:	341*	1477*	1541*	1687*	1729*	1891	1921*	2821	3133	3277*
16:	15*	51	85*	91*	255	341*	435*	451*	561*	595
17:	9*	45	91*	145*	261	781*	1111*	1305*	1729*	1885
18:	25*	49*	65*	85	133*	221	323	325*	343*	425*
19:	9*	15	45*	49*	153	169*	343*	561*	637	889*
20:	21*	57*	133*	231	399	561	671*	861	889*	1281*
21:	55	65*	85	221*	703*	793*	1045*	1105*	2035	2465*
22:	21*	69*	91*	105*	161*	169*	345*	483	485*	645
23:	33*	91	165	169*	265*	341*	385*	451	481*	553*
24:	25*	115	175*	325	553*	575	805*	949*	1105	1541*
25:	39	91	217*	403	451	561*	703	781*	1541*	1729*

* Strong pseudoprime.

You may notice that several numbers appear with greater frequency in Table 8.2. They are the absolute pseudoprimes. Absolute pseudoprimes, also called *Carmichael numbers*, are numbers that satisfy $b^{n-1} \equiv 1 \pmod{n}$ for all $1 < b < n$, where $gcd(b, n) = 1$. Carmichael numbers give fits to probabilistic primality tests and are the bane of cryptologists seeking secure enciphering. The smallest of these insidious numbers is 561.

Carmichael (1912) showed that if p is any prime dividing n and each $p-1$ divides $n-1$, then n is an absolute pseudoprime. Every Carmichael number is odd, square free, and the product of three or more distinct primes. The first 20 Carmichael numbers and their factorizations are listed in Table 8.3.

Table 8.3. The First 20 Carmichael Numbers and Their Factorizations

561	=	$3 \cdot 11 \cdot 17$
1,105	=	$5 \cdot 13 \cdot 17$
1,729	=	$7 \cdot 13 \cdot 19$
2,465	=	$5 \cdot 17 \cdot 29$
2,821	=	$7 \cdot 13 \cdot 31$
6,601	=	$7 \cdot 23 \cdot 41$
8,911	=	$7 \cdot 19 \cdot 67$
10,585	=	$5 \cdot 29 \cdot 73$
15,841	=	$7 \cdot 31 \cdot 73$
29,341	=	$13 \cdot 37 \cdot 61$
41,041	=	$7 \cdot 11 \cdot 13 \cdot 41$
46,657	=	$13 \cdot 37 \cdot 97$
52,633	=	$7 \cdot 73 \cdot 103$
62,745	=	$3 \cdot 5 \cdot 47 \cdot 89$
63,973	=	$7 \cdot 13 \cdot 19 \cdot 37$
75,361	=	$11 \cdot 13 \cdot 17 \cdot 31$
101,101	=	$7 \cdot 11 \cdot 13 \cdot 101$
115,921	=	$13 \cdot 37 \cdot 241$
126,217	=	$7 \cdot 13 \cdot 19 \cdot 73$
162,401	=	$17 \cdot 41 \cdot 233$

From this table, you might infer (correctly) that there are fewer absolute pseudoprimes than there are b-pseudoprimes or strong b-pseudoprimes. In fact, there are merely 43 Carmichael numbers less than 1 million. That is the good news for cryptologists who need to select true primes for secure encryption. The other good news is that Carmichael numbers can be detected using the Miller-Rabin test in multiple bases.

Some Carmichael numbers having three factors can be found employing a formula developed by Chernick (1939). Let $m \geq 1$ and $M_3(m) = (6m + 1)(12m + 1)(18m + 1)$. If m is such that all three factors are prime, then $M_3(m)$ is a Carmichael number. This construction method can be generalized to Carmichael numbers having four or more factors by the following formula:

$$M_k(m) = (6m + 1) \cdot (12m + 1) \cdot \prod_{i=1}^{k-2} (9 \cdot 2^i \cdot m + 1).$$

If m is chosen so that all k factors are prime and $2^{k-4} \mid m$, then $M_k(m)$ is a Carmichael number with k prime factors (Ribenboim, 1991).

Challenges

1. Modify the Miller-Rabin test so that it iterates for ten different bases.

2. Write a program that finds pseudoprimes.

3. Write a program that finds Carmichael numbers. Find the smallest absolute pseudoprime that has five factors. (Ignore: $825{,}265 = 5 \cdot 7 \cdot 17 \cdot 19 \cdot 73$.)

8.7 Gaussian Integers and Primes in the Complex Plane

There is an interesting domain (one of many) that is worth mentioning. If you are familiar with complex numbers, this domain is already familiar to you. Complex numbers are actually two-dimensional numbers (a, b) having a real part a and an imaginary part b. In their usual formulation, a and b are real numbers and are often presented as $a + bi$, where i is $\sqrt{-1}$, the imaginary unit. You may have run across complex numbers before in the form of the remarkable identity involving π, i, and e is the natural logarithm base

$$e^{\pi i} = -1.$$

Graphically, complex numbers can be visualized easily by treating (a, b) as an ordered pair in an xy plane called the complex plane. The complex numbers $2 + i$, $-1 + i$, $-2 - 2i$, and $1 - 2i$ depicted on the complex plane in Figure 8.3.

Figure 8.3. Gaussian integers in the complex plane.

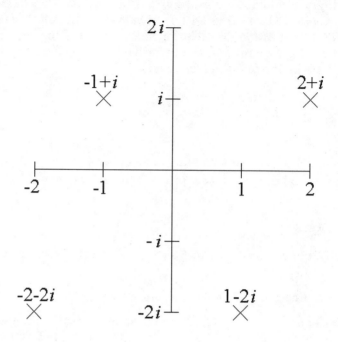

The Gaussian integers are a subset of the set of complex numbers $a + bi$ in which a and b are ordinary integers and again i is $\sqrt{-1}$, the imaginary unit. Even though a and b are integers, a is still called the real part and b is the imaginary part. The domain of Gaussian integers is designated by $Z[i]$. It is one type of quadratic domain, although other quadratic domains are possible, such as $Z[\sqrt{2}]$, $Z[\sqrt{3}]$, and so on.

Arithmetic using complex numbers can be derived using the definition of i. Two Gaussian integers $a + bi$ and $c + di$ have the following arithmetic properties:

Equality:	$(a + bi) = (c + di)$ if and only if $a = c$ and $b = d$
Addition:	$(a + bi) + (c + di) = (a + c) + (b + d)i$
Subtraction:	$(a + bi) - (c + di) = (a - c) + (b - d)i$
Multiplication:	$(a + bi) \cdot (c + di) = (ac - bd) + (ad + bc)i$
Division:	$(a + bi) / (c + di) = \dfrac{(ac + bd) + (bc - ad)i}{(c^2 + d^2)}$

If $z = a + bi$, then the complex conjugate is defined as $z^* = a - bi$. The nonnegative rational integer

$$z \cdot z^* = (a + bi) \cdot (a - bi) = a^2 + b^2$$

is called the *norm* of z, defined by $N(z)$. The norm has some useful properties. For example, for every z in $Z[i]$,

$$N(z_1 \cdot z_2) = N(z_1) \cdot N(z_2).$$

It follows that if $z_1 \mid z_2$, then $N(z_1) \mid N(z_2)$. From the definition of norm, you can see that there are four Gaussian integers that will divide every Gaussian integer: $1, -1, i, -i$. These are the *units* of $Z[i]$. If u is a unit in $Z[i]$ and $z_1 = u \cdot z_2$, then z_1 and z_2 are called *associates*. A Gaussian integer is prime if it has no divisors other than its associates and the units.

It may surprise you to find out that certain numbers that are prime in Z are composite in $Z[i]$. For example, 2 is prime in Z but in $Z[i]$ $2 = (1 + i)^2 \cdot (-i)$. On the other hand, 3 is prime in both Z and $Z[i]$. It makes sense that any composite in Z is composite in $Z[i]$. Of primes in Z, any rational prime that can be represented as the sum of two squares is a composite in $Z[i]$. Put another way, any prime in Z of the form $4k - 1$ is prime in $Z[i]$, any prime in Z of the form $4k + 1$ is composite in $Z[i]$.

If a Gaussian integer is prime, its norm must be prime or a perfect square. However, not all norms that are perfect squares are Gaussian primes. Consider the factorization of the first quadrant Gaussian integers that have a norm equal to 25:

$$3 + 4i = (2 + 1i) \cdot (2 + 1i)$$

$$4 + 3i = (1 + 2i) \cdot (1 + 2i) \cdot (0 + -1i)$$

$$5 + 0i = (1 + 2i) \cdot (2 + 1i) \cdot (0 + -1i)$$

$$0 + 5i = (1 + 2i) \cdot (2 + 1i)$$

Just because two numbers have the same norm does not mean that they have the same factorization. Table 8.4 shows Gaussian primes for $a \leq 10$ and $b \leq 10$; the norms are in parentheses.

Table 8.4. Gaussian Primes for a ? 10 and b ? 10 and their Norms

$1 + 1i$ (2)	$2 + 5i$ (29)	$5 + 6i$ (61)	$3 + 10i$ (109)
$1 + 2i$ (5)	$5 + 2i$ (29)	$6 + 5i$ (61)	$10 + 3i$ (109)
$2 + 1i$ (5)	$1 + 6i$ (37)	$3 + 8i$ (73)	$8 + 7i$ (113)
$3 + 0i$ (9)	$6 + 1i$ (37)	$8 + 3i$ (73)	$7 + 8i$ (113)
$0 + 3i$ (9)	$5 + 4i$ (41)	$5 + 8i$ (89)	$7 + 10i$ (149)
$3 + 2i$ (13)	$4 + 5i$ (41)	$8 + 5i$ (89)	$10 + 7i$ (149)
$2 + 3i$ (13)	$0 + 7i$ (49)	$9 + 4i$ (97)	$9 + 10i$ (181)
$4 + 1i$ (17)	$7 + 0i$ (49)	$4 + 9i$ (97)	$10 + 9i$ (181)
$1 + 4i$ (17)	$7 + 2i$ (53)	$1 + 10i$ (101)	
	$2 + 7i$ (53)	$10 + 1i$ (101)	

Gaussian integers share many of the properties of rational integers. An important property is that each Gaussian integer can be represented as the product of a unique set of Gaussian primes (LeVeque, 1977) (e.g., unique factorization). Likewise, Euclid's greatest common divisor algorithm can be extended to the complex plane.

For one Gaussian integer to divide another without a remainder, it is necessary for the norm of the divisor must be able to divide both the real and imaginary parts of the dividend. The units in the complex plane can be considered purely rotational components. A unit can always be factored out to rotate the complex number into the first (positive) quadrant. For example, to rotate the complex number ($a + bi$) into the first quadrant, multiply by the appropriate unit:

Quadrant	a	b	Unit	
I	$a > 0$	$b > 0$	1	(no rotation)
II	$a < 0$	$b > 0$	$-i$	
III	$a < 0$	$b < 0$	-1	
IV	$a > 0$	$b < 0$	i	

Using these facts about Gaussian integers, a simple recursive factoring algorithm based on trial division can be developed. The routine is recursive since it is possible to find a nonprime factor.

Program 8_7_1.c. Factor Gaussian integers by trial division

```
/*
** Program 8_7_1 - Factor Gaussian integers by trial division
*/
#include "numtype.h"

void main(int argc, char *argv[])
{
  INT re, im;

  printf("Factor Gaussian integers (a + bi)\n\n");
  if (argc == 3) {
    re = atol( argv[1]);          /* real part */
    im = atol( argv[2]);          /* imaginary part */
  } else {
    printf("a = ");
    scanf("%ld", &re);
    printf("b = ");
    scanf("%ld", &im);
  }

  FactorZ(re, im);
}

/*
```

```
** FactorZ - factor Gaussian integers by trial division
*/
#include "numtype.h"

#define N(a,b) (a * a + b * b)          /* re, im input */

FactorZ(INT re, INT im)
{
  INT a, b, d, i, j, n, q, r;

  if (re < 0) re = -re;
  if (im < 0) im = -im;

  n = N(re, im);                        /* norm */

  if (!IsPrime(n))                      /* norm is non-prime */
    for (i = 0; i*i <= n; i++)          /* real part */
      for (j = 0; N(i,j) <= n; j++)     /* complex part */
        if (!((i == re && j == im) ||
              (i == im && j == re)))    /* no associates */
          if ((d = N(i,j)) > 1)
          {
            a = re * i + im * j;
            b = im * i - re * j;
            while (((a % d) == 0) && ((b % d) == 0))
            {
              FactorZ(i,j);
              re = a / d;
              im = b / d;
              a = re * i + im * j;
              b = im * i - re * j;
              n = N(re, im);
            }
          }

  if ((re != 1) || (im != 0))
    printf("%ld + %ldi\n", re, im);

  return(0);
}

/*
** IsPrime - determines if a number is prime by trial division
*/
#include "numtype.h"

int IsPrime(NAT n)
{
  NAT i;

  if (n == 1)     return(0);
  if (n == 2)     return(1);
  if (!(n & 1))   return(0);            /* n is even */

  for (i = 3; i*i <= n; i++, i++)       /* n is odd */
    if ((n % i) == 0) return(0);

  return(1);
}
```

Output is the factorization of the Gaussian integer by Gaussian primes:

```
C> 8_7_1 11 13
Factor Gaussian integers (a + bi)

1 +   1i
2 +   1i
2 +   5i
0 +  -1i
```

Just as we did in Section 7.4, we can plot a picture showing the Gaussian integers in the complex plane. The factoring routine shown above is simplified to return only 0 for composite or 1 for prime, thus making it only a primality test. If the Gaussian integer is prime, a white dot is plotted; otherwise a black dot is plotted.

If a Gaussian integer is prime, its associate is also prime. Because of this fact, only half of the first quadrant needs to be factored: those points on one side of the diagonal $a = b$. If you were to plot all four quadrants of the complex plane, you would observe that each octant is a mirror image of its neighbor. For this reason, the program plots only the points in the first quadrant. As noted above, numbers in other quadrants can be rotated into the first using one of the units.

Program 8_7_2.c. Plot Gaussian integers in the complex plane

```c
/*
** Program 8_7_2 - Plot Gaussian integers in the complex plane
**                 using Microsoft C graphics functions
*/
#include "numtype.h"

#define N(a,b) (a * a + b * b)          /* Norm -- re, im */

void main(void)
{
  struct videoconfig vc;
  INT     a, b;
  NAT     m, np;
  short int k, x0, y0;

/* initialize graphics, compute center of the graph */

  _setvideomode( _MRESNOCOLOR );
  _getvideoconfig( &vc);

  x0 = vc.numxpixels - 1;              /* x origin */
  y0 = vc.numypixels - 1;              /* y origin */
  m = vc.numypixels;                   /* total primes */
  np = 0;

  k = 7;                               /* 7 = white */
```

```
   _setcolor(k);

/* make a picture of primes in the complex plane */

   for (a = 0; a < m; a++)
   {
     for (b = a; b < m; b++)
     {
       if ( IsPrimeZ(a, b) )
       {
         _setpixel( (short) a, (short) (y0 - b));
         _setpixel( (short) b, (short) (y0 - a));
         np++;
       }
     }
     if (kbhit()) break;
   }

   while(!kbhit());
     _setvideomode( _DEFAULTMODE);

   printf("%lu / %lu\n", np, m*m);
}

/*
** IsPrimeZ - determines if a Gaussian integer is prime
*/
#include "numtype.h"

IsPrimeZ(re, im)
INT re, im;
{
   INT a, b, d, i, j, n, q, r;

   if (re < 0) re = -re;
   if (im < 0) im = -im;

   n = N(re, im);                          /* norm */

   if ( !IsPrime(n) )                      /* norm is non-prime */
     for (i = 0; i*i <= n; i++)            /* real part */
       for (j = 0; N(i,j) <= n; j++)       /* complex part */
         if (!((i == re && j == im) ||
              (i == im && j == re)))       /* no associates */
           if ((d = N(i,j)) > 1)
           {
             a = re * i + im * j;
             b = im * i - re * j;
             if (((a % d) == 0) && ((b % d) == 0))
               return(0);
           }
   if (n > 1)
     return(1);
   else
     return(0);
}

/*
** IsPrime - determines if a number is prime by trial division
```

```
*/
#include "numtype.h"

int IsPrime(NAT n)
{
  NAT i;

  if (n == 1)   return(0);
  if (n == 2)   return(1);
  if (!(n & 1)) return(0);              /* n is even */

  for (i = 3; i*i <= n; i++, i++)       /* n is odd */
    if ((n % i) == 0) return(0);

  return(1);
}
```

What you might observe is that although some structures appear, the Gaussian primes appear relatively randomly distributed in the plane (Figure 8.4). Even though their density may appear to be uniform, their occurrence does diminish at roughly the same rate as the rational primes.

Figure 8.4. Gaussian integers in quadrant I of the complex plane.

In the table that follows, the number of Gaussian integers $(a + bi)$ is equal to m^2, where $0 \le a < m$, $0 \le b < m$:

m	Number of Gaussian integers	Number of Gaussian primes	
10	100	15	(15.0%)
100	10,000	756	(7.56%)
1000	1,000,000	48,979	(4.90%)
(ignore: 100,000,000	—	(_._%)	)

Challenges

1. Modify program 8_7_1 to print a table of the first Gaussian primes in which the $N(z) < 1000$.

2. Prove that any rational prime that can be represented as the sum of two squares is a composite in $Z[i]$.

3. Modify Program 8_7_2 to plot all four quadrants of the complex plane with no additional factorizations.

4. Work out the arithmetic laws for the quadratic domain $Z[\sqrt{2}]$, in which each number in the domain is represented as $a + b\sqrt{2}$. Write a program to factor numbers over this domain.

5. Using your result from Challenge 4, write a program to plot primes in the quadratic domain $Z[\sqrt{2}]$.

Diophantine Equations

On two occasions I have been asked [by members of Parliament], "Pray, Mr. Babbage, if you put into the machine wrong figures, will the right answers come out?" I am not able rightly to apprehend the kind of confusion of ideas that could provoke such a question.

CHARLES BABBAGE (1792–1871)

9.0 Introduction

An interesting and widely applied class of equations is called *diophantine equations*. They are the mathematical machines; numerical abstractions of hammers and chisels, gears and axles, and sorters and sifters. Diophantine equations span virtually all scientific disciplines; none can escape their exigent simplicity. Why try?

Diophantine equations bear the name of the great Greek algebraist Diophantus of Alexandria (c.a 250 A.D.), who studied equations with integer solutions extensively (Boyer and Merzbach, 1991). Today the term "diophantine equations" encompasses a broad class of equations having a variety of forms for which integer solutions are sought. You were probably first exposed to "diophantine equations" in elementary school in the form of a story problem, perhaps something along these lines: "A storekeeper has 50 coats to set out. His racks hold either 6 or 8 coats. How many of each rack will be required?"

The methodology you used then to solve this problem probably did not rely on any theory but on a strategy called "guess, test, and revise". You would set up a table and then start guessing at answers, as in Table 9.1.

Unlike most diophantine equations, linear types can be solved explicitly and under well-defined circumstances. Quadratic types have also been studied successfully since antiquity, including the well-known Pythagorean Theorem:

$$x^2 + y^2 = z^2.$$

271

Table 9.1. Guessing a Solution to a Diophantine Equation

Guess #	Number of racks that hold 6 coats	Number of racks that hold 8 coats	Total number of coats
1	1	5	46
2	2	5	52
3	3	3	42
4	3	4	50 success!

Other types are more difficult to analyze. The proof of the most infamous and puzzling of diophantine equations known as Fermat's last theorem — that the equation

$$x^n + y^n = z^n$$

has no nonzero integer solutions for x, y, and z when $n > 2$ — was finally achieved in 1994 by Andrew Wiles (Hogan, 1993), but not without controversy (see Section 11.3).

9.1 Linear Diophantine Equations

Linear diophantine equations have been studied at least since the time of Diophantus, generally with solutions being sought for specific problems. The Hindu mathematician Brahmagupta (c.a 628) apparently was the first one to give a general solution to linear diophantine equations (Boyer and Merzbach, 1991).

Recall Euclid's algorithm (Section 2.2) for finding the greatest common divisor of two (or more) integers. As the algorithm proceeds by dividing out smaller and smaller numbers until the remainder is reduced to zero. The same methodology can be applied to linear diophantine equations.

Linear diophantine equations have the general form

$$a \cdot x + b \cdot y = c,$$

where a, b, and c are integer coefficients. The equation has a solution if and only if $d \mid c$, where $d = gcd(a, b)$.

Consider the following equation from the coat rack example earlier in this chapter:

$$6 \cdot x + 8 \cdot y = 50,$$

where x is the number of racks that hold 6 coats and y is the number of racks that hold 8 coats and together they must hold 50 coats.

Solving for y gives

$$y = 6 + (2 - 6 \cdot x) / 8 = 6 + (1 - 3 \cdot x) / 4.$$

For y to be an integer, the term $(1 - 3 \cdot x) / 4$ must also be an integer, which is possible only if $(1 - 3 \cdot x)$ is a multiple of 4. Therefore,

$$1 - 3 \cdot x = 4 \cdot w.$$

So solving for x gives

$$x = - w + (1 - w) / 3.$$

Similarly, we can solve for w in the expression $(1 - w) / 3$ to get

$$w = 1 - 3 \cdot v.$$

Because the denominator of $3 \cdot v$ is 1, we now have a parametric form for generating solutions to the original equation with any integer value of v. Through backsubstitution, we can arrive at values of x and y. Table 9.2 shows values of v ranging from –5 to 5.

Table 9.2. Solutions to the Diophantine Equation, $6 \cdot x + 8 \cdot y = 50$, through Backsubstitution

v	w	x	y
–5	16	–21	22
–4	13	–17	19
–3	10	–13	16
–2	7	–9	13
–1	4	–5	10
–0	1	–1	7
1	–2	3	4
2	–5	7	1
3	–8	11	–2
4	–11	15	–5
5	–14	19	–8

All the xs and ys satisfy the original equation, but obviously we cannot have a negative number of coat racks. Only two pairs are reasonable solutions given the formulation of the problem: (3, 4) and (7, 1). Many problems in real life have similar implicit constraints.

Writing a program that implements the Euclidean algorithm for finding a single solution to a specific linear diophantine equation is a little tricky. The reason is the backsubstitution step. Given a general problem, the number of actual substitutions required is finite but indeterminate. For this reason, I've used a recursive algorithm that is seeded with a single value v. This is the initial value for the backsubstitution. The value v can be changed if a different solution is desired.

Program 9_1_1.c. Find a single solution to a linear diophantine equation

```
/*
** Program 9_1_1 - Find a single (non-unique) solution to linear
**                 diophantine equations of the form a*x + b*y = c.
**
** input: a, b, c are integer coefficients
**        seed -- the seed value for the computed solution.
*/
#include "numtype.h"

void main(int argc, char *argv[])
{
  INT a, b, c, x, y;

  if (argc == 5) {
    a = atol(argv[1]);
    b = atol(argv[2]);
    c = atol(argv[3]);
    y = atol(argv[4]);          /* store seed in y */
  } else {
    printf("Solve a*x + b*y = c\n\n");
    printf("a, b, c = ");
    scanf("%ld, %ld, %ld", &a, &b, &c);
    printf("seed = ");
    scanf("%ld", &y);           /* store seed in y */
  }

  SolveLinearDioEq(a, b, c, &x, &y);

  printf("(x, y) = (%ld,%ld)\n", x, y);
}

/*
** SolveLinearDioEq - find a single (non-unique) solution to linear
**                    diophantine equations of the form a*x + b*y = c.
**
** input: a, b, c are integer coefficients
**        y contains a seed value for the computed solution
*/
#include "numtype.h"

int SolveLinearDioEq(INT a, INT b, INT c, INT *x, INT *y)
{
  INT d;

/* check for the existence of a solution */

  if (c % (d = GCD(a, b)))
    {
```

```
        printf("no solution\n");
        exit(1);
    }

/* find the next set of remainders */

    if (b > 1)
        SolveLinearDioEq(b / d, (a / d) % b, (c / d) % b, x, y);

/* find the solution by backsubstitution */

    *x = *y;                    /* use seed value now */
    *y = (c - a * *x) / b;

    return(0);
}

/*
** GCD - find the greatest common divisor using Euclid's algorithm
*/
#include "numtype.h"

NAT GCD(NAT a, NAT b)
{
    NAT r;

    while (b > 0)
    {
        r = a % b;
        a = b;
        b = r;
    }

    return(a);
}
```

```
C> 9_1_1
Solve a*x + b*y = c

a, b, c = 6,8,50
seed = 1
(x, y) = (3,4)
```

Finding a single solution to a given linear diophantine equation is actually the first stage in finding the general parametric solution. We know that $a \cdot x + b \cdot y = c$ has a solution if and only if $d \mid c$, where $d = gcd(a, b)$. Suppose that $d \mid c$, let (x_0, y_0) represent an explicit solution; then, of course,

$$a \cdot x + b \cdot y = c = a \cdot x_0 + b \cdot y_0$$

for all solutions (x, y). Since $d \mid c$, we get at once

$$\frac{a}{d} \cdot x + \frac{b}{d} \cdot y = \frac{a}{d} \cdot x_0 + \frac{b}{d} \cdot y_0.$$

Therefore,

$$\frac{a}{d} \cdot (x - x_0) = \frac{b}{d} \cdot (y_0 - y).$$

Since $gcd(a / d, b / d) = 1$, we can conclude that

$$\frac{b}{d} \,\Big|\, (x - x_0).$$

Divisibility means that there exists a t such that $(x - x_0) = t \cdot b / d$. Substituting this result back into the first equation yields a solution for y. To summarize, the set of solutions to the linear diophantine equation is all integer pairs (x, y), where

$$x = x_0 + t \cdot b / d \quad \text{and} \quad y = y_0 - t \cdot a / d$$

for all t, $(t = ..., -2, -1, 0, 1, 2, ...)$.

With this general result, we can now write a program that will compute the parametric form of the solutions to any soluble linear diophantine equation. The great advantage of obtaining the parametric form is that we have the means for finding any solution, as opposed to a single solution. In fact, so many solutions are possible that we'll be swimming in them if we do not introduce some useful and often necessary constraints.

Continuing with the previous example, $6 \cdot x + 8 \cdot y = 50$, and using one of the solutions found, $(3, 4)$, we can write the following equations

$$x = 3 + t \cdot 4$$

$$y = 4 - t \cdot 3$$

for all t, $(t = ..., -2, -1, 0, 1, 2, ...)$. As an aside, t is not the same as the value v that we used to seed the backsubstitution step when finding a solution. They are, however, linearly dependent.

Application: Solving Specific Linear Diophantine Equations

Linear diophantine equations are so common that the procedure for solving them should probably be built into pocket calculators. Using our results from the previous section, we can develop a program that gives the parametric form of the solution to a linear diophantine equation. Also, it prints solution pairs over an interval defined by a minimum t_{min} value and a maximum t_{max} value.

Program 9_1_2.c. Find the parametric form for solutions to linear diophantine equations

```
/*
** Program 9_1_2 - Find the parametric form of a solution to linear
**                 diophantine equations of the form a*x + b*y = c.
**
** input: a, b, c are integer coefficients
** t1, t2 are starting t value and the ending t value
*/
#include "numtype.h"

void main(int argc, char *argv[])
{
   INT a, b, c, d, t, t1, t2, x, y;

   if (argc == 6) {
     a = atol(argv[1]);
     b = atol(argv[2]);
     c = atol(argv[3]);
     t1 = atol(argv[4]);
     t2 = atol(argv[5]);
   } else {
     printf("Solve a*x + b*y = c for t = tmin, ..., tmax\n\n");
     printf("a, b, c = ");
     scanf("%ld, %ld, %ld", &a, &b, &c);
     printf("tmin, tmax = ");
     scanf("%ld, %ld", &t1, &t2);
   }

   y = 0;                        /* solution seed */

   SolveLinearDioEq(a, b, c, &x, &y);

/* print equations */

   d = GCD(a, b);

   printf("\nDiophantine Equation\n--------------------\n");
   printf("%ld * x + %ld * y = %ld\n", a, b, c);

   printf("\nParametric Solution\n-------------------\n");
   printf("x = %ld + %ld * t\n", x, b/d);
   printf("y = %ld - %ld * t\n", y, a/d);

   printf("\nExplicit Solutions\n------------------\n");
   for (t = t1; t <= t2; t++)
     printf("t = %ld => (%ld,%ld)\n", t, x+t*b/d, y - t*a/d);
}

/*
** SolveLinearDioEq - find a single (non-unique) solution to linear
**                    diophantine equations of the form a*x + b*y = c.
**
** input: a, b, c are integer coefficients
**        y contains a seed value for the computed solution
*/
#include "numtype.h"
```

```
int SolveLinearDioEq(INT a, INT b, INT c, INT *x, INT *y)
{
  INT d;

/* check for the existence of a solution */

  if (c % (d = GCD(a, b)))
  {
    printf("no solution\n");
  exit(1);
  }

/* find the next set of remainders */

  if (b > 1)
  SolveLinearDioEq(b / d, (a / d) % b, (c / d) % b, x, y);

/* find the solution by backsubstitution */

  *x = *y;                      /* use seed value now */
  *y = (c - a * *x) / b;

  return(0);
}

/*
** GCD - find the greatest common divisor using Euclid's algorithm
*/
#include "numtype.h"

NAT GCD(NAT a, NAT b)
{
  NAT r;

  while (b > 0)
  {
    r = a % b;
    a = b;
    b = r;
  }

  return(a);
}
```

```
C> 9_1_2
Solve a*x + b*y = c for t = tmin, ..., tmax

a, b, c = 6,8,50
tmin, tmax = 0,5

Diophantine Equation
--------------------
6 * x + 8 * y = 50

Parametric Solution
--------------------
x = -1 + 4 * t
y = 7 - 3 * t
```

```
Explicit Solutions
------------------
t = 0 => (-1,7)
t = 1 => (3,4)
t = 2 => (7,1)
t = 3 => (11,-2)
t = 4 => (15,-5)
t = 5 => (19,-8)
```

Solutions where the values of x and y are both positive are frequently the goal. If the coefficients a and b have different signs, an infinite number of positive solutions will exist. Put more explicitly, if $b > 0$ and $a < 0$, then

$$t > \max(-x_0 \cdot d / b , y_0 \cdot d / a);$$

whereas if $b < 0$ and $a > 0$, then

$$t < \min(-x_0 \cdot d / b , y_0 \cdot d / a).$$

On the other hand, if the coefficients a and b have the same sign, the coefficients in the parametric form of the solution will have opposite signs. The positive solution will exist if and only if there exists an integer t such that

$$-x_0 \cdot d / b < t < y_0 \cdot d / a.$$

Returning one last time to our example, we conclude that $0.25 < t < 2.33$. Therefore, being an integer, $t = 1, 2$ will give us the positive (x, y) pairs. No other pairs can exist.

Congruences and Linear Diophantine Equations

Any linear congruence can be expressed as a linear diophantine equation. As noted in Chapter 3, congruences of the form

$$a \cdot x \equiv b \ (\text{mod } m)$$

have solutions if and only if $gcd(a,m) \mid b$.

The general congruence rewritten as a linear diophantine equation looks like

$$a \cdot x + m \cdot y = b,$$

and, identically, the equation has a solution if and only if $gcd(a, m) \mid b$. Thus, methods used to solve linear congruences can be equally applied to solve linear diophantine equations.

The Extended Euclidean Algorithm

As noted above, the linear diophantine equation

$$a \cdot x + b \cdot y = c$$

has a solution if and only if $d \mid c$, where $d = gcd(a, b)$. I now present an algorithm that solves this equation using a variation on Euclid's GCD algorithm. The extended Euclidean algorithm takes the coefficients a and b, computes $d = gcd(a, b)$, and finds the integers x and y solving the linear diophantine equation. If we rewrite the algorithm as a series of matrix operations, we get

$$\begin{bmatrix} a \\ b \end{bmatrix} = \begin{bmatrix} q_1 & 1 \\ 1 & 0 \end{bmatrix} \cdot \begin{bmatrix} b \\ r_1 \end{bmatrix}$$

$$\begin{bmatrix} b \\ r_1 \end{bmatrix} = \begin{bmatrix} q_2 & 1 \\ 1 & 0 \end{bmatrix} \cdot \begin{bmatrix} r_1 \\ r_2 \end{bmatrix}$$

$$\begin{bmatrix} r_1 \\ r_2 \end{bmatrix} = \begin{bmatrix} q_3 & 1 \\ 1 & 0 \end{bmatrix} \cdot \begin{bmatrix} r_2 \\ r_3 \end{bmatrix}$$

$$\cdots$$

$$\begin{bmatrix} r_{n-1} \\ r_n \end{bmatrix} = \begin{bmatrix} q_{n-1} & 1 \\ 1 & 0 \end{bmatrix} \cdot \begin{bmatrix} r_n \\ 0 \end{bmatrix},$$

where $d = r_n = gcd(a, b)$. Combining the matrix operations we get

$$\begin{bmatrix} a \\ b \end{bmatrix} = \begin{bmatrix} q_1 & 1 \\ 1 & 0 \end{bmatrix} \cdot \begin{bmatrix} q_2 & 1 \\ 1 & 0 \end{bmatrix} \cdots \begin{bmatrix} q_n & 1 \\ 1 & 0 \end{bmatrix} \cdot \begin{bmatrix} d \\ 0 \end{bmatrix}.$$

If we let

$$M = \begin{bmatrix} s & t \\ u & v \end{bmatrix} = \begin{bmatrix} q_1 & 1 \\ 1 & 0 \end{bmatrix} \cdot \begin{bmatrix} q_2 & 1 \\ 1 & 0 \end{bmatrix} \cdots \begin{bmatrix} q_n & 1 \\ 1 & 0 \end{bmatrix},$$

then

$$M^{-1} = \begin{bmatrix} a \\ b \end{bmatrix} = \begin{bmatrix} d \\ 0 \end{bmatrix}.$$

Because det $M_{n-1} = (-1)^n$

$$M^{-1} = (1 - 1)^n \begin{bmatrix} v & -t \\ -u & s \end{bmatrix}$$

Therefore, $x = (-1)^n \cdot v \cdot c / d$ and $y = (-1)^{n+1} \cdot t \cdot c / d$ to properly scale the solution.

Program 9_1_3.c. Finds a single solution to a linear diophantine equation using the extended Euclidean algorithm

```
/*
** Program 9_1_3 - Find a single (non-unique) solution to linear
**                 diophantine equations of the form a*x + b*y = c
**                 using the extended Euclidean algorithm
*/
#include "numtype.h"

void main(int argc, char *argv[])
{
  INT a, b, c, d, n, p, q, r, x, y;
  INT M[2][2], T[2][2];

  if (argc == 4) {
    a = atol(argv[1]);
    b = atol(argv[2]);
    c = atol(argv[3]);
  } else {
    printf("Solve a*x + b*y = c (extended Euclidean algorithm)\n\n");
    printf("a, b, c = ");
    scanf("%ld, %ld, %ld", &a, &b, &c);
  }

/* initialize matrix, iteration counter */

  M[0][0] = M[1][1] = 1;
  M[0][1] = M[1][0] = 0;
  p = 1;                             /* parity */

/* loop until remainder is zero */

  while (b != 0)
  {
    q = a / b;
    r = a % b;
    a = b;
    b = r;
    if (p > 0) p = -1;
    else p = 1;
    T[0][0] = M[0][0] * q + M[0][1];
    T[0][1] = M[0][0];
    T[1][0] = M[1][0] * q + M[1][1];
    T[1][1] = M[1][0];
    memcpy(M, T, sizeof(M));
  }
```

```
/* and the solution is... */

x = p * M[1][1] * c / a;
y = -p * M[0][1] * c / a;

printf("(x, y) = (%ld,%ld) gcd = %ld", x, y, a);
}
```

Like Program 9_1_1, this method only finds a single nonunique solution.

C> **9_1_3**
Solve a*x + b*y = c (extended Euclidean algorithm)

a, b, c = **16,-3,9**
(x, y) = (9,45) gcd = 1

Challenges

1. Modify Program 9_1_1 to solve linear congruences.

2. Modify Program 9_1_2 to generate only positive solutions (x, y). If there are more than ten positive solutions, compute only ten solutions where $(x + y)$ is minimal.

3. Modify Program 9_1_3 to find the parametric form of the solution.

9.2 Simultaneous Linear Diophantine Equations

Very often, real-world problems develop into systems of equations that require a simultaneous solution. In two dimensions this amounts to the intersection of two lines. In three, this is the intersection of three planes. With respect to real numbers, if the lines or the planes are not parallel to each other, the system of equations will have a single solution, the point of intersection. This concept is easily extrapolated to an n-dimensional space, although the geometric visualization becomes more difficult.

In expanded notation, a system of equations looks like this:

$$a_{1,1} \cdot x_1 + a_{1,2} \cdot x_2 + a_{1,3} \cdot x_3 + \ldots + a_{1,n} \cdot x_n = b_1$$

$$a_{2,1} \cdot x_1 + a_{1,2} \cdot x_2 + a_{1,3} \cdot x_3 + \ldots + a_{1,n} \cdot x_n = b_2$$

$$a_{3,1} \cdot x_1 + a_{1,2} \cdot x_2 + a_{1,3} \cdot x_3 + \ldots + a_{1,n} \cdot x_n = b_3$$

$$\ldots$$

$$a_{n,1} \cdot x_1 + a_{n,2} \cdot x_2 + a_{n,3} \cdot x_3 + \ldots + a_{n,n} \cdot x_n = b_n$$

although we prefer the more compact matrix notation

$$A \cdot x = b,$$

where

$$A = \begin{vmatrix} a_{1,1} \, a_{1,2} \, a_{1,3} \, \cdots \, a_{1,n} \\ a_{2,1} \, a_{1,2} \, a_{1,3} \, \cdots \, a_{1,n} \\ a_{3,1} \, a_{1,2} \, a_{1,3} \, \cdots \, a_{1,n} \\ \cdots \quad \cdots \quad \cdots \quad \cdots \\ a_{n,1} \, a_{n,2} \, a_{n,3} \, \cdots \, a_{n,n} \end{vmatrix}, \, x = \begin{vmatrix} x_1 \\ x_2 \\ x_3 \\ \cdots \\ x_n \end{vmatrix}, \text{ and } b = \begin{vmatrix} b_1 \\ b_2 \\ b_3 \\ \cdots \\ b_n \end{vmatrix}.$$

What we are seeking is the solution vector $x = (x_1, x_2, x_3, \ldots, x_n)$. Our problem is only slightly more difficult than the general problem when restricting our equations to integer coefficients and integer solutions. As before, a single equation has a solution if $gcd(a_{i,1}, a_{i,2}, a_{i,3}, \ldots, a_{i,n})$ $\mid b_i$ for $i = 1$ to n. However, this does not guarantee the simultaneity.

Using row reductions, we will create a diagonal matrix, D, and a transformed b vector, b'. What must be true is that $d_{i,i} \mid b_i'$ for $i = 1$ to n. If any one fails, the intersection occurs at a nonlattice point and no integer solution exists.

The procedure is similar to traditional Gaussian elimination found in any good book on numerical analysis (Conte and DeBoor, 1980). The difference here is that instead of dividing by the diagonal element, we multiply by the least common multiple for the column. Consider the following example. We begin with an augmented matrix and perform row operations until only a diagonal matrix remains:

$$\begin{array}{ccc} A & & b \\ 3 & 4 & 18 \\ 2 & 2 & 10 \end{array}.$$

1. The $lcm(a_{1,1}, a_{2,1}) = 6$. Multiply row 1 by $6/a_{1,1}$ and multiply the remaining rows by $6/a_{i,1}$:

$$\begin{array}{ccc} 6 & 8 & 36 \\ 6 & 6 & 30 \end{array} \qquad \begin{array}{c} \text{multiply by 2} \\ \text{multiply by 3} \end{array}.$$

2. Subtract row 1 from each of the remaining rows:

$$\begin{array}{ccc} 6 & 8 & 36 \\ 0 & -2 & -6 \end{array} \qquad \text{row 2 } - \text{ row 1.}$$

3. The $lcm(a_{2,1}, a_{2,2}) = 8$. Multiply row 2 by $8/a_{2,2}$ and multiply the remaining rows by $8/a_{i,2}$:

| | 6 | 8 | 36 | | multiply by 1
| | 0 | 8 | 24 | | multiply by –4.

4. Subtract row 2 from each of the remaining rows:

| | 6 | 0 | 12 | | row 1 – row 2.
| | 0 | 8 | 24 | |

5. Each element of the solution vector is determined by b_i' / d_i:

| | 1 | 0 | 2 | | divide by 6
| | 0 | 1 | 3 | | divide by 8.

Therefore, the solution vector $x = (2, 3)$.

The procedure is implemented in the next program for solving matrices up to 20×20 in size. However, the larger the matrix the greater the risk of overflow.

Program 9_2_1.c. Solve a system of linear diophantine equations using Gaussian elimination

```
/*
** Program 9_2_1 - Solve a system of linear diophantine equations
*/
#include "numtype.h"

#define MAXDIM 20

void main(void)
{
   INT a[MAXDIM][MAXDIM], b[MAXDIM], n;
   int i, j;

   printf("Solve a system of linear diophantine equations A*x = b\n\n");
   printf("n = ");
   scanf("%ld", &n);
   for (i = 0; i < n; i++)
   {
      for (j = 0; j < n; j++)
      {
         printf("a[%d][%d] = ", i, j);
         scanf("%ld", &a[i][j]);
      }
      printf("b[%d] = ", i);
      scanf("%ld", &b[i]);
   }

   printf("\ninput matrix\n------------\n");
   for (i = 0; i < n; i++)
   {
      printf("1", b[i]);
      for (j = 0; j < n; j++)
         printf("%6ld", a[i][j]);
```

```
      if (i == n / 2)
        printf(" 1 * 1x[%d]1 = 1%61d 1\n", i, b[i]);
      else
        printf(" 1   1x[%d]1   1%61d 1\n", i, b[i]);
    }

  if (SolveLinearDioEqN(a, b, n))
  {
    printf("\n=> not solvable\n");
  }
  else
  {
    printf("\nsolution vector\n---------------\n");
    for (i = 0; i < n; i++)
      printf("x[%d] = %61d\n", i, b[i]);
  }
}

/*
** SolveLinearDioEqN - solve a system of linear diophantine equations
*/
#include "numtype.h"

#ifndef MAXDIM
#define MAXDIM 20
#endif

int SolveLinearDioEqN(INT a[MAXDIM][MAXDIM], INT b[MAXDIM], INT n)
{
  int i, j, k;
  INT d, r, lcm;

  for (i = 0; i < n; i++)
  {
    lcm = labs(a[i][i]);
    if (lcm == 0) return(1);                   /* singular matrix */
    for (j = 0; j < n; j++)
      if (a[j][i])
        lcm = LCM(lcm, labs(a[j][i]));         /* column LCM */
    d = lcm / a[i][i];                         /* diagonal entry */

    for (j = 0; j < n; j++)
      a[i][j] *= d;
    b[i] *= d;

    for (j = 0; j < n; j++)
    {
      if ((i != j) && a[j][i])
      {
        r = lcm / a[j][i];
        for (k = 0; k < n; k++)
          a[j][k] = a[j][k] * r - a[i][k];
        b[j] = b[j] * r - b[i];
      }
    }
  }

/* reduce the solution vector if possible */
```

```
   for (i = 0; i < n; i++)
     if ( (b[i] % a[i][i]) != 0)
       return(2);                          /* not solvable */
     else
       b[i] /= a[i][i];

   return(0);                              /* no errors */
}

/*
** LCM -  finds the least common multiple for two integers
**        LCM = a * b / GCD(a, b)
*/
#include "numtype.h"

NAT LCM(NAT a, NAT b)
{
  b /= GCD(a,b);

  if ( (0xffffffff / b) < a )
    return(0);                             /* overflow */
  else
    a *= b;                                /* ok to multiply */

  return(a);
}

/*
** GCD - find the greatest common divisor using Euclid's algorithm
*/
#include "numtype.h"

NAT GCD(NAT a, NAT b)
{
  NAT r;

  while (b > 0)
  {
    r = a % b;
    a = b;
    b = r;
  }

  return(a);
}
```

The example takes an augmented 3 × 3 matrix

$$
A = \begin{vmatrix} 2 & 2 & 5 \\ 3 & 2 & 1 \\ 1 & 3 & 2 \end{vmatrix}, b = \begin{vmatrix} 7 \\ 2 \\ 7 \end{vmatrix}
$$

and solves for x in $A \cdot x = b$. Also, I used the more natural programming subscripts that start at 0 rather than 1.

```
C> 9_2_1
Solve a system of linear diophantine equations A*x = b

n = 3
a[0][0] = 2
a[0][1] = 2
a[0][2] = 5
b[0] = 7
a[1][0] = 3
a[1][1] = 2
a[1][2] = 1
b[1] = 2
a[2][0] = 1
a[2][1] = 3
a[2][2] = 2
b[2] = 7

input matrix
------------
|  2   2   5  |   |x[0]|       | 7 |
|  3   2   1  | * |x[1]|  =    | 2 |
|  1   3   2  |   |x[2]|       | 7 |

output matrix
------------
| -4212     0      0 |   |x[0]|   | 4212  |
|     0  2592      0 | * |x[1]| = | 5184  |
|     0     0  16848 |   |x[2]|   | 16848 |

solution vector
---------------
x[0] = -1
x[1] =  2
x[2] =  1
```

9.3 Quadratic Diophantine Equations

Quadratic diophantine equations are another very broad and important class of equations that generally involve one or more squared terms. The two most well-known types are the Pythagorean Theorem and Pell's Equation, each of which can be solved under certain conditions.

The Pythagorean Theorem and Related Forms

Undoubtedly the most famous of the quadratic diophantine equations is the Pythagorean Theorem.

Pythagorean Theorem (9.1). $x^2 + y^2 = z^2$

The solutions of the Pythagorean Theorem correspond to the lengths of the sides of a right triangle (Figure 9.1).

Figure 9.1. The right triangle.

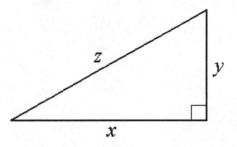

This equation has enjoyed extensive study by many mathematicians over the centuries. Although universally known as the Pythagorean equation, you may find that some authors from mainland China refer to this equation as the Shang-Gao (580–568 B.C.) equation (Hsiung, 1992).

It is possible to resolve the Pythagorean equation into parametric form. Once that form is obtained, we can crank out solutions all day long. Most of us know the solution is

$$3^2 + 4^2 = 5^2$$

and I think we can agree that

$$6^2 + 8^2 = 10^2$$

is not a new solution. Our goal will be to create the parametric form for generating solutions that are not mere multiples of other solutions. We will call the solutions we find "primitive" if x, y, and z have no common factor. In fact, if any two of the variables have a common factor, so must the third. Therefore, x, y, and z must be pair-wise relatively prime.

The first thing we can notice is that both x and y cannot be even. If they were, then z would be even also. The solution would not be primitive because of the common factor 2.

Also, both x and y cannot be odd. From congruence theory we know that all odd numbers are congruent to ± 1 (mod 4). Hence all odd squares are congruent to 1 (mod 4). If x and y were both odd then z^2 would have to be congruent to 2 (mod 4). This is impossible because all even squares have 4 as a factor and thus are congruent to 0 (mod 4).

With those possibilities out of the way, we can suppose that x is odd and y is even. Now we can introduce a new variable $u = y/2$:

$$x^2 + 4 \cdot u^2 = z^2.$$

Next we rearrange the terms to get:

$$4 \cdot u^2 = z^2 - x^2$$
$$= (z + x) \cdot (z - x)$$

Clearly, both $(z + x)$ and $(z - x)$ will be even. Next we introduce two new variables, s and r, such that

$$2 \cdot s = z + x$$
$$2 \cdot r = z - x.$$

After substitution we get

$$4 \cdot u^2 = 2 \cdot s \cdot 2 \cdot r$$

or

$$u^2 = s \cdot r.$$

You already may have noticed that $z = s + r$ and $x = s - r$. From this you can conclude that r and s have no common factors because then x and z would; a possibility we have already ruled out. This means that for $u^2 = s \cdot r$, s and r must each be perfect squares. Put another way, $s = m^2$ and $r = n^2$, so finally

$$x = m^2 - n^2$$
$$y = 2 \cdot m \cdot n$$
$$z = m^2 + n^2.$$

With regard to m and n, we can say that m must be greater than n, they both cannot be odd, and they cannot share a common factor. Now we can write a program (linked to GCD()) that will print a table of primitive solutions based on our analysis. Right triangles with the sides being each of integral length are called Pythagorean triangles, the sides of which form Pythagorean triples (x, y, z).

Program 9_3_1.c. Prints a table of primitive Pythagorean triples

```
/*
** Program 9_3_1 - Print a table of primitive solutions to the
**                 Pythagorean Theorem: x^2 + y^2 = z^2
*/
#include "numtype.h"
```

```
void main(void)
{
  INT m, n, x, y, z;

  printf("Table of primitive Pythagorean triples\n\n");
  for ( m = 2; m < 10; m++)
    for (n = (m & 1) + 1; n < m; n += 2)
      if (GCD(m,n) == 1)
        {
          x = m * m - n * n;
          y = 2 * m * n;
          z = m * m + n * n;
          printf(" (%ld, %ld) => %ld^2 + %ld^2 = %ld^2\n",
                 m, n, x, y, z);
        }
}

/*
** GCD - find the greatest common divisor using Euclid's algorithm
*/
#include "numtype.h"

NAT GCD(NAT a, NAT b)
{
  NAT r;

  while (b > 0)
    {
      r = a % b;
      a = b;
      b = r;
    }

  return(a);
}
```

```
C> 9_3_1
Table of primitive Pythagorean triples

(2, 1) => 3² + 4² = 5²
(3, 2) => 5² + 12² = 13²
(4, 1) => 15² + 8² = 17²
(4, 3) => 7² + 24² = 25²
(5, 2) => 21² + 20² = 29²
(5, 4) => 9² + 40² = 41²
(6, 1) => 35² + 12² = 37²
(6, 5) => 11² + 60² = 61²
(7, 2) => 45² + 28² = 53²
(7, 4) => 33² + 56² = 65²
(7, 6) => 13² + 84² = 85²
(8, 1) => 63² + 16² = 65²
(8, 3) => 55² + 48² = 73²
(8, 5) => 39² + 80² = 89²
(8, 7) => 15² + 112² = 113²
(9, 2) => 77² + 36² = 85²
(9, 4) => 65² + 72² = 97²
(9, 8) => 17² + 144² = 145²
```

The Congruum

In the Middle Ages, a certain amount of intellectual effort was spent trying to find what were called *congruent numbers*. A positive integer k is said to be a congruent number if there exists a rational number x such that both $x^2 + k$ and $x^2 - k$ are rational squares. Although this problem is quite difficult, a related and simpler problem asks us to find nontrivial solutions to the system of diophantine equations (Shockley, 1967):

$$x^2 + k = y^2$$
$$x^2 - k = z^2.$$

The number k in the solution is called the *congruum*.

The integer solution to the congruum problem lies in the solution to the Pythagorean equation. If (a, b, c) is a Pythagorean triple, then $k = 2ab$ is a congruum. The converse is also true: if k is a congruum, then there exists a Pythagorean triple (a, b, c) such that $k = 2ab$. Also, if k is a congruum, then $k \equiv 0 \pmod{24}$ (see Table 9.3).

Table 9.3. Congrua Where x, y, and $z < 100$ and gcd $(x, y, z) = 1$

x	y	z	k
5	7	1	24
13	17	7	120
17	23	7	240
25	31	17	336
29	41	1	840
37	47	23	840
41	49	31	720
53	73	17	2520
61	71	49	1320
65	79	47	2016
65	89	23	3696
85	97	71	2184

Application: The Integer Square Root Function

A new function was introduced with Program 6_3_1 called iSquareRoot() — the integer square root. By way of definition, iSquareRoot(n) = $[\sqrt{n}]$ is the greatest integer less than or equal to the square root of n. Although we could employ the C/C++ function sqrt(), iSquareRoot() has the feature of returning the remainder:

the difference between the number and its integer root squared. Also the algorithm presented is easily extended for multiple-precision arithmetic (see Chapter 12).

To find integer square roots, we use Newton's method. Newton's method is a general iterative method used for finding a solution to the equation

$$f(x) = 0,$$

Other iterative methods in the same vein include the bisection method and the secant method. A survey of these methods can be found in any book on elementary numerical analysis (Conte and de Boor, 1980).

Although Newton's method is usually applied to continuously differentiable functions, it can be applied to certain integer functions as well. In general, given $f(x)$ and a point x_0, for $n = 0, 1, 2, \ldots$, we calculate

$$x_{n+1} = x_n - f(x_n) / f'(x_n)$$

until $x_n = x_{n+1}$.

For the iSquareRoot() function based on Newton's method, we are given n and are seeking x, the square root of n. With regard to the integers, to find the square root $x = \sqrt{n}$ we need to find the zero for

$$f(x) = x^2 - n = 0.$$

This is a continuously differentiable function and

$$f'(n) = 2 \cdot x,$$

giving the following iteration recursive relation:

$$x_{n+1} = (x_n + n / x_n) / 2$$

For a given $n = 15$, the procedure begins by assigning $x_0 = n$. The result of this process is shown in Table 9.4.

Table 9.4. Values Obtained from $x_{n+1} = (x_n + n / x_n) / 2$

i_i	x_i	n / x_i
0	15	1
1	8	1
2	4	3
3	3	5

As noted above, for continuously differentiable functions when $x_{n+1} = x_n$, then we have found x. However, for integer solutions, Newton's method acts differently. Once a solution is reached, further iteration gives values that oscillate between x and $x + c$, where c is some constant. This behavior is typical when using Newton's method for finding integer solutions to functions.

From Table 9.4, the solution x is 3, the first instance where $x_i < n$ / x_i. Further iteration gives values that oscillate between 3 and 4. In the case of the integer square root, we want to make an additional computation for the remainder:

$$r = n - x^2.$$

If n is a perfect square, then the remainder is 0.

Program 9_3_2.c. Finds integer square roots using Newton's method

```c
/*
** Program 9_3_2 - Find the integer square root of a number
*/
#include "numtype.h"

void main(int argc, char *argv[])
{
    NAT a, q, r;

    if (argc == 2) {
        a = atol(argv[1]);
    } else {
        printf("Find integer square root\n\n");
        printf("a = ");
        scanf("%lu", &a)
    }

    iSquareRoot(a, &q, &r);

    printf("['%lu] = %lu, r = %lu\n", a, q, r);
}

/*
** iSquareRoot - finds the integer square root of a number
**               by solving: a = q² + r
*/
#include "numtype.h"

NAT iSquareRoot(NAT a, NAT *q, NAT *r)
{
    *q = a;

    if (a > 0)
        while (*q > (*r = a / *q))      /* integer root */
            *q = (*q + *r) >> 1;        /* divided by 2 */

    *r = a - *q * *q;                   /* remainder */

    return(*q);
}
```

```
C> 9_3_2
Find integer square root

a = 1000
[√Λ?1000] = 31, r = 39
```

Challenges

1. Prove that a circle inscribed inside any Pythagorean triangle has a radius of integer length and that the radius $r = n \cdot (m - n)$.

2. Write a program that finds Pythagorean triangle pairs that have equal hypotenuses but legs of different lengths (e.g., $33^2 + 56^2 = 63^2 + 16^2 = 65^2$). Write a program that finds Pythagorean triangle pairs that have equal areas with legs of different lengths (e.g., $1/2 \cdot 21 \cdot 20 = 1/2 \cdot 35 \cdot 12 = 210$). Can you find two triangles that have equal hypotenuses and area but legs of different lengths?

3. An alternative parametric form for generating Pythagorean triangles is

$$x = 2 \cdot t + 1$$
$$y = 2 \cdot t^2 + 2 \cdot t$$
$$z = 2 \cdot t^2 + 2 \cdot t + 1$$

for all t ($t = 1, 2, ...$). Find the relationship between t and m and n.

4. Write a program to find congrua less than 10,000 and where $gcd(x, y, z) = 1$.

5. Write a program that finds integer cubic roots using Newton's method.

Pell's Equation

The following type of equation has as rich a history as the Pythagorean theorem owing to its wide application in the solution of quadratic diophantine equations. The diophantine equation commonly known as Pell's equation has the form

$$x^2 - d \cdot y^2 = n,$$

with a coefficient and solutions in integers. Interestingly, the equation was never studied by John Pell (1611–1685). Pell's remote relation to the problem is that he revised a translation of someone else's discussion of Wallis and Brouncker's solution. His name became attached to the equation because of a mistake of attribution made by Euler, and thus was laid the foundation of Pell's fame. In fact, Pell's equation made its first known appearance long before John Pell.

Pell's equation often pops up in problems requiring rational approximations, although it also appears unexpectedly. It is unlikely that Archimedes (c.a. 287 B.C. – 212 B.C.) could solve the "Pellian," even though it figures prominently in his famous "cattle problem." A very readable account of the cattle problem and its solution can be found in Hoffman (1988). Instances of Pell's equation also were known to have been studied by other Greek and Indian mathematicians before receiving systematic treatment by Fermat.

Finding solutions to Pell's equation is facilitated by continued fractions. Without a loss of generality, consider the more specific case

$$x^2 - d \cdot y^2 = \pm 1,$$

where d, x, and y are integers and d is not a perfect square.

In the case that $d < -1$, no solutions exist. If $d = -1$, then the only trivial solutions are $(\pm 1, 0)$ and $(0, \pm 1)$. You also can observe that if d is a perfect square, the only solution is $(\pm 1, 0)$, the only integers whose squares differ by 1. The case when $d > 0$ and d is not a perfect square is not trivial, and this is what I will examine next. One final note: If (x, y) is a solution to Pell's equation, then so are $(-x, y)$, $(x, -y)$ and $(-x, -y)$. Therefore, only positive solutions need to be determined since the others are readily obtainable.

Let a / b be a rational approximation to the irrational number $\sqrt{d}$. If this is the case, then a^2 / b^2 is an approximation to d. Put another way, we are looking for c in

$$a^2 \backslash b^2 - d = c,$$

where c is a small integer. If it happens that $c = \pm 1$, we have the solution to Pell's equation. Clearly then, the problem of solving Pell's equation is closely associated with finding rational approximations to $\sqrt{d}$.

In Chapter 6 we found rational approximations to $\sqrt{d}$ in the form of infinite periodic continued fractions. For example, the continued fraction for $\sqrt{2} = [\ 1, 2, 2, 2, \ldots]$ yielding convergents p and q shown in Table 9.5:

$$\sqrt{2} \approx p_k / q_k, k = 0, 1, 2, \ldots$$

Whenever $p^2 - 2q^2 = 1$, then a solution has been found for $x^2 - 2y^2 = 1$.

Table 9.5. Convergents to $\sqrt{2} = [1, 2, ...]$ (Period = 1)

N	p	q	p/q	$p^2 - 2q^2$
0	1	1	1.000000000000	−1
1	3	2	1.500000000000	1
2	7	5	1.400000000000	−1
3	17	12	1.416666666667	1
4	41	29	1.413793103448	−1
5	99	70	1.414285714286	1
6	239	169	1.414201183432	−1
7	577	408	1.414215686275	1
8	1,393	985	1.414213197970	−1
9	3,363	2,378	1.414213624895	1
10	8,119	5,741	1.414213551646	−1
11	19,601	13,860	1.414213564214	1
12	47,321	33,461	1.414213562057	−1
13	114,243	80,782	1.414213562427	1
14	275,807	195,025	1.414213562364	−1
15	665,857	470,832	1.414213562375	1
16	1,607,521	1,136,689	1.414213562373	−1

In all cases the p and q give "good" approximations to $\sqrt{2}$. You can see that every (p, q) is a solution to the equation

$$x^2 - 2y^2 = \pm 1,$$

although we may be interested only in the solution to $x^2 - 2y^2 = 1$.

The first solution $(3, 2)$ solving $x^2 - 2y^2 = \pm 1$ is called the fundamental solution and is easily found from the continued fraction expansion of $\sqrt{d}$. The following theorem gives us all positive solutions from the fundamental solution.

Theorem 9.2. If (x_0, y_0) is the fundamental solution of Pell's equation $x^2 - d \cdot y^2 = \pm 1$, all positive solutions are of the form (x_n, y_n), where

$$x_n + y_n \cdot \sqrt{d} = (x_0 + y_0 \cdot \sqrt{d})^{n+1};$$

(x_n, y_n) can be found using recursive formulas

$$(x_n = x_0 \cdot x_{n-1} + d \cdot y_0 \cdot y_{n-1}$$

$$y_n = x_0 \cdot y_{n-1} + x_{n-1} \cdot y_0.$$

Clearly, if one solution can be found, then an infinite number of solutions exist. Although the first known examples of continued fractions came from Rafael Bombelli (ca. 1526–1573), it is known that these formulas were used by the Hindu mathematician Bhascara (ca. 1140 A.D.). However, it is not thought that Bhascara used continued fractions to find solutions of quadratic diophantine equations.

The next program finds solutions to $x^2 - d \cdot y^2 = \pm 1$ by first finding the fundamental solution to the equation and then using the recursion formulas above to generate additional solutions.

Program 9_3_3.c. Solve the basic Pell's equation $x^2 - d \cdot y^2 = \pm 1$

```
/*
** Program 9_3_3  - Solve the basic Pell's equation
**                      x^2 - d*y^2 = ñ1
*/
#include "numtype.h"

#define MAXB 100

void main(int argc, char *argv[])
{
   INT b[MAXB], d, i, m;
   INT p, pm1, pm2, q, qm1, qm2;
   NAT x, x0, xm1, y, y0, ym1;

   if (argc == 2) {
     d = atol(argv[1]);
   } else {
     printf("Solve basic Pell's equation: x^2 - d*y^2 = ñ1\n\n");
     printf("d = ");
     scanf("%ld", &d);
   }

/* find the CF representation of û'd (note: period = m) */

   m = cfSquareRoot(d, b, (INT) MAXB);

/* evaluate the CF using the forward recursion algorithm */

   pm2 = 0;                 /* p[-2] (next to next to last) */
   pm1 = 1;                 /* p[-1] (next to last) */
   qm2 = 1;                 /* q[-2] (next to next to last) */
   qm1 = 0;                 /* q[-1] (next to last) */

   for (i = 0; i < m; i++)
   {
     p = b[i] * pm1 + pm2;
     q = b[i] * qm1 + qm2;
     pm2 = pm1;
     pm1 = p;
     qm2 = qm1;
     qm1 = q;
   }

/* get the fundamental solution (i.e., (x0, y0) => ñ1) */

   x0 = xm1 = p;
   y0 = ym1 = q;
   printf("(%ld, %ld) (fundamental solution)\n", x0, y0);
```

```
/* generate other solutions using recursion formula */

  for (i = 0; i < 8; i++)
  {
    x = x0 * xm1 + d * y0 * ym1;
    y = x0 * ym1 + xm1 * y0;
    xm1 = x;
    ym1 = y;
    printf("(%lu, %lu)\n", x, y);
  }
}

/*
** cfSquareRoot —  find a continued fraction representation
**                 of a square root of an integer
*/
#include "numtype.h"

int cfSquareRoot(INT n, INT b[], INT maxb)
{
  INT i, p, q, r, s;
  int k = 0;

  b[k] = iSquareRoot(n, &s, &r);

  if (r > 0)
  {
    p = 0;
    q = 1;
    do
    {
      p = b[k] * q - p;
      q = (n - p * p) / q;
      if (k == maxb) return(k); else k++;
      b[k] = (p + s) / q;
    } while (q != 1);
  }

  return(k);                            /* return period */
}

/*
** iSquareRoot - finds the integer square root of a number
**               by solving: a = q² + r
*/
#include "numtype.h"

NAT iSquareRoot(NAT a, NAT *q, NAT *r)
{
  *q = a;

  if (a > 0)
    while (*q > (*r = a / *q))          /* integer root */
      *q = (*q + *r) >> 1;              /* divided by 2 */

  *r = a - *q * *q;                     /* remainder */

  return(*q);
}
```

```
C> 9_3_3
Solve basic Pell's equation: x^2 - d*y^2 = ±1

d = 2
(1, 1) (fundamental solution)
(3, 2)
(7, 5)
(17, 12)
(41, 29)
(99, 70)
(239, 169)
(577, 408)
(1393, 985)
```

The function cfSquare root() finds one period of the continued fraction expansion of $\sqrt{d}$ using the method described in Chapter 4. Its return value is the length of the period m of the continued fraction for $\sqrt{d}$. A solution to $x^2 - d \cdot y^2 = \pm 1$ can always be found in m iterations, as confirmed in the following theorem:

Theorem 9.3. Let d be a positive nonsquare integer and let m be the period of the expansion of $\sqrt{d}$. Then

Case 1. If m is even, the positive solutions of $x^2 - d \cdot y^2 = 1$ are

$$x = p_{km-1}, y = q_{km-1}, k = 1, 2, 3, \ldots$$

and $x^2 - d \cdot y^2 = -1$ has no solutions.

Case 2. If m is odd, the positive solutions of $x^2 - d \cdot y^2 = 1$ are

$$x = p_{km-1}, y = q_{km-1}, k = 2, 4, 6, \ldots$$

and the positive solutions of $x^2 - d \cdot y^2 = -1$ are

$$x = p_{km-1}, y = q_{km-1}, k = 1, 3, 5, \ldots$$

The proof of this is given in Niven and Zuckerman (1980).

When solving specific problems, it turns out that when the period m is odd, the fundamental solution yields -1. To obtain solutions to the equation $x^2 - d \cdot y^2 = 1$, we need to square the solution according to Theorem 9.2 using the recursion formula.

In our example $x^2 - 2y^2 = 1$, $\sqrt{2} = [1, 2, \ldots]$, the period is 1; therefore, we need to compute the next solution from the fundamental solution $(x_0, y_0) = (1, 1)$

$$x_1 = 1 \cdot 1 + 2 \cdot 1 \cdot 1$$

$$y_1 = 1 \cdot 1 + 1 \cdot 1$$

or (3, 2). A quick check shows that this is the correct solution to the original equation

$$3^2 - 2 \cdot 2^2 = 1 \ \sqrt{\ }.$$

It is interesting to observe the seemingly random nature of solutions to $x^2 - d \cdot y^2 = 1$, which can be seen in Table 9.6. What is most striking is the unpredictability in the magnitude of x and y. For some ds, x and y are small and, at the same time, the very next solution is quite large. For example, $x^2 - 60 \, y^2 = 1$ is solved by (31, 4), whereas the least solution for $x^2 - 61 \, y^2 = 1$ is (1,766,319,049, 226,153,980).

Table 9.6. Least Solutions to $x^2 - d \cdot y^2 = 1$, for $d < 100$ and d is Nonsquare

d	x	y
2	3	2
3	2	1
5	9	4
6	5	2
7	8	3
8	3	1
10	19	6
11	10	3
12	7	2
13	649	180
14	15	4
15	4	1
17	33	8
18	17	4
19	170	39
20	9	2
21	55	12
22	197	42
23	24	5
24	5	1
26	51	10
27	26	5
28	127	24
29	9,801	1,820
30	11	2
31	1,520	273
32	17	3
33	23	4
34	35	6
35	6	1
37	73	12
38	37	6
39	25	4

Table 9.6. (continued)

d	x	y
40	19	3
41	2,049	320
42	13	2
43	3,482	531
44	199	30
45	161	24
46	24,335	3,588
47	48	7
48	7	1
50	99	14
51	50	7
52	649	90
53	66,249	9,100
54	485	66
55	89	12
56	15	2
57	151	20
58	19,603	2,574
59	530	69
60	31	4
61	1,766,319,049	226,153,980
62	63	8
63	8	1
65	129	16
66	65	8
67	48,842	5,967
68	33	4
69	7,775	936
70	251	30
71	3,480	413
72	17	2
73	2,281,249	267,000
74	3,699	430
75	26	3
76	57,799	6,630
77	351	40
78	53	6
79	80	9
80	9	1
82	163	18
83	82	9

Table 9.6. (continued)

d	x	y
84	55	6
85	285,769	30,996
86	10,405	1,122
87	28	3
88	197	21
89	500,001	53,000
90	19	2
91	1,574	165
92	1151	120
93	12151	1,260
94	2,143,295	221,064
95	39	4
96	49	5
97	62,809,633	6,377,352
98	99	10
99	10	1

I now return to the more general problem of finding solutions to

$$x^2 - d \cdot y^2 = n.$$

The basic theorem with respect to solutions to Pell's equation is as follows:

Theorem 9.4. Let d be a positive nonsquare integer and let n be such that $0 < |n| < \sqrt{d}$. If (x_0, y_0) is a solution of $x^2 - d \cdot y^2 = n$, then x_0 / y_0 is a convergent to $\sqrt{d}$ (Shockley, 1967).

For our programming efforts, this means that the equation is not solvable unless one of the convergents to $\sqrt{d}$ solves it. What is interesting about solutions derived this way is that no statement can be made about which convergent will solve it. Even so, by Theorem 9.3 we know that by cycling through the periodic continued fraction expansion, we are certain to arrive at a solution (if it exists) in either m or $2m$ iterations, depending on the parity of m.

Consider the following equation:

$$x^2 - 29 \cdot y^2 = 4.$$

The continued fraction expansion of $\sqrt{29} = [\,5, 2, 1, 1, 2, 10, \ldots\,]$ with a period $m = 5$. Since m is odd, if a solution exists we can expect the solution within ten iterations of Theorem 4.2. If it is not found within ten then the equation is insoluble. Computing the convergents derived from the continued fraction then, the following results in the series, of values shown in Table 9.7.

Table 9.7. Convergents to the Pell's Equation $x^2 - 29 \cdot y^2 = 4$

n	pn	qn	$p^2 - 29q^2$
0	5	1	-4
1	11	2	5
2	16	3	-5
3	27	5	4

Therefore, the fundamental solution is $(x_0, y_0) = (27, 5)$. The solution (x_0, y_0) said to be a *primitive solution* if $gcd(x_0, y_0) = 1$.

The fundamental solution to the equation $x^2 - 29 \cdot y^2 = 1$ is $(s, t) = (9801, 1820)$. Therefore, additional solutions to $x^2 - 29 \cdot y^2 = 4$ can be found using

$$x_n + y_n \cdot \sqrt{d} = (x_0 + y_0 \cdot \sqrt{d}) \cdot (s + t \cdot \sqrt{d})^n.$$

Substituting to get the next solution (x_1, x_1):

$$x_1 + y_1 \cdot \sqrt{d} = (27 + 5 \cdot \sqrt{29}) \cdot (9801 + 1820 \cdot \sqrt{29})$$
$$= 528{,}527 + 98{,}145 \cdot \sqrt{29},$$

so $(x_1, y_1) = (528{,}527, 98{,}145)$.

Another solution to the equation is $(727, 135)$. This is obtained by multiplying two solutions — the first to $x^2 - 29 \cdot y^2 = -4$, which is $(5, 1)$ (see Table 9.7). The second is the solution to $x^2 - 29 \cdot y^2 = -1$, namely, $(70, 13)$. Multiplying these gives

$$x + y \cdot \sqrt{d} = (5 + 1 \cdot \sqrt{29}) \cdot (70 + 13 \cdot \sqrt{29})$$
$$= 727 + 135 \cdot \sqrt{29}.$$

Program 9_3_4. Solves the general Pell's equation $x^2 - d \cdot y^2 = n$.

```
/*
** Program 9_3_4 - Solve the general Pell's equation
**                 x^2 - d*y^2 = n
*/
#include "numtype.h"

#define MAXB 200

void main(int argc, char *argv[])
{
   INT b[MAXB], d, i, m, n;
   INT p, pm1, pm2, q, qm1, qm2;
   NAT x, x0, xm1, y, y0, ym1;
```

```
  if (argc == 3) {
    d = atol(argv[1]);
    n = atol(argv[2]);
  } else {
    printf("Solve Pell's equation: x^2 - d*y^2 = n\n\n\n");
    printf("d, n = ");
    scanf("%ld, %ld", &d, &n);
  }

/* find the CF representation of √d (note: period = m) */

  m = cfSquareRoot(d, b, (INT) MAXB / 2);
  for (i = m + 1; i <= 2*m; i++)
    b[i] = b[i-m];
  m += m;

/* evaluate the CF using the forward recursion algorithm */

  pm2 = 0;                    /* p[-2] (next to next to last) */
  pm1 = 1;                    /* p[-1] (next to last) */
  qm2 = 1;                    /* q[-2] (next to next to last) */
  qm1 = 0;                    /* q[-1] (next to last) */

  for (i = 0; i < m; i++)
  {
    p = b[i] * pm1 + pm2;
    q = b[i] * qm1 + qm2;
    pm2 = pm1;
    pm1 = p;
    qm2 = qm1;
    qm1 = q;
    if ((p * p - d * q * q) == n)
    {
      printf("(%lu, %lu)\n", p, q);
      break;
    }
  }

  if (i == m)
    printf("no solution\n");
}


/*
** cfSquareRoot -- find a continued fraction representation
**                 of a square root of an integer
*/
int cfSquareRoot(INT n, INT b[], INT maxb)
{
  INT i, p, q, r, s;
  int k = 0;

  b[k] = iSquareRoot(n, &s, &r);

  if (r > 0)
  {
    p = 0;
    q = 1;
```

```
    do
    {
      p = b[k] * q - p;
      q = (n - p * p) / q;
      if (k == maxb) return(k); else k++;
      b[k] = (p + s) / q;
    } while (q != 1);
  }

  return(k);                            /* return period */
}

/*
** iSquareRoot - finds the integer square root of a number
**               by solving: a = q² + r
*/
NAT iSquareRoot(NAT a, NAT *q, NAT *r)
{
  *q = a;

  if (a > 0)
    while (*q > (*r = a / *q))          /* integer root */
      *q = (*q + *r) >> 1;              /* divided by 2 */

  *r = a - *q * *q;                     /* remainder */

  return(*q);
}
```

```
C> 9_3_4
Solve Pell's equation: x^2 - d*y^2 = n

d, n = 29,4
(27, 5)
```

Pell's equation is important to number theory for a number of reasons. Under our consideration here, Pell's equation is a form that can be obtained by the algebraic manipulation of the general quadratic equation

$$a \cdot x^2 + b \cdot x \cdot y + c \cdot y^2 + d \cdot x + e \cdot y + f = 0$$

with integral coefficients and solutions. For example, rewriting the equation so that it is a polynomial in x, we get

$$a \cdot x^2 + (b \cdot y + d) \cdot x + c \cdot y^2 + e \cdot y + f = 0.$$

This is solvable if the discriminant is a perfect square:

$$z^2 = (b \cdot y + d)^2 - 4 \cdot a \cdot (c \cdot y^2 + e \cdot y + f).$$

Rewriting the equation for the discriminant gives

$$z^2 = (b^2 - 4 \cdot a \cdot c) \cdot y^2 + (2 \cdot b \cdot d - 4 \cdot a \cdot e) \cdot y + d^2 - 4 \cdot a \cdot f.$$

If we let

$$p = b^2 - 4 \cdot a \cdot c$$
$$q = 2 \cdot b \cdot d - 4 \cdot a \cdot e$$
$$r = d^2 - 4 \cdot a \cdot f,$$

then

$$p \cdot y^2 + q \cdot y + r - z^2 = 0.$$

As above, the discriminant of this quadratic in y also must be a perfect square. Therefore,

$$w^2 = q^2 - 4 \cdot p \cdot (r - z^2).$$

Knowing the solution to this equation gives the solution to our original, general quadratic.

Application: Irrational Gear Ratios

Although gears have an integral number of teeth, it is possible to create gear assemblages that have irrational gear ratios. To do this, we use a rational approximation to a quadratic irrational ratio sought using Pell's equation.

Suppose we seek a gear ratio of $1 : \sqrt{5}$. Program 9_3_3 gives the solution to

$$x^2 - 5y^2 = 1$$

as 2889/1292, which agrees with $\sqrt{5}$ to seven decimal places. To create the gear assembly, we need to factor

$$\frac{2889}{1292} = \frac{27}{17} \cdot \frac{107}{76}$$

to obtain more useful gear ratios. Our final gear assembly looks something like Figure 9.2.

Using continued fractions and their convergents, the concept of approximating irrationals can be extended well beyond quadratic irrationals.

Figure 9.2. Gears with an irrational ratio $1 : \sqrt{5}$.

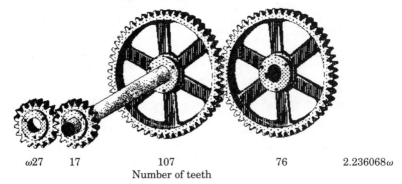

| $\omega 27$ | 17 | 107 | 76 | 2.236068ω |

Number of teeth

Challenges

1. Using Newton's method, write a program that computes integer cube roots, that is, iCubeRoot$(n) = [\sqrt[3]{n}]$.

2. Design gear assemblies with gear ratios of $1{:}\phi$ and $1{:}\pi$.

9.4 Exponential Diophantine Equations

It is generally true that a certain amount of algebraic manipulation is often required to solve exponential diophantine equations. The simplest of these have only one unknown and have the following form:

$$a^x = b,$$

where a and b are integer constants greater than 0 and x is the integer we are seeking. The most efficient algorithm runs in $O(\log_a(b))$ time and is defined by a simple recursive relationship

$$s_{x+1} = s_x \cdot a$$

until $s_{x+1} = b$ (where $s_0 = 1$). Obviously, if s_x is less than b and s_{x+1} is greater than b, the equation cannot be solved. In practice, even the largest bs and the smallest as will not require a prohibitive number of multiplications. In the following program, note how the overflow condition is protected against to avoid an infinite loop.

Program 9_4_1.c. Solves exponential diophantine equations using the trial method

```
/*
** Program 9_4_1 - Solve the exponential diophantine equation
**                 a^x = b by trial method
```

```
*/
#include "numtype.h"

void main(int argc, char *argv[])
{
  NAT a, b, s, x;

  if (argc == 3) {
    a = atol(argv[1]);
    b = atol(argv[2]);
  } else {
    printf("Solve exponential diophantine equation: a^x = b\n\n");
    printf("a, b = ");
    scanf("%lu, %lu", &a, &b);
  }

  printf("%lu^x = %lu => ", a, b);
  x = 0;
  s = 1;
  while (s < b)
  {
    if (s > 0xffffffff / a)
      break;              /* overflow */
    s *= a;
    x++;
  }

  if (s == b)
    printf("x = %lu\n", x);
  else
    printf("no solution\n");
}
```

```
C> 9_4_1
Solve exponential diophantine equation: a^x = b

a, b = 13,62748517
13^x = 62748517 => x = 7
```

Exponential diophantine equations having more than one unknown require more powerful techniques. Overall they do not present a problem that is significantly more difficult than are linear diophantine equations, in theory. However, computationally there are several complications. Consider the following exponential diophantine equations with two unknowns:

$$a^x + b^y = c,$$

where a, b, and c are integer constants and (x, y) is the solution sought. Primitive roots and the index (discrete logarithm), discussed in Chapter 4, provide a means to attack problems in this class. Consider the following example:

$$3^x = 5^y + 2.$$

By inspection we can ascertain that there is at least one solution, (3, 2), but from a software-development point of view that is cheating. Putting it another way, how can we find solutions algorithmically? The answer is: We can create a series of exponential congruences of the form

$$a^x \equiv c \pmod{b^y}.$$

Returning to the example, we have

$3^x \equiv 2 \pmod 5$ for $y = 1$

$3^x \equiv 2 \pmod{25}$ for $y = 2$

$3^x \equiv 2 \pmod{125}$ for $y = 3$

$3^x \equiv 2 \pmod{625}$ for $y = 4$

and so on. You can use Program 5_6_2 to solve each of these specific congruences.

Challenges

1. Modify Program 9_4_1 to work for $a < 0$ and $b < 0$.

2. Modify Program 5_6_2 to solve exponential diophantine equations.

Chapter

10

Number Curios

It is a Law of Nature with us that a male child shall have one more side than his father, so that each generation shall rise one step in the scale of development and nobility. Thus the son of a Square is a Pentagon; the son of a Pentagon, a Hexagon; and so on.

EDWIN A. ABBOTT, *FLATLAND*

10.0 Introduction

Number curios can be described as unusual numbers displaying special structures or properties. Aside from their amusing recreational aspects, some of the curios have interesting and deep historical roots, such as the figurate numbers. In other instances, they are pathological examples of particular problems, such as Mersenne primes and perfect numbers. Certainly one of the earliest uses that number-theoretic problems were put to were puzzles and amusements. Number curios derived from problems having interesting associations of digits and operators or digit patterns have always stirred controversy. Do these problems constitute "serious" mathematics? My feeling is that there are many ways to look at this question, but there really is only one answer.

What is considered "serious" mathematics seems to depend largely on its applicability. If a particular analysis has no applicability beyond the problem's constraints, it is less "serious" than one that is immediately useful. Must we wait for applied technology to validate our lines of research? I think not or we deny the existence of mathematical beauty. I believe that there is as much aesthetic beauty in an elegantly engineered program as in any painting hanging in a museum.

Even in the cases of so-called recreational mathematics, the tools needed for solutions tend to draw from deeper results in related areas of investigation. Some of the "simplest" problems auger into the nature of numbers so profoundly that they expose the black abyss of intractability. Clearly, curios and novelty problems should not be dismissed as not being "serious" problems. They serve as a vital link in the chain of mathematical thought extending between creative and pragmatic thought.

Lastly, I would point out that many of the examples in this chapter bear the names of famous mathematicians who first delineated some of their attributes. And why shouldn't they? Their curios, unlike some so-called serious work, are at once both entertaining and provocative. If our greatest mathematical minds did not find number curios too trivial for their consideration, how can we?

10.1 Triangular, Square, and Pentagonal Numbers

It is hard to imagine that the types of numbers studied 2500 ago would be of interest to modern mathematicians and programmers. But they are. The figurate numbers of Pythagoras of Samos (ca. 580–500 B.C.) are applicable in modern everyday life as well as in number-theoretic investigations. Some of the figurate numbers you may already be familiar with, perhaps because of their geometrical interpretation. We'll look at the first three kinds of these old and venerable numbers.

Triangular numbers are constructed by making a pyramid of points each row increasing by one point (Figure 10.1). The number of rows in the pyramid is called its rank. A number generated so constructed is equal to the sum of the integers up to its rank. Triangular numbers exhibit an odd-odd, even-even pattern.

The square numbers are simply that, perfect squares (Figure 10.2). The rank r of a square number is the number of points on a side.

Figure 10.1. The geometry of triangular numbers reveals their recursive nature.

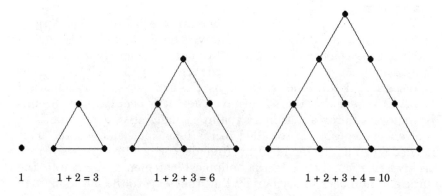

$$1 \qquad 1 + 2 = 3 \qquad 1 + 2 + 3 = 6 \qquad 1 + 2 + 3 + 4 = 10$$

Another way of looking at it is that a square number is equal to the sum of the first r odd integers. Unlike triangular numbers, square numbers alternate odd and even, and the difference between any two is always an odd number.

Figure 10.2. Square numbers have a familiar geometry and recursion.

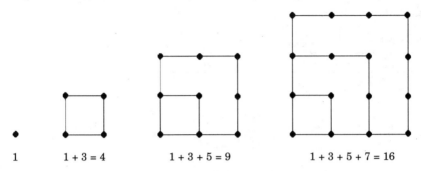

| 1 | 1 + 3 = 4 | 1 + 3 + 5 = 9 | 1 + 3 + 5 + 7 = 16 |

Pentagonal numbers are less common but are developed in a manner similar to their more common cousins. The points are arranged in ever-growing pentagons have rank r points on each side (Figure 10.3). Unfortunately, there is no convenient formulaic way to think of pentagonal numbers. However, pentagonal numbers are commonplace in works dealing with additive number theory and partitioning functions. A pentagonal number of rank r is equal to the sum of a rank r square number and a rank $(r - 1)$ triangular number.

Figure 10.3. Pentagonal numbers are more unusual in their geometry but the recursion is still evident.

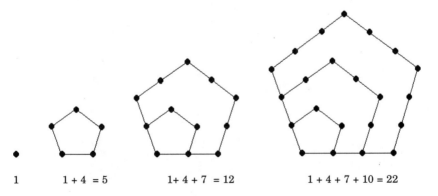

| 1 | 1 + 4 = 5 | 1 + 4 + 7 = 12 | 1 + 4 + 7 + 10 = 22 |

Figurate numbers were the darlings of the ancient Pythagoreans. Their enchantment with these numbers was so great that the triangular pattern for the number 10, the holy tetractys, competed with the pentagon (and pentagram) for adoration (Boyer and Merzbach,

1991). Although the pentagon won out, one must wonder what effect the greater veneration of the holy tetractys by the Pythagoreans might have had on the present-day sport of bowling.

Writing a program to compute figurate numbers is no harder than knowing the formula for the numbers desired. In Table 10.1 are the formulas for each of the figurate numbers described, where k = the kth figurate number (rank) and n = number of sides.

Table 10.1. Formulas for Figurate Numbers

Type of number	Formula
Triangular	$k \cdot (k + 1) / 2$
Square	k^2
Pentagonal	$k \cdot (3 \cdot k - 1) / 2$
n–gonal	$k \cdot ((n - 2) \cdot k - (n - 4)) / 2$

Program 10_1_1. Create a table of figurate numbers

```
/*
** Program 10_1_1 - Create a table of figurate numbers
*/
#include "numtype.h"

void main(int argc, char *argv[])
{
  INT i, k, n;

  if (argc == 2) {
    k = atol(argv[1]);
  } else {
    printf("Table of figurate numbers\n\n");
    printf("starting rank = ");
    scanf("%ld", &k);
  }

  printf("\n  rank   triangular      square   pentagonal");
  printf("\n------   ----------   ----------   ----------\n");

  for (i = 0; i < 10; i++, k++)
  {
    printf("%6ld", k);
    for (n = 3; n < 6; n++)
      printf("  %10ld", k * ((n - 2) * k - (n - 4)) / 2 );
    printf("\n");
  }
}
```

```
C> 10_1_1
Table of figurate numbers

starting rank = 1
```

rank	triangular	square	pentagonal
1	1	1	1
2	3	4	5
3	6	9	12
4	10	16	22
5	15	25	35
6	21	36	51
7	28	49	70
8	36	64	92
9	45	81	117
10	55	100	145

The problem of representing numbers as the sum of square numbers dates back at least to Diophantus of Alexandria. Diophantus apparently knew that every integer is the sum of two, three, or four square numbers. However, although first explicitly stated by Bachet in 1621, the theorem remembered today is that of J. L. Lagrange (1736–1813).

Fermat, most famous for the fact that his copy of *Diophantus* had margins too narrow to hold a certain proof, made other noteworthy claims. He wrote in a letter to Pascal on September 25, 1654, that every number is expressible as the sum of n n-gonal numbers. The proof of this statement was ultimately provided by Cauchy in 1813 (Guy, 1994).

While the world was waiting for Cauchy to prove Fermat's statement, Gauss made his famous "Eureka!" discovery. As Gauss wrote in his diary on July 10, 1796:

$$\text{Eureka! } num = \Delta + \Delta + \Delta,$$

meaning that any integer can be represented as the sum of three triangular numbers. Interestingly, all even perfect numbers are also triangular numbers (see below).

Application: A Probabilistic Square Number Test

It is possible to determine probabilistically if a number is a square without extracting the square root. This can be done by checking the remainders of the number modulo one or more numbers. The greater the number of successes, the greater the likelihood that the number is a square.

For example, the quadratic residues mod 16 are 0, 1, 4, 9. Any number mod 16 that does not have one of these remainders cannot be a square. Given a random number, there is a 12/16 chance that the number can be shown to not be a square. In the spirit of most probabilistic tests, having a quadratic residue does not guarantee that it is a square. For example, 33 is not a square but 33 mod 16 = 1. Looking at it another way, there is a 4/16 chance that a nonsquare will pass the test. Similarly, there are 22 quadratic residues mod 100, giving a 78/100 chance of proving that a given number is not a square. Fermat used this fact when looking for squares using his $x^2 - y^2$ factoring method (Chapter 5).

We can apply a variation on the casting out 9s rule discussed in Chapter 3. The number $999999 = 3^3 \cdot 7 \cdot 11 \cdot 13 \cdot 37$. And of course $1000000 \equiv 1 \bmod 999999$. We may take any number and break it up into blocks of six digits. After summing the blocks, we can test the sum for quadratic residues for 7, 11, 13, 27, and 37 shown in Table 10.2.

Table 10.2. Shows All Possible Quadratic Residues for Various Moduli

m	Number of quadratic residues	Quadratic residues
7	4	0, 1, 2, 4
11	6	0, 1, 3, 4, 5, 9
13	7	0, 1, 3, 4, 9, 10, 12
16	4	0, 1, 4, 9
27	11	0, 1, 4, 7, 9, 10, 13, 16, 19, 22, 25
37	19	0, 1, 3, 4, 7, 9, 10, 11, 12, 16, 21, 25, 26, 27, 28, 30, 33, 34, 36
100	22	0, 1, 4, 9, 16, 21, 24, 25, 29, 36, 41, 44, 49, 56, 61, 64, 69, 76, 81, 84, 89, 96

Suppose we want to test a large number for "squareness." Consider the number Fibonacci number F_{150}:

$$9,969,216,677,189,303,386,214,405,760,200.$$

Break the number into blocks of six digits and add them up:

$$
\begin{array}{r}
9 \\
969,216 \\
677,189 \\
303,386 \\
214,405 \\
+\ 760,200 \\
\hline
2,924,405
\end{array}
$$

Now check the sum for each of the factors of 999,999 until we have a quadratic nonresidue:

$$2{,}924{,}405 \quad \equiv 1 \bmod 7$$

$$2{,}924{,}405 \quad \equiv 0 \bmod 11$$

$$2{,}924{,}405 \quad \equiv 3 \bmod 13$$

$$2{,}924{,}405 \quad \equiv 8 \bmod 27$$

Since 8 is a quadratic nonresidue mod 27, the number F_{150} is not a square.

The probability p of a nonsquare passing this test is

$$p = \frac{4}{7} \cdot \frac{6}{11} \cdot \frac{7}{13} \cdot \frac{11}{27} \cdot \frac{19}{37} = 0.0351.$$

Recall from Chapter 8 how Fermat used the mod 100 test to help him detect squares. If this test is combined with the mod 100 test, then the probability falls to 0.00772.

The next program implements the squareness test. I am using the Jacobi symbol to identify quadratic residues for each of the factors of 999,999. If any are a quadratic nonresidue, then the number is definitely not a square. If it passes all the tests, then it is probably a square.

Program 10_1_2. Test for squareness using a probabilistic test

```
/*
** Program 10_1_2 - A probabilistic squareness test
*/
#include "numtype.h"

void main(int argc, char *argv[])
{
  INT n;

  if (argc == 2) {
    n = atol(argv[1]);
  } else {
    printf("Probabilistic squareness test\n\n");
    printf("n = ");
    if (scanf("%ld", &n) != 1) exit(1);
  }

  if (SquareTest(n))
    printf("%ld is probably a square\n", n);
  else
    printf("%ld is not a square\n", n);
}

/*
** SquareTest - performs a probabilistic squareness test
**
** returns: 0 = not square, 1 = probably square
*/
#include "numtype.h"
```

```
int SquareTest(INT n)
{
  INT i, r, s;
  NAT p[5] = { 7, 11, 13, 27, 37 };      /* factors of 999999 */

  s = 0;
  while (n > 0)
  {
    s += n % 1000000;                    /* extract 6 digits */
    n /= 1000000;
  }

  for (i = 0; i < 5; i++)
    if (JacobiSymbol(s, p[i]) < 0)
      return(0);

  return(1);
}

/*
** JacobiSymbol - evaluates the Jacobi Symbol (generalized Legendre
**                Symbol) using Euler's criterion and the Law of
**                Quadratic Reciprocity
*/
#include "numtype.h"

int JacobiSymbol(INT a, INT n)
{
  INT e, t;

  e = 1;
  while (a != 0)
  {
    while (!(a & 1))                  /* a is even */
    {
      a >>= 1;
      if ((n % 8) == 3 || (n % 8) == 5) e = -e;
    }

    t = a; a = n; n = t;             /* swap */

    if ((a % 4) == 3 && (n % 4) == 3) e = -e;

    a %= n;
  }

  if (n == 1)
    return(e);
  else
    return(0);
}
```

```
C> 10_1_2
Probabilistic squareness test

n = 100000000
100000000 is probably a square
```

Challenges

1. Prove that the sum of any two successive triangular numbers is a square number.

2. Write a program for computing hexagonal, heptagonal, and octagonal numbers.

3. Write a program for computing tetragonal and cubic numbers.

4. Design a squareness test based on 1001 similar to the one shown above. Determine the probability of a nonsquare passing this test.

5. Modify Program 10_1_2 to include the mod 100 test.

10.2 Mersenne Primes

There is a certain type of prime number called a Mersenne, so named in honor of the Franciscan friar Marin Mersenne (1588–1648). In spite of their namesake, these numbers have been studied by many cultures since the time of the ancient Greeks. A Mersenne number is a number generated by the following expression:

$$M(p) = 2^p - 1.$$

If $M(p)$ happens to be prime, it is called a Mersenne prime. A brief list of the first 22 Mersenne numbers showing the relative occurrence of primes among them appears in Table 10.3.

Table 10.3. The First 22 Mersenne Numbers and Their Primality

p	$M(p)$	$M(p)$ prime?
2	3	Yes
3	7	Yes
4	15	No
5	31	Yes
6	63	No
7	127	Yes
8	255	No
9	511	No
10	1,023	No
11	2,047	No
12	4,095	No
13	8,191	Yes
14	16,383	No

Table 10.3. (continued)

p	M(p)	M(p) prime?
15	32,767	No
16	65,535	No
17	131,071	Yes
18	262,143	No
19	524,287	Yes
20	1,048,575	No
21	2,097,151	No

You can see that although that although a Mersenne prime must have p prime, not all prime ps yield a Mersenne prime (e.g., 11). Fermat and Euler both worked out that any factor of $M(p)$ must have the forms $4 \cdot n + 1$ and $2 \cdot k \cdot p + 1$. This information can be used to search for factors of $M(p)$ that are spaced by $4 \cdot p$ (Schroeder, 1997, Section 3.5).

At this time there are only 37 known Mersenne primes. The last was discovered by Roland Clarkson on January 27, 1998, using a program written by George Woltman. Spence's success came after many months and with the aid of more than 2000 volunteers worldwide in the Great Internet Mersenne Prime Search GIMPS (website: http://www.mersenne.org/prime.htm). The final validation took 15 days on a 100 MHz Pentium. Table 10.4 shows the presently known Mersenne primes and the number of digits they contain.

Table 10.4. The Known Mersenne Primes

n	v	$p_i + 1/p_i$	Number of digits in $M(p)$
1	2	1.5000	1
2	3	1.6667	1
3	5	1.4000	2
4	7	1.8571	3
5	13	1.3077	4
6	17	1.1176	6
7	19	1.6316	6
8	31	1.9677	10
9	61	1.4590	19
10	89	1.2022	27
11	107	1.1869	33
12	127	4.1024	39
13	521	1.1651	157
14	607	2.1071	183

Table 10.4. (continued)

n	p	$p_i + 1/p_i$	Number of digits in $M(p)$
15	1,279	1.7224	386
16	2,203	1.0354	664
17	2,281	1.4103	687
18	3,217	1.3220	969
19	4,253	1.0400	1,281
20	4,423	2.1906	1,332
21	9,689	1.0260	2,917
22	9,941	1.1280	2,993
23	11,213	1.7780	3,376
24	19,937	1.0885	6,002
25	21,701	1.0695	6,533
26	23,209	1.9172	6,987
27	44,497	1.9382	13,395
28	86,243	1.2813	25,962
29	110,503	1.1950	33,265
30	132,049	1.6364	39,751
31	216,091	3.5029	65,050
32	756,938	1.1354	227,862
33	859,433	1.4635	258,716
34	1,257,787	1.1117	378,632
35	1,398,269	2.1285	420,921
36	2,976,221	1.0152	895,932
37	3,021,377	–	909,526

What is interesting about the known Mersenne primes is their regularity and the possibility that they are governed by a Poisson process. Gillies (1963) was probably the first to use statistical analysis to aid in the discovery of the 21st, 22nd, and 23rd Mersenne primes. Subsequently refined by Wagstaff (1983), Gillies suggested that the density of primes near p that yield Mersenne primes should be asymptotic to

$$\frac{e^\gamma}{p \cdot \log 2},$$

where e is the natural logarithm base and γ is Euler's constant ($\gamma \approx 0.577216$). Comparing this to the general density of primes near p, $1/\log p$ (see Chapter 4.1). Based on this, one can make the prediction that the ratio on any two successive primes should be roughly equal to $2^{1/e^\gamma} \approx 1.47576$.

Although there are some notable gaps, a striking trend appears if one plots the $\log_2$ of the values of the primes that generate the Mersenne prime. Figure 10.4 shows $\log_2 p_i$ plotted against $\log_2 p_{i+1}$. This graph may give us some insight into their relative occurrence.

Figure 10.4. Distribution of Mersenne primes.

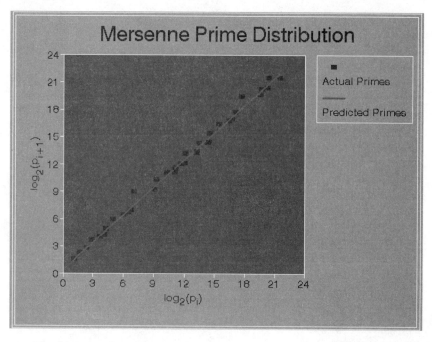

The line on the plot can serve as a heuristic predicting approximately where Mersenne primes may fall either in the gaps or after extrapolation. If the gaps are overlooked (after $p = 127, 607, 4423,$ and $216{,}091, 1{,}398{,}269$), then the ratio of $p_{i+1}/p_i = 1.3799$ gives a significantly better match than Wagstaff's ratio.

All of the values under $p = 365{,}000$ and most under $472{,}000$ have been checked so the first 31 numbers are the first 31 Mersenne primes (Klee and Wagon, 1991). However there is a notable gap after $M(216{,}091)$. Using the heuristic ratio of 1.3799 we might expect Mersenne primes in the vicinity of $p = 298187, 411469,$ and 567779.

Lucas Sequences and Their Relationship to Mersenne Primes

You may well wonder—if it is so difficult to factor large numbers (> 1000 digits) how the largest known Mersenne prime has 420,921 digits. The reason is because it was never actually factored. A primality test was developed by Edouard Lucas (1842–1891) in 1876 that is particularly well suited for testing the primality of Mersenne primes. In fact, Lucas verified *by hand* that $M(127)$, a 39-digit number, is indeed prime.

The test, called the Lucas-Lehmer test, is based on the Lucas sequence defined as follows. Let $L_0 = 4$ and let $L_{n+1} = L_n^2 - 2$. $M(p)$ is prime if and only if $M(p)$ divides L_{p-2} or, put another way, $L_{p-2} \equiv 0 \pmod{M(p)}$. Look at Table 10.5 for the first few numbers in the Lucas sequence.

Table 10.5. Lucas Numbers 0 Through 6

n	L_n
0	4
1	14
2	194
3	37,634
4	1,416,317,954
5	2,005,956,546,822,746,114
6	4,023,861,667,741,036,022,825,635,656,102,100,994

You can see that the sequence grows very large very fast. However, you can test without ever actually calculating the complete value of the Lucas number. For example, $M(5)$ divides L_3 (i.e., $37,634 \equiv 0 \pmod{31}$); therefore, $M(5)$ is a Mersenne prime. On the other hand, $M(6)$ does not divide L_4 (i.e., $1,416,317,954 \equiv 23 \pmod{31}$), so $M(6)$ is not a Mersenne prime.

What is so handy about the Lucas-Lehmer test is that the test can be implemented so that only p multiplications are needed to determine if $2^p - 1$ is a Mersenne prime (Rosen, 1988). The reason is that we are testing divisibility, so we can use multiplication and reduction by the modulus at every iteration of the sequence. We could perform the modulus operation at the same time we multiply using the line

```
1 = (1 * 1 - 2) % m,
```

and this would work fine. But because of the special form of the Mersenne number, the product (mod $M(p)$) can be accomplished using a logical AND rather than modular arithmetic. This gives us an important computational edge.

Remember Program 5_5_1 from Chapter 5. Divisibility of a number by 7, was checked using the "casting out 9s" principle. In that case, the low-order bits were masked and summed. If the sum was divisible by

7 then the number was divisible by 7. Here we have an identical situation that will allow us to optimize the Lucas-Lehmer test. We are testing the divisibility of a number by $M(p) = 2^p - 1$, where 2^p is the radix. The Lucas number is masked by the Mersenne number twice, first for the low-order "digit" and then for the high order "digit." There can never be more than two digits since on each iteration the Lucas number is squared and reduced.

In the program, the operation is implemented as a while loop so that the sum of the digits is always less than or equal to the Mersenne number. Thus, if the sum of the digits is equal to the Mersenne number, then the Mersenne number divides the Lucas number. This method is shown in the next program:

```
/*
** Program 10_2_1 - Test Mersenne numbers for primality
**                  using the Lucas test
*/
#include "numtype.h"

void main(void)
{
  NAT i, l, m, p, s;

  printf("Test Mersenne numbers for primality\n\n");

  for (p = 3; p < 16; p++)
  {
    m = (1 << p) - 1;              /* the mersenne number */
    l = 4;                        /* initialize lucas sequence */

    for (i = 1; i < p - 1; i++)
    {
      l = (l * l - 2);
      while (l > m)
        l = (l & m) + (l >> p);
    }

    if (l == m)
      printf("M(%lu) = %lu ==> prime\n", p, m);
    else
      printf("M(%lu) = %lu ==> not prime\n", p, m);
  }
}
```

```
C> 10_2_1
Test Mersenne numbers for primality

M(3)  = 7 ==> prime
M(4)  = 15 ==> not prime
M(5)  = 31 ==> prime
M(6)  = 63 ==> not prime
M(7)  = 127 ==> prime
M(8)  = 255 ==> not prime
M(9)  = 511 ==> not prime
M(10) = 1023 ==> not prime
M(11) = 2047 ==> not prime
M(12) = 4095 ==> not prime
```

```
M(13) = 8191  ==> prime
M(14) = 16383 ==> not prime
M(15) = 32767 ==> not prime
```

Challenge

1. Check the primality of the number of the form $2p + 1$.

10.3 Abundant, Deficient, and Perfect Numbers

Consider the divisors of a particular number and their sum indicated by the function $\sigma(n)$ (see Chapter 2.4). If the sum of the divisors, excluding the number itself, is less than the number, then it is said to be *deficient*. On the other hand, if the sum of the divisors is greater than the number, it is called *abundant*. A number is *perfect* when the sum of its divisors is equal to itself. Constructing a short table of the first 30 numbers (Table 10.6), we can see that these characteristics have a somewhat random occurrence, although there are clearly more deficient numbers than abundant numbers.

Table 10.6. Abundant, Deficient, and Perfect Numbers for n Between 2 and 30

n	Sum of divisors	abu/def/per
2	1	Deficient
3	1	Deficient
4	3	Deficient
5	1	Deficient
6	6	Perfect
7	1	Deficient
8	7	Deficient
9	4	Deficient
10	8	Deficient
11	1	Deficient
12	16	Abundant
13	1	Deficient
14	10	Deficient
15	9	Deficient
16	15	Deficient
17	1	Deficient
18	21	Abundant
19	1	Deficient

Table 10.6. (continued)

n	Sum of divisors	abu/def/per
20	22	Abundant
21	11	Deficient
22	14	Deficient
23	1	Deficient
24	36	Abundant
25	6	Deficient
26	16	Deficient
27	13	Deficient
28	28	Perfect
29	1	Deficient
30	42	Abundant

It turns out as one might hope; perfection in numbers, as in most things, is the rarest of attributes. Perfect numbers, at least those that are known, are directly related to Mersenne primes. Since ancient times, the prime numbers that have the special form

$$M(p) = 2^p - 1$$

have been used to construct perfect number that have the form

$$P_p = 2^{p-1} \cdot M(p) = 2^{p-1} \cdot (2^p - 1).$$

This fact was actually known to the ancient Greeks (Proposition 36, Book IX of Euclid's *Elements*). It also was known then that this is the only form that *even* perfect numbers can take. You can see without much effort that every perfect number is a triangular number as well:

$$P_p = 2^{p-1} \cdot (2^p - 1) = 2^{p-1} \cdot (2^p - 1) = \Delta_{2^p-1}.$$

With respect to the sum of the divisors function introduced in Chapter 2, a number n is perfect if and only if $\sigma(n) = 2 \cdot n$.

Given a particular number, it is not possible to know if it is abundant or deficient without recognizing its special form. Of particular special forms it is interesting to note the percentage of abundant versus deficient numbers generated ($1 \leq n \leq 1,000$):

Form	% Abundant	% Deficient	% Perfect
n	24.6	75.1	0.3
$3 \cdot n$	50.3	49.6	0.1
$5 \cdot n$	38.3	61.7	0.0
n^2	65.3	34.7	0.0
$n^2 + 1$	1.6	98.4	0.0

Clearly, some forms of numbers generate more of one type of number than another. It is an open question whether there is a form that generates nothing but numbers of strictly one type.

Challenge

1. Write a program that classifies numbers according to whether they are abundant, perfect, or deficient and that produced output like Table 10.6. Check numbers of the various forms discussed in this chapter.

10.4 Fibonacci Numbers

During the Middle Ages little was added to the body of mathematical thought. One noteworthy exception to this were the contributions of Leonardo of Pisa (ca. 1200–1256), better know as Fibonacci, or "son of Bonaccio." Fibonacci was an Italian merchant of means whose book, *Liber Abaci* (or *Book of the Abacus*) did much to advance algebraic methods and the use of Hindu-Arabic numerals (Boyer, 1991).

Among the various problems presented in his book, one has deservedly received more far more attention than any of the others:

> How may pairs of rabbits will be produced in a year, beginning with a single pair, if in every month each pair bears a new pair which becomes productive from the second month on?

The solution to this problem gives rise to the celebrated "Fibonacci sequence." The sequence is developed such that each term is the sum of the two preceding terms:

$$F_n = F_{n-1} + F_{n-2}.$$

Given $F_0 = 0$, $F_1 = 1$, this yields 0, 1, 1, 2, 3, 5, 8, 13, 21, 34, ... The rate at which this sequence grows is quite impressive, as can be observed in the following program.

```
/*
** Program 10_4_1 - Print a list of Fibonacci numbers
*/
#include "numtype.h"

#define MAXF 50

void main(void)
{
   NAT   f[MAXF], n;
```

```
printf("Fibonacci numbers\n\n");

f[0] = 0;
f[1] = 1;

for (n = 0; n < MAXF; n++)
{
  if (n > 1)
  {
    f[n] = f[n-1] + f[n-2];
    if (f[n] < f[n-1]) break;        /* overflow */
  }
  printf("F[%lu] = %lu\n", n, f[n]);
}
}
```

```
C> 10_4_1
Fibonacci numbers

F[0] = 0
F[1] = 1
F[2] = 1
F[3] = 2
F[4] = 3
F[5] = 5
F[6] = 8
F[7] = 13
F[8] = 21
F[9] = 34
F[10] = 55
F[11] = 89
F[12] = 144
F[13] = 233
F[14] = 377
F[15] = 610
F[16] = 987
F[17] = 1597
F[18] = 2584
F[19] = 4181
...
F[46] = 1836311903
F[47] = 2971215073
```

The Fibonacci sequence seems to possess almost magical qualities. The ratio of any two successive terms rationally approximates the golden ratio, the limit of which is the golden ratio:

$$\phi = \lim_{n \to \infty} \frac{F_{n+1}}{F_n} = \frac{(1 + \sqrt{5})}{2}.$$

It has long been know that every third F_n is divisible by F_3, every fourth F_n is divisible by F_4, every fifth F_n is divisible by F_5, and so on. Put another way,

$$Fn \mid F_{k \cdot n}, \text{ for } k > 1.$$

Thus, any composite n leads to a composite Fibonacci number. You might be tempted to speculate that if n is prime, then F_n is prime, since $F_3 = 2$, $F_5 = 5$, $F_7 = 13$, $F_{11} = 89$, $F_{13} = 233$, and $F_{17} = 1597$. But this sequence comes to an abrupt end with $F_{19} = 4181 = 37 \cdot 113$. However, it is a remarkable fact that for *every* prime p there exists an F_n that p divides.

Because of the way Fibonacci numbers are constructed, certain characteristics emerge. For example, if u and v are integers such that $a > b > 0$ and the Euclidean algorithm computes the greatest common divisor $d = gcd(a, b)$ in n steps, then $a \geq F_{n+2}$ and $b \geq F_{n+1}$.

It seems that Fibonacci numbers exhibit unusual properties in just about every way they can be looked at. The GCD of any two Fibonacci numbers is equal to the Fibonacci number indexed by the greatest common divisor of the indexes of the Fibonacci numbers:

$$gcd(F_m, F_n) = F_{gcd(m, n)}.$$

For example.

$$gcd(F_{26}, F_{39}) = gcd(121393, 63245986) = 233 = F_{13}.$$

Now that's something one would not expect!

There are a number of interesting identities relating to Fibonacci numbers. One that can be used to factor a number n is

$$F_{2n+1} = F_{n+1}^2 + F_n^2.$$

If n divides F_{2n+1}, then

$$F_{n+1}^2 \equiv - F_n^2 \ (\text{mod } n).$$

As you will recall from Chapter 8, this give us a chance to find a factor for n by evaluating the GCD $gcd(F_{n+1}^2, n - F_n^2)$.

Fibonacci numbers seem as prolific in nature as the rabbit populations they were invented to study. They are applicable to many regenerative processes, such as the number of seeds, leaves, and petals of many plants; the curvature of certain cephalopod shells, and bumble bee progeny. Fibonacci numbers also have been used to model successfully the population growth of "exotic" species that invade natural habitats. It should come as no surprise that serious research continues into the properties of these remarkable numbers even though 750 years have passed since their development.

The Golden Ratio

Since the time of the ancient Greeks, the golden ratio, also known as the divine proportion, has been regarded with special reverence. So much can be said about it that entire books are dedicated to the study of its nature (Huntley, 1970; Boles, 1987). Although derived geometrically, the golden ratio's beauty derives from its application in the visual arts and its occurrence in the natural sciences. Consider the line segment of length l:

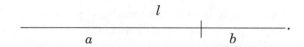

We seek a cut that will divide the line segment into two segments having lengths a and b, such that

$$\frac{l}{a} = \frac{a + b}{a} = \frac{a}{b}.$$

If the length of b is taken to be 1, then we get

$$a^2 - a - 1 = 0,$$

the solutions of which are the golden ratio

$$\phi = (1 + \sqrt{5})/2 = 1.61803\ 39887\ 49894\ 84820\ 45868\ 34365\ 63811\ 77203\ ...$$

and its negative reciprocal

$$\phi' = (1 - \sqrt{5})/2 = -0.61803\ 39887\ 49894\ 84820\ 45868\ 34365\ 63811\ 77203\ ...$$

That is,

$$\phi \cdot \phi' = -1.$$

ϕ and ϕ' are the only numbers that when diminished by 1 are their own reciprocals:

$$\phi - 1 = 1 / \phi \quad \text{and} \quad \phi' - 1 = 1 / \phi'$$

Another remarkable feature of ϕ is its behavior when squared:

$$\phi^2 = \phi + 1.$$

In fact,

$$\phi^3 = \phi^2 + \phi$$

$$\cdots$$

$$\phi^{n+1} = \phi^n + \phi.$$

Is there anything that ϕ can't do?

This ratio itself can be derived from a variety of construction methods. A rectangle whose sides are in the golden ratio $\phi: 1$ is called a golden rectangle. Such a rectangle is easily constructed:

1. Start with a square (Figure 10.5).

2. Bisect the square (Figure 10.5).

3. Draw a diagonal from the intersection of the bisector to the opposite corner (Figure 10.5).

Figure 10.5. Steps 1-3 in the geometric construction of a golden rectangle.

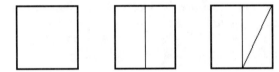

4. Using the diagonal as a radius, draw an arc from the square's corner to the base (Figure 10.6).

5. Form a rectangle with the extended base and the top of the square (Figure 10.6).

Figure 10.6. Steps 4 and 5 in the geometric construction of a golden rectangle.

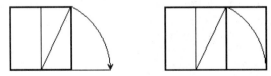

6. Now we have the golden rectangle (Figure 10.7).

Figure 10.7. Completed geometric construction of a golden rectangle.

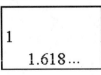

In fact, not only is this a golden rectangle but the rectangular part that was added on to the square is a golden rectangle as well. There is also a golden triangle with similar properties (Figure 10.8). See if you can create a golden triangle by construction.

Figure 10.8. The golden triangle.

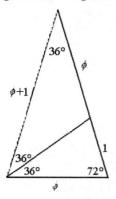

The numerous and complex connections between the golden mean and the Fibonacci sequence has been and continues to be subject of investigation (Vajda, 1989; Rosen, 1991). We know that the ratio of any two successive terms of the Fibonacci sequence rationally approximates the golden ratio

$$\phi = \lim_{n \to \infty} \frac{F_{n+1}}{F_n} = \frac{(1 + \sqrt{5})}{2}$$

Also, the continued fraction expansion for $\phi = [\ 1, 1, 1, 1, 1, ...]$ is the slowest to converge whereas the forward recursion formula yields nothing but Fibonacci numbers.

The famous spirals that have the same curvature as nautilus shells and elephant tusks never cease to fascinate. Starting with a golden rectangle, a sequence of golden rectangles are nested one inside the other. The spiral that can be constructed by drawing a curve through the corners of the squares is called a logarithmic spiral.

The squares themselves are linear combinations of ϕ in the form $a + b\phi$, where a and b are successive Fibonacci numbers. The logarithmic spiral is called equiangular because any line drawn through the center of the spiral will cut the spiral at a constant angle (~73°). Unlike other geometric forms, the logarithmic spiral occurs in nature; the most famous example being in the shape of the shell of the nautilus. So fascinated was Jakob Bernoulli (1654–1705) with its beauty that he called it *spira mirabilis* (Figure 10.9) and asked that it be engraved on his tombstone (Huntley, 1970). From its formal definition (in polar coordinates)

$$r = e^{\theta \cdot \frac{2}{\pi} \cdot \log \phi}$$

you can see that the ratio of the sides of the squares that this spiral circumscribes is $1:\phi$.

Figure 10.9. The *Spira Mirabilis:* $r = e^{\theta \cdot \frac{2}{\pi} \cdot \log \phi}$.

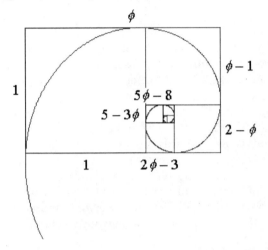

As a final note on the golden ratio, you may be surprised to know that there is an infinite radical expansion for ϕ:

$$\phi = \sqrt{1+ \sqrt{1+ \sqrt{1+ \sqrt{1+ \sqrt{1+ \sqrt{1+ \ldots}}}}}}$$

Its resemblance to the continued fraction is more than uncanny. Later I will show how general infinite radical expressions can actually converge to integers under certain circumstances.

Challenges

1. Show that Fibonacci numbers $Fn+2$ and $Fn+1$ will yield the worst–case for the Euclidean algorithm (i.e., taking the n number of steps).

2. Show that $\phi^n = \phi^{n-1} + \phi^{n-2}$.

3. Show that $\phi = \phi^{-1} + \phi^{-2} + \phi^{-3} + \ldots$

4. Write a program that evaluates the infinite radical representation for ϕ and compare its rate of convergence to the continued fraction.

10.5 Fermat Numbers

Fermat numbers, like Fibonacci numbers, have a long association with prime numbers, in part because of a mistaken notion. Fermat numbers have the form

$$F(n) = 2^{2^n} + 1.$$

Based on the fact that for $n = 0, 1, 2, 3, 4$, $F(n)$ is prime, Fermat conjectured that all numbers of this type would be prime. Euler answered his conjecture one hundred years after it was proposed with the factorization of $F(5) = 641 \cdot 6,700,417$. He obtained this factorization by first proving that any prime factor of a Fermat number must have the form $2^{n+1} \cdot k + 1$. For $n = 5$, a prime factor will have the form $p = 64 \cdot k + 1$, yielding the candidates 193, 257, 449, 577, and 641.

Fermat numbers grow extremely rapidly. Owing to their large size, ease of construction, and (supposed) composite nature, Fermat numbers serve as the acid test for factoring techniques. Notwithstanding Euler's discovery, authors of factoring algorithms earn major bragging rights when their shiny new method factors a Fermat number in record time. The first few Fermat numbers are shown in Table 10.7.

Table 10.7. Fermat Numbers $F(0)$ Through $F(7)$

n	$F(n)$
0	3
1	5
2	17
3	257
4	65,537
5	4,294,967,297
6	18,446,744,073,709,551,617
7	340,282,366,920,938,463,463,374,607,431,768,211,457

It is now known that $F(5)$ through $F(21)$ are composite. It is now widely believed that there may not be any more Fermat primes; yet another conjecture based on only scant more evidence than Fermat's initial inference. Will mathematicians ever learn?

Pepin's Test for the Primality of Fermat Numbers

The test for the primality of Fermat numbers (Pepin, 1877) is really a corollary to the Lucas test for primality. The test is as follows. Given the Fermat number F, then F is prime if and only if

$$k^{(F-1)/2} \equiv -1 \pmod{F},$$

where $k > 3$.

Challenges

1. Analyze the bit patterns of Fermat numbers.

2. Write a program that computes Fermat numbers and tests them using Pepin's test.

10.6 Armstrong Numbers

Armstrong numbers are venerable number curios; the stuff that mathematical lore is made of. Over the centuries, numerologists have sought meaning in curios such as these through creative interpretation. These are n-digit numbers whose digits, when raised to the nth power and summed equal that number (e.g., the digits of 407 cubed and summed equal 407, a third order Armstrong number). The number 153 has a special place in history because of its many "mystical" properties. Not only is it the smallest Armstrong number, but it is also the sum of the integers from 1 to 17.

An elementary analysis of the ranges of Armstrong numbers shows that although any order is possible, there are no second order Armstrong numbers. The following program will find Armstrong numbers up to the seventh order. This may take a couple of minutes to run depending on the speed of your computer.

Program 10_6_1.c. Makes a table of Armstrong numbers

```
/*
** Program 10_6_1 - Print a list of Armstrong numbers
*/
#include "numtype.h"

void main(void)
{
  NAT a, b, n, s, x;

  printf("Armstrong numbers\n\n");

  for (n = 2; n < 8; n++)                 /* 1 - 7 digit numbers */
  {
    for (a = IPOW(10L, n - 1); a < IPOW(10L, n); a++)
    {
      b = a;
      s = 0;
```

```
      while (b)
      {
        s += IPOW(b % 10, n);
        b /= 10;
      }
      if (s == a)
        printf("%lu\n", a);
    }
    printf("\n");
  }
}

/*
** IPOW -- compute the integer power of an integer
*/
#include "numtype.h"

NAT IPOW(NAT m, NAT n)
{
  NAT p;

  if (n == 0) {
    p = 1;
  } else {
    p = m;
    while (--n) p *= m;
  }

  return(p);
}
```

```
C> 10_6_1
Armstrong numbers

153
370
371
407

1634
8208
9474

54748
92727
93084

548834

1741725
4210818
9800817
9926315
```

Challenges

1. Prove that there cannot be a two digit Armstrong number.

2. Show that Armstrong numbers of any order can exist. Can you predict if there are any orders > 7 for which Armstrong numbers do not exist?

3. Make your own "numbers" based on a relationship between the numbers and their digits.

10.7 Bailey-Borwein-Plouffe π Digit Extraction Algorithm

Until very recently it was believed that one could not calculate the nth digit of π, or any irrational number for that matter, without first calculating the preceding $n - 1$ digits. David Bailey, Peter Borwein, and Simon Plouffe astounded mathematicians everywhere when they announced that they had computed the 10-billionth digit of π. Their remarkable formula has been called nothing less than miraculous. The following infinite series

$$\pi \sum_{n=0}^{\infty} \left(\frac{4}{8n+1} - \frac{2}{8n+4} - \frac{1}{8n+5} - \frac{1}{8n+6} \right) \cdot \left(\frac{1}{16} \right)^n$$

can be evaluated up to a given hexadecimal digit. As can be seen in the program that follows, to compute to a specific hexadecimal digit n of π, we evaluate the first n terms plus a few extra. Clearly, the larger the value of n, the more terms that need to be evaluated. However, because we are only interested in a specific digit, we can discard all the extraneous digits to the left using modular arithmetic.

Program 10_7_1.c. Evaluate the nth hexadecimal digit of π

```
/*
** Program 10_7_1 - Evaluate the nth hexadecimal digit of pi
*/
#include "numtype.h"
#include <math.h>

#define EPSILON 1e-16

void main(int argc, char *argv[])
{
   INT     i, n;
   double  s, t;
   char    str[8];

   if (argc == 2) {
      n = atol(argv[1]);
   } else {
      printf("Evaluate the nth hexadecimal digit of pi\n\n");
      printf("n = ");
```

```
        scanf("%lu", &n);
    }

/* compute the digits up to n */

    for (i = 0, s = 0; i < n; i++)
        s += 4.0 * (double) PowMod(16L, n - i, (8*i + 1)) / (8*i + 1) -
             2.0 * (double) PowMod(16L, n - i, (8*i + 4)) / (8*i + 4) -
                   (double) PowMod(16L, n - i, (8*i + 5)) / (8*i + 5) -
                   (double) PowMod(16L, n - i, (8*i + 6)) / (8*i + 6);

/* compute additional terms until they are too tiny to matter */

    for (t = 1.0; t > EPSILON; i++, t /= 16.0)
        s += 4.0 * t / (8*i + 1) -
             2.0 * t    / (8*i + 4) -
                   t    / (8*i + 5) -
                   t    / (8*i + 6);

/* print the result */

    ConvertHexFraction(s, sizeof(str), str);
    printf("[%ld] (%lf) = 3.%s%8.8s\n", n, s, (n ? " ... " : ""), str);
}

/*
** Convert the decimal fraction of a double to hexadecimal fraction
*/

void ConvertHexFraction(double n, int nc, char str[])
{
    char hex[] = "0123456789abcdef";
    int  i;

    for (i = 0; i < nc; i++)
    {
        n = (n - floor(n)) * 16;
        str[i] = hex[(int) n];
    }
}

/*
** PowMod - computes r for aü a^n ≡ r (mod m) given a, n, and m
*/
#include "numtype.h"

NAT PowMod(NAT a, NAT n, NAT m)
{
    NAT r;

    r = 1;
    while (n > 0)
    {
        if (n & 1)                     /* test lowest bit */
            r = MulMod(r, a, m);       /* multiply (mod m) */
        a = MulMod(a, a, m);           /* square */
        n >>= 1;                       /* divided by 2 */
    }
```

```
        return(r);
}

/*
** MulMod - computes r for a * b ≡ r (mod m) given a, b, and m
*/
#include "numtype.h"

NAT MulMod(NAT a, NAT b, NAT m)
{
    NAT r;

    if (m == 0) return(a * b);          /* (mod 0) */

    r = 0;
    while (a > 0)
    {
        if (a & 1)                      /* test lowest bit */
            if ((r += b) > m) r %= m;   /* add (mod m) */
        a >>= 1;                        /* divided by 2 */
        if ((b <<= 1) > m) b %= m;      /* times 2 (mod m) */
    }

    return(r);
}
```

The digits themselves are counted from the right of the decimal point starting with 0. In the `printf` statement I handle the special case $n = 0$; otherwise the skipped digits are indicated with an ellipsis.

```
C> 10_7_1
Evaluate the nth hexadecimal digit of pi

n = 0
[0] (3.141593) = 3.243f6a88

C> 10_7_1
Evaluate the nth hexadecimal digit of pi

n = 1000
[1000] (-58.711155) = 3. ... 49f1c09b
```

10.8 Other Curios

Surely there is no end to the number of puzzles and numerical oddities that can be invented or discovered. Permutations and manipulations of a number's digits are interesting, if not useful, types of problems. Remarkable patterns and seemingly incongruous relationships can arise. Take, for example, the rather odd diophantine equation

$$a^b \cdot c^d = abcd,$$

where a, b, c, and d are the digits 1 through 9. Analytic methods for problems such as these necessarily rely on ingenious algebraic manipulations. But suppose we do not feel so clever today. We can still find the only solution,

$$2^5 \cdot 9^2 = 2592,$$

by brute force. Simply program your computer to try all digit combinations and you're done; Proof by Exhaustion. Try other combinations for four- and six-digit numbers and see if you can find other expressions that behave similarly.

Illegal Cancellation

If you ever want to undermine your child's confidence in the laws of mathematics completely, show him or her this:

$$\frac{16}{64} = \frac{1\cancel{6}}{\cancel{6}4} = \frac{1}{4}, \qquad \text{or} \qquad \frac{19}{95} = \frac{1\cancel{9}}{\cancel{9}5} = \frac{1}{5}$$

Clearly, the correct answer was arrived at, but the math police may have already been summoned. This type of puzzle is an indeterminate diophantine equation. To find other examples of fractions behaving badly, we look to find the xs, ys, and zs that solve

$$\frac{10x + y}{10y + z} = \frac{x}{z},$$

with the additional constraints that x, y, and z are positive integers less than 10 and $x \neq z$.

At first blush, it would seem that there are 729 possible solutions (i.e., 9^3). However, since $x \neq z$, we have $8 \cdot 9^2 = 846$. Further, with algebraic manipulations, the expression can be transformed into

$$9xz = (10x - z)y$$

If $9 \mid (10x - z)$, then

$$10x \equiv z \pmod 9,$$

which is the same as

$$x = z,$$

a disallowed solution. Nontrivial solutions can be arrived at only when $3 \mid y$. Now we only have to try $y = 3$, 6, or 9. We should also recognize that $x < z$, again reducing the number of possible solutions to 108. At this point, we can attempt to solve the problem, by exhaustion.

If we want three-digit examples we need to vary slightly the constraints on y or x and z. The following fractions also have the property of allowing illegal double-digit cancellations: 166/664 = 1/4, 266/665 = 2/5, 199/995 = 1/5, 484/847 = 4/7, and 499/998 = 4/8.

Infinite Radicals

We first encountered this strange beast in the discussion of the golden ratio, where

$$\phi = \sqrt{1+\sqrt{1+\sqrt{1+\sqrt{1+\sqrt{1+\sqrt{1+\cdots}}}}}}$$

You will probably be surprised to learn that this type of expression can actually converge to an integer. Consider the more general expression

$$x = \sqrt{n+\sqrt{n+\sqrt{n+\sqrt{n+\sqrt{n+\sqrt{n+\cdots}}}}}},$$

where the limit of the expression is x. Although it looks a bit scary, it can be simplified very easily. Since the iteration is infinite, we can substitute x under the first radical to get

$$x = \sqrt{n+x},$$

a remarkable simplification. Now squaring both sides yields

$$x^2 = n + x,$$

so

$$n = x\,(x - 1).$$

Now we can produce infinite radicals that evaluate integers all day long for $x = 2$, $n = 2$; $x = 3$ $n = 6$; $x = 4$, $n = 12$; and so on.

A program for evaluating infinite radicals is fairly simple. You can use any n as input, but only those special ns just discussed will converge to an integer limit.

Program 10_8_1.c. Evaluate an infinite radical

```
/*
** Program 10_8_1 - Evaluate an infinite radical
*/
#include "numtype.h"

void main(int argc, char *argv[])
```

```
{
  NAT n;
  double x, y;

  if (argc == 2) {
    n = atol(argv[1]);
  } else {
    printf("Evaluate an infinite radical\n\n");
    printf("n = ");
    scanf("%lu", &n);
  }

  x = 0.0;
  do
  {
    y = x;
    x = sqrt(n + x);
    printf("%.6lf\n", x);
  } while (x - y >= 0.000001);
}
```

```
C> 10_8_1
Evaluate an infinite radical

n = 6
2.449490
2.906801
2.984426
2.997403
2.999567
2.999928
2.999988
2.999998
3.000000
3.000000
```

If you don't want the result printed on every iteration, simply move the print statement outside the do while loop. I have included it inside so that the rate of convergence can be observed.

Magic Squares

Magic squares have been known for centuries. They are the arrangement of unique numbers in a grid such that the rows, columns, and diagonals add up to the same value. The value to which they sum is usually called the magic number. The order of a square is the number of rows and columns. The simplest case is a 3×3 square (order 3):

8	1	6
3	5	7
4	9	2

Each row, column, and diagonal of this magic square sum to 15. The magic number m can be easily determined by

$$m = \Delta_o2 \,/\, o,$$

where o is the order and Δ is the triangle number.

Using a general method called the De la Loubere method (Spencer, 1982) one can construct odd-order squares of any size. To create an order o magic square, the method proceeds as follows:

1. Place the number 1 at the top center cell.

2. Move up and right to place the next number. Whenever the digit would be placed above the magic square, move to the bottommost cell in that column and place it there. If the digit falls to the right of the square, place it in the first cell in that row.

3. When o digits have been written, drop down one cell and place the next number. Go to step 2 and continue until the square has been completely filled. Even-order magic squares are a bit different. Here is a 4×4 magic square whose magic number is 34:

16	2	3	13
5	11	10	8
9	7	6	12
4	14	15	1

The method for construction presented here (Spencer, 1982) for a doubly even (i.e., $4 \mid o$)-order magic square is a bit more interesting. Using a method called the double diagonal method:

1. Start numbering with the number 1 in the lower right corner.

2. If the cell lies on either diagonals of the 4×4 subsquare, number it; otherwise skip it. Move to the left and continue to increment the numbers. When the left edge is reached, go up one row and continue at the rightmost column.

3. Continue until the square in the upper left corner is reached.

4. Starting with the upper left and counting from 1, number the squares, moving to the right. Do not fill in any subdiagonal that

has already been assigned a number. Advance to the next row, first column, when the end of a row is reached. Continue until the doubly even magic square is completely filled out.

The next program will create an odd-order magic square or a doubly even magic square. An even magic square whose order is not divisible by 4 is left as a challenge to you (see below).

Program 10_8_2.*c*. Create odd-order and doubly even-order magic squares

```
/*
** Program 10_8_2.c - Make magic squares
**
** notes: if an odd order is entered, use the De la Loubere method
**        else if doubly even use double diagonal method
*/
#include "numtype.h"

#define MAXO 20

void main(int argc, char *argv[])
{
    int i, c, r, o, s[MAXO][MAXO];

    if (argc == 2) {
        o = atol(argv[1]);
    } else {
        printf("Create Magic Squares\n\n");
        printf("order = ");
        scanf("%d", &o);
    }

    if (o > MAXO) o = 3;

    printf("magic number = %d\n\n", (o*o*o + o) / 2);

/* odd or even order */

    if (o & 1)                          /* odd order */
    {
        r = 0;
        c = o / 2;

        for (i = 1; i <= o*o; i++)
        {
            s[r][c] = i;
            if (i % o) { r--; c++; } else r++;
            if (r < 0) r = o - 1;
            if (c == o) c = 0;
        }
    }
    else if (o % 4 == 0)                 /* doubly even order */
    {
        r = c = o - 1;
```

```
    for (i = 1; i <= o*o; i++)        /* diagonals */
    {
      if ((r %4 == c % 4) ||
        (r % 4 + c % 4 == (o - 1) % 4))
        s[r][c] = i;

      if (c == 0) { c = o - 1; r--; } else c--;
    }

    r = c = 0;

    for (i = 1; i <= o*o; i++)        /* off-diagonals */
    {
      if (!((r %4 == c % 4) ||
        (r % 4 + c % 4 == (o - 1) % 4)))
        s[r][c] = i;

      if (c == o - 1) { c = 0; r++; } else c++;
    }
  }
  else                              /* singly even order */
  {
    printf("Sorry, I don't know how to do that.\n");
    o = 0;
  }

/* now print it */

  for (r = 0; r < o; r++, putchar('\n'))
    for (c = 0; c < o; c++)
      printf("%4d", s[r][c]);
}
```

```
C> 10_9_2
Create Magic Squares

order = 7
magic number = 175

  30   39   48    1   10   19   28
  38   47    7    9   18   27   29
  46    6    8   17   26   35   37
   5   14   16   25   34   36   45
  13   15   24   33   42   44    4
  21   23   32   41   43    3   12
  22   31   40   49    2   11   20

C> 10_9_2
Create Magic Squares

order = 8
magic number = 260

  64    2    3   61   60    6    7   57
   9   55   54   12   13   51   50   16
  17   47   46   20   21   43   42   24
  40   26   27   37   36   30   31   33
```

32	34	35	29	28	38	39	25
41	23	22	44	45	19	18	48
49	15	14	52	53	11	10	56
8	58	59	5	4	62	63	1

Magic squares of any order can be created using relatively simple techniques. However, the magic hexagon is a different bird altogether. The magic hexagon is a hexagonal array of cells and each "row" sums to the magic number, even though the rows are of varying length. The number of cells in the shortest row is the order. The order 3 magic hexagon is shown in Figure 10.10.

Figure 10.10. A magic hexagon of order 3. (Copyright 1963, *Scientific American*.)

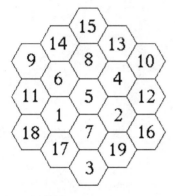

The story associated with this magic hexagon is even more remarkable than the figure itself. Clifford Adams, a railroad clerk, began searching for an order 3 magic hexagon in 1910. After 47 years he found just one and then, to his great dismay, lost the paper that contained the solution. After Mr. Adams spent another 5 years trying to rediscover the solution, he located the lost paper. He then sent it to Martin Gardner, editor of the popular "Mathematical Games" column in *Scientific American*. Gardner sent it to the mathematician C. W. Trigg, who proved that, excluding reflections and rotations, Mr. Adams's hexagon is the one and only magic hexagon of any order that exists. Wow!

Challenges

1. Write a program that finds fractions that permit illegal single-digit cancellation in three-digit numbers, such as 217/775 = 21/75.

2. Modify your program to find cancellations of the type 143185/ 1701856 = 1435/17056.

3. Modify Program 10_8_2 to work for any-order magic squares.

11

Unsolved Problems

What force is in relation to our will,
the impenetrable opacity of mathematics is
in relation to our intelligence.

SIMON WEIL, *NOTEBOOKS*, (VOL.2)

11.0 Introduction

History is replete with instances of unsolved problems and conjectures in number theory. In this chapter I will look at only a few. From the extensive number of problems, I have tried to select a few that are both infamous and undemanding (from a number theoretical point of view). I also have attempted to harvest from various branches of number theory, acknowledging that there are many more delicacies that have gone unmentioned.

Number-theoretic problems arise all the time. Some have been solved; others will be solved in the future; many will never be. Broadly speaking, there are two approaches to attacking these open questions: intellectual force and brute force. Conjectures, statements we suspect are true, remain undecided because the number-theoretic tools to resolve them have not been developed yet and not enough brute force is available. In the quest for truth, if the human mind is like a surgeon's tool, then the computer is like a sledgehammer.

Conjectures often spring to mind when considering difficult (if not intractable) related problems. After considering Fermat's last theorem and proving that it was true when the exponent is equal to 3; that is, that

$$x^3 + y^3 = z^3$$

cannot be solved in integers; Euler made the bold conjecture that if n ≥ 3, then fewer than n nth powers cannot sum up to an nth power. Although it seems plausible, nobody has been able to prove or disprove the statement using analytical methods. Enter the computer. In 1966 L. J. Lander and T. R. Parker found

$$27^5 + 84^5 + 110^5 + 133^5 = 144^5$$

using a computer (Guy, 1981). In 1996, Colvin, Scher, and Seidl (1996), reported that they had found a second solution to the fifth-order equation

$$14132^5 + 220^5 - 14068^5 - 6237^5 = 5027^5.$$

The search for this solution required a week of computer time on a massively parallel 72-node Intel Paragon computer.

Not so long ago, R. Frye discovered the first instance where three numbers raised to the fourth power sum to a fourth power (Elkies, 1988):

$$95,800^4 + 217,519^4 + 414,560^4 = 422,481^4,$$

again using a computer. Although unproved for sixth powers and higher, the future for Euler's conjecture does not appear to be very promising. As it sometimes turns out in number theory, conjectures founded on scant evidence later appear to be diametrically true.

These days the great interest in number-theoretic conjectures comes not from renewed intellectual vigor but from advances in computing machinery. Though not very bright, computers do work hard. Can computers help us? Yes, on both sides of the coin of truth. Although a daunting task, humans helped by computers can prove conjectures true. On the other hand, the exciting possibility exists that we can prove long-standing conjectures false simply by programming our tireless electronic slaves to find but one single counterexample.

11.1 Primes and Their Allies

Many of the unsolved problems in number theory have to do with primes and factoring. These can be grouped generally as problems from multiplicative number theory. Part of the reason for their appeal is the simplicity with which they can be stated. It is fair to say that until new analysis tools are developed, general results regarding primes will be limited.

Areas where research is needed were published by Lucas (1891). In these areas, not much progress has been made:

1. finding the prime that follows a given prime

2. finding a prime greater than a given prime

3. computing $p(n)$, the number of primes less than n

4. computing p_n, the nth prime

5. finding a function or family of functions that generate only primes.

Prime-generating functions are especially curious since it is possible to create functions that produce primes in significantly greater numbers than they naturally occur. Some of the reasons why this is a far from trivial task is discussed in the next section.

Prime-Generating Sequences

In Chapter 4, we looked at what is probably the most famous of the prime generating sequences, first proposed by Euler:

$$\{ a_n \} = \{ n^2 + n + 41 \}$$

Euler's polynomial will produce an uninterrupted sequence of 80 primes for $n = -40, -39, ..., 39$. It is unknown even today if even the simple expression $n^2 + 1$ will yield an infinite number of primes. Like Euler's prime generating polynomial, it certainly yields a respectable number of primes.

Other variations are more difficult to analyze. For example, is there an n, other than 1, 2, or 4, that makes $n^n + 1$ a prime number? And there are other famous prime-generating sequences. Consider the following: does the sequence

$$c_0 = 2$$

$$c_1 = c_0{}^{c_0} - 1$$

$$c_2 = c_0{}^{c_1} - 1$$

$$...$$

$$c_i = c_0{}^{c_{i-1}} - 1$$

generate nothing but primes? The difficulty is attempting to find a counterexample because the numbers grow extremely rapidly:

$$c_0 = 2$$

$$c_1 = 3$$

$$c_2 = 7$$

$$c_3 = 127$$

$$c_4 = 170{,}141{,}183{,}460{,}469{,}231{,}731{,}687{,}303{,}715{,}884{,}105{,}727$$

$$c_5 > 10^{51{,}217{,}599{,}719{,}369{,}681{,}879{,}879{,}723{,}386{,}331{,}576{,}246.}$$

The number $c_4 = 2^{127} - 1$ was proved to be prime by Catalan in 1876. Even if c_5 is shown to be prime by a large supercomputer manufacturer with extra CPU cycles on its hands, it is unlikely that we could ever determine the primality of the number c_6 or those that follow it.

Twin Primes

It has long been observed that in sequences of primes, some occur in pairs. That is, if p and $p + 2$ are prime, they are called twin primes. Consider the first few pairs less than 100: (3, 5), (5, 7), (11, 13), (17, 19), (29, 31), (41, 43), (59, 61), (71, 73).

Are there infinitely many twin primes? Most believe that there are. However, there appear to be no analytic tools that can be brought to bear on this problem at this time. For this reason, nobody has yet been able to prove that there are infinitely many prime pairs. Consider Table 11.1 showing the ratio of the number of twin primes $\pi_t(n)$ to the number of primes $\pi(n)$.

Table 11.1. Ratio of Twin Primes to Primes

n	$\pi(n)$	$\pi_t(n)$	$\pi_t(n) / \pi(n)$
10	5	2	0.4000
100	26	8	0.3077
1,000	168	35	0.2093
10,000	1,229	205	0.1668
100,000	9,592	1,224	0.1276
1,000,000	78,498	8,169	0.1041
10,000,000	664,579	58,980	0.0887

We can say that overall, roughly 1 out of every 10 primes has a twin. Of course this depends largely on the interval we are considering. Although the ratios appear to be smoothly asymptotic, if we look at the number of twin primes over intervals of 100,000, we see the erratic behavior more typical of number theoretic functions. This is what is plotted in Figure 11.1.

Figure 11.1. Graph of average twin prime to prime ratio and over intervals of 100,000.

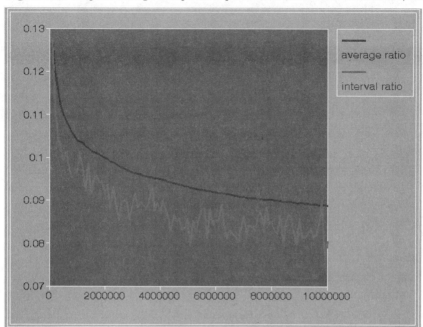

If it can be shown that the function $f(n) = \pi_t(n) / \pi(n)$ is asymptotic to some constant, say 0.08, then the number of twin primes is indeed infinite. Like Goldbach's conjecture described below, this is a problem that probably needs to be addressed from a sequence-density perspective.

Twin primes can be found by trial division using the simple program that follows. For this program, we seek the two numbers n and $n - 2$ that are prime using the IsPrime function created alone.

Program 11_1_1.c. Find twin primes

```
/*
** Program 11_1_1 - Find twin primes
*/
#include "numtype.h"

void main( int argc, char *argv[])
{
  NAT n0, n1;

  n0 = 3;
  n1 = 5;
  while (!kbhit())
  {
  if (IsPrime(n1))
    {
```

```
        if (n1 == n0 + 2)
            printf("%lu, %lu\n", n0, n1);
        n0 = n1;
        }
        n1 += 2;
    }
}

/*
** IsPrime - determines if a number is prime by trial division
*/
int IsPrime(NAT n)
{
  NAT i;

  if (n == 1)    return(0);
  if (n == 2)    return(1);
  if (!(n & 1))  return(0);                /* n is even */

  for (i = 3; i*i <= n; i++, i++)          /* n is odd */
     if ((n % i) == 0) return(0);

  return(1);
}
```

Program 11_1_1 continues executing until any key is hit. The output is a list of twin primes. In just a few seconds, the twin primes less than 1,000,000 can be found.

```
C> 11_1_1
3, 5
5, 7
11, 13
17, 19
29, 31
41, 43
59, 61
71, 73
101, 103
107, 109
137, 139
149, 151
179, 181
191, 193
197, 199
. . .
```

Mersenne Primes

Are there infinitely many Mersenne primes? The answer is probably yes owing to the divergence of the harmonic series. Of Mersenne composites, Euler showed that if $k > 1$ and $p = 4k + 2$ is prime, then $2p+1$ is prime if and only if $2p \equiv 1 \pmod{2^p + 1}$. Therefore, if $p = 4k+3$ is prime and $2p+1$ is prime the Mersenne number $2^p - 1$ is composite.

Because there are probably infinitely many twin primes, we can conjecture that there are infinitely many prime pairs such p, $2p+1$.

Bateman, Selfridge, and Wagstaff (1989) put forward a new conjecture regarding Mersenne primes. Let p be any odd natural number. If two of the following conditions hold, then so does the third:

1. $p = 2^k \pm 1$ or $p = 4^k \pm 3$

2. $2^p - 1$ is a Mersenne prime

3. $(2^p + 1)/3$ is a prime.

This conjecture has been verified for all primes $p \leq 100{,}000$.

We have previously observed that Mersenne numbers are either prime or composite. Of the composite Mersenne numbers, is every Mersenne number square-free? Although this is not really a conjecture (which we guess to be true), it is an open question that should be answered if the apparatus to do so can be developed.

Challenges

1. Investigate various prime generating sequences and try to develop some of your own. See if you can create a piecewise collection of prime-generating sequences.

2. Write a program that counts twin primes over specific intervals.

11.2 A Perfect Number That Is Odd

This may well be the holy grail of number theory. A perfect number is a number that is equal to the sum of its divisors (excluding itself). All of the known perfect numbers are even, but is there an odd perfect number? This age-old question remains unanswered; however, empirical calculations relating to abundant and deficient numbers suggest that one may exist. Not only are all perfect numbers even, but they can be formed from any Mersenne prime, a fact known at least since the time of Euclid.

Euclid-Euler Formula. For every Mersenne prime $2p - 1$, the even number of the form

$$2^{p-1} \cdot (2^p - 1)$$

is perfect.

Recall the sum of the divisors function σ and its multiplicative nature. If $2^p - 1$ is prime, then $\sigma(2^p - 1) = 2^p$. Further, $\sigma(2^{p-1}) = 2^p - 1$, since

$$1 + 2 + 4 + 8 + \ldots + 2^{p-1} = 2^p - 1.$$

This gives $\sigma(2^{p-1} \cdot (2^p - 1)) = \sigma(2^{p-1}) \cdot \sigma(2^p - 1) = 2^p \cdot (2^p - 1)$. This formula demonstrates one more fact: that there are infinitely many even perfect numbers if and only if there are infinitely many Mersenne numbers.

But as for odd perfect numbers, we do not have such a tidy formulation. However, if such a number does exist, it must be a perfect square times an odd power of a single prime. Additionally, it must divisible by at least eight primes and have at least 29 nondistinct prime factors (Sayers, 1986). And by the way, did I mention that it has at least 300 decimal digits (Brent, Cohen, and de Riele, 1991)?

With all of these constraints it seems to be a formidable task to find an odd perfect number. However, it is precisely this parameterization of the problem that may lead us to that number.

Challenge

1. Try to develop some computational machinery to find an odd perfect number that does not rely on factoring. You will probably need to use long double-type data or multiple precision integers to avoid an immediate overflow condition.

11.3 Fermat's last Theorem

Many problems in number theory remain unsolved in spite of hundreds of years of diligent effort on the part of gifted mathematicians. Among the most widely known is Fermat's last theorem. Its beauty flows from the simplicity with which it can be stated.

Fermat's last theorem. The equation

$$x^n + y^n = z^n$$

has no nonzero integer solutions for x, y, and z when $n > 2$.

The sense of frustration felt by mathematicians attempting this proof was intensified by a tantalizing note written by Fermat in the margin of his copy of *Diophantus's Arithmetic*. He wrote the immortal words "*I have discovered a truly remarkable proof which this margin is too small to contain.*" It is almost certain that Fermat later realized that his "remarkable proof" was flawed since he issued challenges to his contemporaries to prove special cases including $n = 3$. In fact, the proof for $n = 3$ was not known until 1753 (94 years later) when Euler was able to verify this case. On top of that even Euler's proof contained a flaw that had to be reconciled later.

Many notable mathematicians have attempted Fermat's last theorem through the centuries: Dirichlet, Legendre, Kummer, Lehmer, and others (Klee and Wagon, 1991). Each added to the body of knowledge by increasing the size of the allowable exponents, eliminating certain types of exponents, and showing that there exist only a finite number of primitive solutions. But each of these falls short of proof.

To encourage serious work on the problem, as if it were necessary, the German mathematician P. Wolfskehl bequeathed 100,000 marks to the Academy of Science in Göttingen as a prize for the first complete proof. The promise of money *plus* glory was enough to encourage a deluge of fallacious proofs from all quarters. Fermat's last theorem is the undisputed champion of mathematical problems having the greatest number of incorrect proofs published.

Interest in the problem flagged and inflation eroded the prize to a mere 10,000 marks. However serious mathematicians, needing no remunerative encouragement, continued working. In 1993, after seven years of work, Andrew Wiles of Princeton University Geometrically proved Fermat's last theorem. Of his proof Wiles says "it is unlikely to have any applications" (Hogan, 1993). In a final ironic twist Wiles had to withdraw his claim temporarily until certain gaps in his proof were filled. These gaps were finally filled in 1994 by restructuring the approach; a paradigm shift from geometric to algebraic that ultimately simplified it. The proof is now generally accepted.

11.4 The 3*n*+1 Problem

It is a remarkable fact that many problems in number theory are easily stated but difficult, if not impossible, to prove. This problem goes by many names, including the Collatz problem, Ulam's problem, the hailstone problem, the Syracuse problem, Kakutani's problem, and probably more. It is another intriguing problem developed from some of the simplest arithmetic concepts. This problem was first conceived in 1952 by B. Thwaites (Lagarias, 1985) and in spite of an onslaught of computer analysis remains unanswered (Hayes, 1984). Consider the following function f:

$$f(n) = \begin{cases} n/2, & \text{if } n \text{ is even} \\ 3n + 1, & \text{if } n \text{ is odd} \end{cases},$$

where the range and domain are the natural numbers. This function can be programmed easily, as is shown in the following example.

Program 11_4_1.c. The hailstone function

```
/*
** Program 11_4_1 - Print values of the 3n+1 function
*/
#include "numtype.h"

void main(int argc, char *argv[])
{
  NAT k, n;

  if (argc == 2) {
    n = atol(argv[1]);
  } else {
    printf("Print values of the 3n+1 function\n\n");
    printf("n = ");
    scanf("%lu", &n);
  }

  for(k = 0; n > 1; k++)
  {
    if (n & 1)
      n = 3 * n + 1;                 /* odd n */
    else
      n /= 2;                        /* even n */
  printf("%8lu", n);
  }

  printf("\ntotal number of values = %lu", k);
}
```

The program terminates once $n = 1$. At that point, the function
yields the values 1, 4, 2, 1, 4, 2, 1.... But will it always reach a point
where $n = 1$ starts with an arbitrary n? This is the question that the
$3n + 1$ function asks, the answer to which has escaped mathematicians
for decades (Legarias, 1985).

Consider the output of Program 11_4_1, when using a starting value
of 41, this sequence is unusually long.

```
C> 11_4_1
Print values of the 3n+1 function

n = 41
      124      62      31      94      47     142      71     214     107     322
      161     484     242     121     364     182      91     274     137     412
      206     103     310     155     466     233     700     350     175     526
      263     790     395    1186     593    1780     890     445    1336     668
      334     167     502     251     754     377    1132     566     283     850
      425    1276     638     319     958     479    1438     719    2158    1079
     3238    1619    4858    2429    7288    3644    1822     911    2734    1367
     4102    2051    6154    3077    9232    4616    2308    1154     577    1732
      866     433    1300     650     325     976     488     244     122      61
      184      92      46      23      70      35     106      53     160      80
       40      20      10       5      16       8       4       2       1
total number of values = 109
```

The $3n+1$ function, more colorfully dubbed the "hailstone function," produces sequences of numbers during successive applications of $f(n)$ that drift up and down like hailstones carried on turbulent winds inside storm clouds. And those numbers, like hailstones, eventually fall to earth (i.e., the integer "1"). But will they in all cases? This is the question that remains open, even though all integers less than 2^{40} have been checked (Legarias, 1985).

There are several ways to look at the problem. A straightforward approach is to consider the sequence to be a random walk where each iteration represents a single step. An odd step is 3/2 since every odd number is followed by an even number. An even step is 1/2. To use this thinking, we convert the products to sums by taking the logarithm base 2 of each step. Each step therefore adds either 0.585 or −1. If odd and even steps occur with equal probability, then the mean step size is −0.208. Therefore, the overall tendency of any arbitrary sequence is to head toward 1.

Of course the statistical tendency says very little about a specific sequence that is a nontrivial loop. Terras (1976) and Garner (1981) showed that any nontrivial loop must have a minimum of a half-million numbers. This fact alone makes it unlikely that any such loop will be discovered by accident. And with all numbers less than 2^{40} having been checked, it will take some serious computer power to answer this question.

Challenges

1. Find the starting number n that produces the longest sequence for $n < 1,000,000$. What is the average sequence length?

2. Attempt to characterize the 10 starting numbers that produce the longest sequences found in Challenge 1.

3. Try other functions such as $3.5n + 1$ or $2.5n + 1$. First make predictions about their behavior and then experiment with different starting values.

11.5 Additive Number Theory

Whereas multiplicative number theory considers divisibility and products, additive number theory is concerned with representability and representations of integers, partitions, and combinatorics. Fundamental to the theory is the concept of partitions for which a definition is given.

Definition. A *partition* of a nonnegative integer n is a representation of n as a sum of positive integers.

The partition function $p(n)$ is defined as the number of ways the positive integer n can be represented as the sum of positive integers. The order of the summands is not considered to be relevant. For convenience, $p(0)$ is defined as 1. Euler was the first to investigate some of the important properties of $p(n)$ in his 1748 book *Introductio in Analysin Infinitorum*, a book that is considered to be the keystone of modern analysis.

By way of example, consider the partitions of 5. They are 5, 4+1, 3+2, 3+1+1, 2+2+1, 2+1+1+1, 1+1+1+1+1. Therefore $p(5) = 7$. These partitions can be represented graphically using rows of equally spaced dots. The graphical representation, called a Ferrer graph, of the partitions of 5 is shown in Figure 11.2.

Figure 11.2. Graphical representation of the seven partitions of 5.

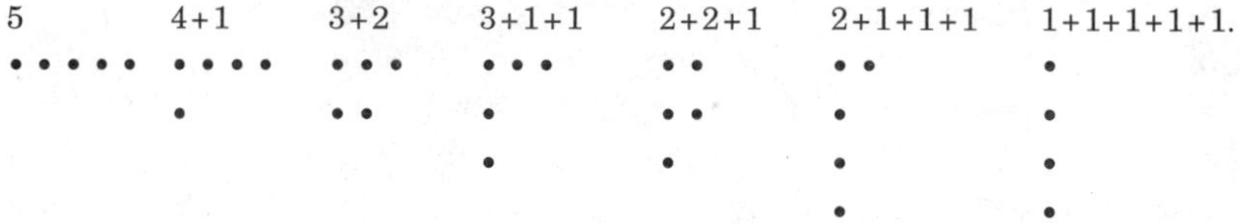

Using the concept of a graphical representation, we can define a conjugate partition. This is the partition resulting from counting the dots vertically rather than horizontally. The conjugate of 4+1 is therefore 2+1+1+1.

We can similarly define other partitioning functions as well.

$p_m(n)$ = the number of partitions of n into summands less than or equal to m

$p^e(n)$ = the number of partitions of n into even summands

$p^o(n)$ = the number of partitions of n into odd summands

$p^d(n)$ = the number of partitions of n into distinct summands

$p^e(n)$ = the number of partitions of n into summands less than or equal to m.

And for distinct summands:

$q(n)$ = the number of partitions of n into distinct summands

$q^e(n)$ = the number of partitions of n into distinct even summands

$q^o(n)$ = the number of partitions of n into distinct odd summands.

Many problems exist relating to partition functions and their relationships. Here I will consider only a few of the basic problems in additive number theory.

Goldbach's Conjecture

This conjecture dates from 1742 and was discovered in correspondence between Goldbach and Euler. It falls under the general heading of partitioning problems in additive number theory. Goldbach made the conjecture that every odd number ≥ 6 is equal to the sum of three primes. Euler replied that Goldbach's conjecture was equivalent to the statement that every even number ≥ 4 is equal to the sum of two primes. Because proving the second implies the first, but not the converse, most attention has been focused on the second representation.

The smallest numbers can be verified easily by hand:

$6 = 3 + 3$	$8 = 3 + 5$	$10 = 3 + 7$	$12 = 5 + 7$
$14 = 3 + 11$	$16 = 3 + 13$	$18 = 5 + 13$	$20 = 3 + 17$
$22 = 3 + 19$	$24 = 5 + 19$	$26 = 3 + 23$	$28 = 5 + 23$
$30 = 7 + 23$	$32 = 3 + 29$	$34 = 3 + 31$	$36 = 5 + 31$
$38 = 7 + 31$	$40 = 3 + 37$	$42 = 5 + 37$	$44 = 3 + 41$
$46 = 3 + 43$	$48 = 5 + 43$	$50 = 3 + 47$	$52 = 5 + 47$

Of course all the examples in the world do not a proof make.

As a partitioning problem it is worth noting that as the numbers get larger, the number of representations grows as well:

$$12 = 5 + 7$$

$$24 = 5 + 19 = 7 + 17 = 11 + 13$$

$$48 = 5 + 43 = 7 + 41 = 11 + 37 = 17 + 31 = 19 + 29.$$

This would suggest that the likelihood of finding that exceptional even number that is not the sum of two primes diminishes as one searches in ever larger even numbers.

Euler was convinced that Goldbach's conjecture was true but was unable to find any proof (Ore, 1948). The first conjecture has been proved for sufficiently large odd numbers by Hardy and Littlewood (1923) using an "asymptotic" proof. They proved that there exists an n_0 such that every odd number $n \geq n_0$ is the sum of three primes. In 1937 the Russian mathematician Vingradov (1937, 1954) again proved the first conjecture for a sufficiently large (but indeterminate) odd number using analytic methods. Calculations of n_0 suggest a value of $3^{3^{15}}$, a number having 6,846,169 digits (Ribenboim, 1988, 1995a).

In 1966 Chen Jing-Run (1966) proved that every sufficiently large even number can be expressed as the sum of a prime and a number with no more than two prime factors (reprinted in Chen, 1973, 1978).

One can verify Goldbach's conjecture by brute force, up to a point. Using about 130 CPU-hours on an IBM 3083 Sinisalo (1993) verified the conjecture up to $4 \cdot 10^{11}$. Although Sinisalo used a bit array and Eratosthenes sieve, the program that follows uses a similar strategy while employing trial division. The procedure is to take an odd number and then finds small primes starting with 3. Next, the difference is tested for primality. If the difference is prime, we are done; if not, select the next larger prime and repeat the process.

If the Goldbach conjecture is true, then for any even n there exists a prime p for which the complementary number $q = n - p$ also is prime. The *Goldbach partition* shall be denoted by the representation $n = p + q$, where p and q are prime. The smallest prime in the Goldbach partition is indicated by partition function $g(n)$. Table 11.2 shows the minimal values for $n \leq 1,000,000$. The last column in the table is the ratio $g(n)/n$. The ratio is especially interesting because it shows the largest values of $g(n)$, and we can see that in almost every case it is less than its predecessor.

Table 11.2. Minimal Goldbach Partition Values for n ? 1,000,000

n	$g(n)$	$n - g(n)$	$g(n) / n$
6	3	3	0.500000
12	5	7	0.416667
30	7	23	0.233333
98	19	79	0.193878
220	23	197	0.104545
308	31	277	0.100649
556	47	509	0.084532
992	73	919	0.073589
2,642	103	2,539	0.038986
5,372	139	5,233	0.025875
7,426	173	7,253	0.023297
43,532	211	43,321	0.004847
54,244	233	54,011	0.004295
63,274	293	62,981	0.004631
113,672	313	113,359	0.002754
128,168	331	127,837	0.002583
194,428	359	194,069	0.001846
194,470	383	194,087	0.001969
413,572	389	413,183	0.000941
503,222	523	502,699	0.001039

The next program finds the first pair of primes in the Goldbach partition for even numbers and the first triplet for odd numbers.

Program 11_5_1.c. Find two primes whose sum is an even number

```
/*
** Program 11_5_1 - Find the two primes whose sum is an even number
**                  or three primes whose sum is an odd number
*/
#include "numtype.h"

void main( int argc, char *argv[])
{
  NAT n, p;

  if (argc == 2) {
    n = atol(argv[1]);
  } else {
    printf("Check Goldbach's conjecture\n\n");
    printf("n = ");
    scanf("%lu", &n);
  }


  printf("%lu = ", n);
  if (n & 1)
  {
    printf("3 + ");                  /* n is odd */
    n -= 3;
  }

  if (n < 4) exit(1);                /* too small */

  if ( IsPrime(n - 2))
    p = 2;
  else
    for (p = 3; p < n; p++, p++)
      if ( IsPrime(p) && IsPrime(n - p)) break;

  printf("%lu + %lu\n", p, n - p);
}

/*
** IsPrime - determines if a number is prime by trial division
*/
int IsPrime(NAT n)
{
  NAT i;

  if (n == 1)    return(0);
  if (n == 2)    return(1);
  if (!(n & 1)) return(0);          /* n is even */

  for (i = 3; i*i <= n; i++, i++)    /* n is odd */
    if ((n % i) == 0) return(0);

  return(1);
}
```

```
C> 11_5_1
Check Goldbach's conjecture

n = 63274
63274 = 293 + 62981
```

An approach that may prove fruitful is the application of the concept of the density of a sequence first proposed in a theorem by Schnirelmann (1930) (translated in Gelfond and Linnik, 1965). These and other lines of attack are assembled in a book by Wang (1984), which is a collection of selected papers on this problem.

Gauss's Circle Problem

The prolific mathematician Gauss gives us a partitioning problem with a direct geometric interpretation called Gauss's circle problem. The partition function $r_2(n)$ is defined as the number of ways a number n can be represented as the sum of two squares

$$n = a^2 + b^2.$$

For example, $r_2(1) = 4$ since $1 = 1^2 + 0 = 0 + 1^2 = (-1)^2 + 0 = 0 + (-1)^2$. Although it has an irregular behavior, its average value tends toward π:

$$\lim_{n \to \infty} \frac{1}{n} \sum_{t=0}^{n} r_2(i) = \pi.$$

Because of symmetry, $r_2(n)$ will always be a multiple of 4, with the exception of the degenerate case of $n = 0$. This can be seen in Table 11.3, which lists the first few values for $r_2(n)$.

Table 11.3. Value and Average Value for $r^2(n)$, where $n < 20$

n	$r_2(n)$	$\Sigma r_2(n) / n$
0	1	—
1	4	5.000
2	4	4.500
3	0	3.000
4	4	3.250
5	8	4.200
6	0	3.500
7	0	3.000
8	4	3.125
9	4	3.222
10	8	3.700

Table 11.3. (continued)

n	$r_2(n)$	$\Sigma r_2(n) / n$
11	0	3.364
12	0	3.083
13	8	3.462
14	0	3.214
15	0	3.000
16	4	3.063
17	8	3.353
18	4	3.389
19	0	3.211
20	4	3.250

Graphically these are the lattice points that lie on a circle with an integral radius, as shown in Figure 11.3.

Figure 11.3. Lattice point on a circle with an integral radius.

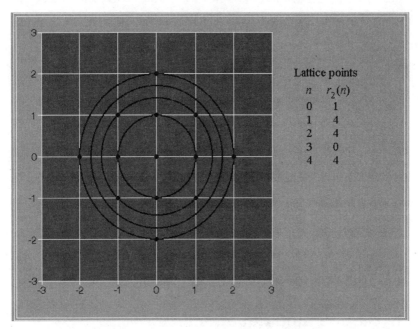

A number of problems have arisen from the analysis of this geometric problem. Gauss's circle problem is to find the best estimate of $\frac{1}{n} \sum_{t=0}^{n} r_2(i)$ for a given n. Gauss was able to show that

$$\sum_{t=0}^{n} r_2(i) = \pi n + \varepsilon,$$

where $\varepsilon = k \cdot \sqrt{n}$ and k is a constant. In 1906 Sierpinski showed that $\varepsilon = k \cdot n^{1/3}$. Since then the exponent 1/3 has been been replaced with the slightly smaller 27/32, and it is known that it is false for 1/4 (Andrews, 1971). This is an instance where further meaningful refinement of this exponent will probably come only through theoretical analysis rather than from computer simulation.

The Four-Color Map Theorem

One of the more important recent advancements in additive number theory came with the proof of the four-color map theorem (Appel and Haken, 1976; Saaty, 1977). Stated formally:

Four-Color Map Theorem. Every loopless planar graph admits a vertex colored with, at most, four colors.

To prove the theorem it is assumed that there exist counterexamples. By using a computer, the 1936 basic forms were identified. By then exhausting all counterexamples by showing that they must contain a subgraph from the basic set proved the theorem. However, no proof that does not rely on the assistance of a computer has yet been discovered.

Challenges

1. Write a program that will print dot graphs of the partition functions presented in this section.

2. Modify Program 11_5_1 to verify Goldbach's conjecture (page 367) for even numbers up to 1,000,000.

3. Write a program that will print the prime pairs whose smallest summand is larger than any previous smallest summand, as in Table 11.2, for $n < 1,000,000$.

4. Modify Program 11_5_1 to find *all* prime pairs that sum to the input number n, i.e., the Goldbach partition.

11.6 Diophantine Equations

To this day there is no finite method for determining whether a particular diophantine equation has solutions. Even so, many specific problems can be solved or at least simplified significantly by first performing a certain amount of analysis. The remaining unsolved part can then be attempted by brute force.

The Discrete Logarithm Problem

The discrete logarithm problem has two parts: finding a primitive root and finding a member of a particular residue system. Many Diophantine equations, as well as certain problems relating to prime numbers, would fall if this problem could be solved. It is not that discrete logarithms cannot be found, it is just that there does not yet exist an efficient algorithm for finding them. Those that are not trial methods rely on an element of probability of success.

This situation does not exist for want of brain power. It has vexed the greatest mathematical minds throughout history. This fact testifies to the difficulty of finding discrete logarithms efficiently.

Pell's Equation

Although I previously looked at Pell's equation from a point of how to solve these, many open questions remain regarding their character and solvability. One condition for solving

$$x^2 - d \cdot y^2 = -1$$

is that it is known that $d \not\equiv 0 \pmod 4$. Also, d cannot be divisible by any prime p such that $p \equiv 3 \pmod 4$. However, in the following Pell's equation, it is conjectured that if p is a prime and $p \equiv 3 \pmod 4$ and

$$x^2 - p \cdot y^2 = 1$$

and if u is the least positive value of y, then

$$u \not\equiv 0 \pmod p.$$

This conjecture has been verified by K. Goldberg for $p < 18{,}000$. Further, it has been shown (Mordell, 1962) that $u \not\equiv 0 \pmod p$ if and only if the Euler number

$$E_{(p-3)/4} \not\equiv 0 \pmod p$$

where E_n is defined as

$$\sec t = \sum_{n=0}^{\infty} \frac{E_n t^{2n}}{(2n)!}.$$

A similar conjecture persists for when $p \equiv 1 \pmod 4$ (Mordell, 1969).

The Perfect Box

The perfect box is one that has integer-length sides, side diagonals, and a main diagonal (Figure 11.4). In Chapter 5, parametric solutions to the Pythagorean theorem were found after some algebraic manipulation. This problem is an extension into three dimensions of the Pythagorean theorem discussed in Chapter 5.

Figure 11.4. A box and its diagonals.

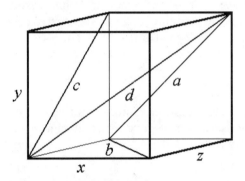

We are seeking a solution to the system of equations:

$$x^2 + y^2 = a^2$$

$$x^2 + z^2 = b^2$$

$$y^2 + z^2 = c^2$$

$$x^2 + y^2 + z^2 = d^2$$

Brocard, in 1895s, believed that he had proved no perfect box can exist but assumed incorrectly that x, y, and z are, pairwise, relatively prime. What he did show is that, for a perfect box to exist, at least two of its side lengths are not relatively prime. To date, the existence of a perfect box has neither been established nor ruled out.

Klee and Wagon (1991) point out that to have a perfect box the four numbers $(xd)^2$, $(xc)^2$, $(cd)^2$, and $(bc)^2$ must be such that their differences are all perfect squares. Searching for these four numbers offers a viable alternative to a trial-and-error methodology using Pythagorean triples.

The Cannonball Problem

The cannonball problem, where cannonballs are arranged in a pyramid and the total number of cannonballs is a perfect square ($1^2 + 2^2$

$+ 3^2 + \ldots + k^2 = n^2$). It has long been known that the only solution to this problem is $k = 24$, $n = 70$; or 4900 cannonballs in total.

However, it remains undecided if a related problem, $1^n + 2^n + 3^n + \ldots + k^n = (k+1)^n$, has any solutions other than the trivial one: $1 + 2 = 3$. It has been proved that if any other solution exists, it must be for $k > 10^{1,000,000}$ (Ogilvy and Anderson, 1966).

Simple Cubics

Although number theoretical results involving cubics are scarce, they have received some attention. Unlike quadratics, numbers that are cubed retain their sign, and this fact leads to far more complications in their analysis. Fermat's last theorem, the godfather of unsolved problems, illustrates this fact. As mentioned earlier, Fermat asserted that

$$x^n + y^n = z^n$$

has a nontrivial integer solution only when $n = 2$, and he proved this for $n = 4$ in 1659 (Ore, 1948). However, mathematics had to wait 94 years for Euler to prove that this was true for $n = 3$. One can also assume that cubics by their nature generate larger products and commensurately require more effort to manipulate.

The cubic diophantine equation

$$x^3 + y^3 = 1$$

only has the integer solutions $(1, 0)$ and $(0, 1)$, no other rational solutions, and is asymptotic to the line $x + y = 0$. If it had a rational solution, the fractions could be cleared and another integer solution would be obtained. Look at its graph in Figure 11.5 and you will see the curve pass through its only two integer solutions, the infinity of other points being irrational.

The situation where a simple diophantine equation has only a very small number of rational solutions is not that uncommon. It was George Cantor who showed that the field of irrational points is so much more dense (a higher order of infinity) than the rational field that the phenomenon was explained. It can be shown by analytic geometry that solutions to the equation

$$x^3 + y^3 = z$$

must lie between

$$\frac{3 - \sqrt{12z - 3}}{6} \leq x, y$$

and

$$x, y \leq \frac{3 + \sqrt{12z - 3}}{6}$$

Figure 11.5. Graph of the equation $x^3 + y^3 = 1$.

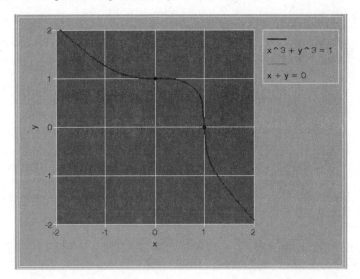

This is true because, when the curve is less than the line $x + y = 1$, no integral solutions are possible. Therefore, all cubics of the form $x^3 + y^3 = z$ have at most a finite number of integer solutions or none at all. Of course finding them is another matter. Without attempting to solve the cubic generally, solutions can be found using enumeration of xs over the interval and solving for y using Newton's method.

Recall from Chapter 6, where the integer square root for n was found by iterating

$$x_{n+1} = x_n - f(x_n) / f'(x_n),$$

where

$$f(x) = x^2 - n = 0.$$

If we let

$$f(x) = x^3 - n = 0,$$

then we will be able to find integer cube roots. The following program implements such a strategy.

Program 11_6_1. Solves the equation $x^3 + y^3 = z$ given z

```
/*
** Program 11_6_1 - Solve the equation x^3 + y^3 = z, given z
*/
#include "numtype.h"
```

```
main( int argc, char *argv[])
{
  INT min, max, x, y, z, r;

  if (argc == 2) {
    z = atol(argv[1]);
  } else {
    printf("Solve x^3 + y^3 = z\n\n");
    printf("enter z = ");
    scanf("%ld", &z);
  }

  min = (3 - sqrt(12*z - 3)) / 6;
  max = (3 + sqrt(12*z - 3)) / 6;

  for (x = min; x <= max; x++)
  {
    iCubeRoot(z - x*x*x, &y, &r);
    if (r == 0)
      printf("%8ld %8ld %8ld\n", x, y, z);
  }
}

/*
** iCubeRoot - finds the integer cube root of a number
**             by solving: a = q*q*q + r
*/
#include "numtype.h"

INT iCubeRoot(INT a, INT *q, INT *r)
{
  int s;

  if (a < 0) { s = -1; a = -a; } else s = 1;

  *q = a;

  if (a > 0)
    while (*q > (*r = a / *q / *q))
    {
      *q = (2 * *q + *r) / 3;
      if (*q == 0) break;
    }

  *r = a - *q * *q * *q;            /* remainder */

  *q *= s;
  *r *= s;

  return(*q);
}
```

```
C> 11_6_1
Solve x^3 + y^3 = z

enter z = 728
      -10         12        728
       -1          9        728
```

6	8	728
8	6	728
9	-1	728
12	-10	728

One of the simplest cubic diophantine equations is known to have an infinite number of solutions (Lehmer, 1956; Payne and Vaserstein, 1991). Any number of solutions to the equation

$$x^3 + y^3 + z^3 = 1$$

can be produced through the use of the algebraic identity

$$(9t^3 + 1)^3 + (9t^4)^3 + (-9t^4 - 3t)^3 = 1$$

by substituting in values of t. This can be verified by multiplying out the terms. By assigning values t for $1 \geq t \leq 10$, we can generate Table 11.4.

Table 11.4. Integer Solutions to the Cubic Diophantine Equation $x^3 + y^3 + z^3 = 1$

t	x	y	z
1	10	9	-12
2	73	144	-150
3	244	729	-738
4	577	2,304	-2,316
5	1,126	5,625	-5,640
6	1,945	11,664	-11,682
7	3,088	21,609	-21,630
8	4,609	36,864	-36,888
9	6,562	59,049	-59,076
10	9,001	90,000	-90,030

Although these are certainly solutions, the identity generates only one family of solutions. Other solutions such as (94, 64, −103), (235, 135, −249), (438, 334, −495), (729, 244, −738), ... can be found. What is not known is if it is possible to parameterize all solutions for this equation. Put another way, are there an infinite number of families of solutions? Probably yes, but that too remains to be shown.

Consider next the slightly more general case of

$$x^3 + y^3 + z^3 = t^3,$$

which also has an infinite number of solutions. A parameterized integer solution has been provided by Ramanujan for this equation:

$$x = 3n^2 + 5n \cdot m - 5m^2$$

$$y = 4n^2 - 4n \cdot m + 6m^2$$

$$z = 5n^2 - 5n \cdot m - 3m^2$$

$$t = 6n^2 - 4n \cdot m + 4m^2,$$

where m and n are integers. But obviously, this too will not produce all possible solutions owing to the dependence of t on m and n. If we let $t = 1$, as in the original equation, then a different set of solutions is required.

Interestingly, the cubic does have a rational solution:

$$x = c \cdot (1 - (a - 3b) \cdot (a^2 + 3b^3))$$

$$y = c \cdot ((a + 3b) \cdot (a^2 + 3b^2) - 1)$$

$$z = c \cdot ((a^2 + 3b^2)^2 - (a + 3b))$$

$$t = c \cdot ((a^2 + 3b^2)^2 - (a - 3b)),$$

where a, b, c, t, x, y, z are any rational numbers. As before, this system of equations cannot be used to generate all known integer solutions either.

The problem of developing integer solutions rests on the analysis of the equation

$$x^3 + y^3 + z^3 = s$$

for some arbitrary integer s. Only one nontrivial solution to the equation:

$$x^3 + y^3 + z^3 = 3$$

is known (4, 4, –5). Are there more? For this case and for cubics in general, the analytic tools that are available are limited.

Other questions regarding cubics remain open. Does the equation $(x + y + z)^3 = x \cdot y \cdot z$ have any solutions in integers?

Euler's Conjecture Revisited

We now know that Euler's conjecture is generally false. However, counterexamples for the myriad special cases remain to be found, if they exist. It is known that

$$1^3 + 12^3 = 9^3 + 10^3$$

and

$$133^4 + 134^4 = 59^4 + 158^4.$$

However, a solution in positive integers to

$$a^5 + b^5 = c^5 + d^5$$

is not known (Guy, 1994).

Consecutive Perfect Powers

A perfect power is defined as a number having the form m^n, where m and n are integers and $n > 1$. $8 = 3^2$ and $9 = 2^3$. The consecutive power problem asks if 8 and 9 are the only consecutive integers that are perfect powers (Ribenboim, 1994). Catalan's conjecture is that 8 and 9 are it—there are no more. Here is a problem made to order for solution by computer.

Challenges

1. Recognizing the symmetry of $x^3 + y^3 = z$ about the line $y = x$, improve Program 11_6_1 by finding one solution and printing both without duplications.

2. There is a problem that is simpler variation of the perfect box. Given a square in a plane, is there a point that is an integer distance from all four corners?

12

Multiple-Precision Arithmetic

Multiplication is vexation,
Division is as bad,
The Rule of Three perplexes me,
And practice drives me mad.

ANONYMOUS (C. 1570)

12.0 Introduction

It is an unfortunate fact that for our number-theoretic needs, general purpose computers offer only fixed precision. Although computations can be performed expediently, the data range is limited. All CPU's support the 1-, 2-, or 4-byte integers data type. Table 12.1 shows the unsigned data ranges available using standard integer precision:

Table 12.1. Standard Integer Precision for Unsigned Data

Data type	Number of bytes	Unsigned data range
Char	1	$0 - 255$
Short int	2	$0 - 65,535$
Long int	4	$0 - 4,294,967,295$

That's fine for most applications, but for many number-theoretic applications it is useful (and often necessary) to have much higher precision. This is especially true for applications involving encryption algorithms based on factoring large numbers. For these, operations on numbers having a hundred digits are not uncommon. Also, there are the more recreational uses mentioned in Chapters 7 and 8. For this reason, I have included this chapter on multiple-precision arithmetic.

A multiple precision integer, as used here, is an unsigned integer having arbitrary length. Recalling the discussion of radix from Chapter 2, the base 256 has this representation:

$$a = c_0 + c_1 \cdot 256 + c_2 \cdot 256_2 + \ldots + c_n \cdot 256^n.$$

By definition, each base 256 MP digit has the range 0 to 255. This conveniently fits the standard unsigned char data type. Digits are stored in the usual computer ordering convention, in which the first digit ($c[0]$) is the least significant and the last digit ($c[n]$) is the most significant. This arrangement makes indexing a bit easier when performing operations result in numbers greater than the operands, in particular multiplication.

The library of multiple-precision routines allows operations on numbers up to 1000 decimal digits (415 base 256 digits). This limit is arbitrary and can be modified easily by changing the constants ADIG-IT and IDIGIT in the include file, mp.h, for the library of routines:

Program mp.h. Header file for multiple-precision library

```
/*
** mp.h - Header file for multiple precision library
*/

#ifndef MP_H
#include <stdio.h>
#include <stdlib.h>
#include <string.h>
#include <memory.h>

#define ADIGIT   1000   /* maximum ASCII digits */
#define IDIGIT   415    /* maximum multiple precision digits */

#define MIN(a,b) a < b ? a : b
#define MAX(a,b) a > b ? a : b

/* accumulator */

union {
  unsigned int x;
  unsigned char b[2];
} a;

/* prototypes */

int AddMP( unsigned char [], unsigned char [], unsigned char []);
int AddMPShort( unsigned char [], unsigned char, unsigned char []);
int DivMP( unsigned char [], unsigned char [], unsigned char [],
unsigned char []);
int DivMPShort( unsigned char [], unsigned char, unsigned char [],
unsigned char []);
int MulMP( unsigned char [], unsigned char [], unsigned char []);
```

```
int MulMPShort( unsigned char [], unsigned char, unsigned char []);
int SubMP( unsigned char [], unsigned char [], unsigned char []);
int SubMPShort( unsigned char [], unsigned char, unsigned char []);
int CompareMP( unsigned char [], unsigned char []);
int IsZeroMP( unsigned char []);

int AndMP( unsigned char [], unsigned char [], unsigned char []);

unsigned char *LeftByteShiftMP( unsigned char [], unsigned int);
unsigned char *LeftBitShiftMP( unsigned char [], unsigned int);
unsigned char *RightByteShiftMP( unsigned char [], unsigned int);
unsigned char *RightBitShiftMP( unsigned char [], unsigned int);

unsigned char *ScanMP( unsigned char *, unsigned char * );
unsigned char *CopyMP( unsigned char *, unsigned char * );
unsigned char *ConvertASCIItoMP( unsigned char *, unsigned char *
);
unsigned char *ConvertMPtoASCII( unsigned char *, unsigned char *
);

char *PackASCIINumber( char * );
char *PrintASCIINumber( char * );

void DumpMP( unsigned char[]);

#define MP_H
#endif
```

The number of ASCII digits permitted for input determines what the maximum number of base 256 digits are necessary. Because

$$10^{\text{ADIGIT}} = 256^{\text{IDIGIT}},$$

the limit is easily computed as

$$\text{IDIGIT} = \text{ADIGIT} / \log_{10}256 = \text{ADIGIT} / 2.408.$$

All the routines use a union to facilitate carries. Rather than trying to extract the high byte after the operation, I simply reference the high byte a.b[1] in the accumulator union as shown above in the header file. This saves us from having to perform bit masking and rotation.

The MP routines presented here are nondestructive. That is to say that the numbers passed into the routines are left intact rather than being overwritten. This feature comes at the expense of a small amount of bookkeeping at the beginning and end of each routine. Multiple precision is notoriously slow. For convenience I have used base 256 (2^8), although better performance can be obtained using base 65,536 (2^{16}).

As suggested by Knuth (1981), I have implemented a "short" version for addition, subtraction, multiplication, and division. If the second operand has only one digit (i.e., <256), the short operation is performed. Otherwise the "long" operation is executed.

You may well ask why I have chosen not to include multiple-precision arithmetic in the programs described in the preceding chapters. The first reason is speed. Some of the applications perform millions of arithmetic operations and the performance penalty would be too great.

Second, and more important, my feeling was that a series of function calls is much less instructive than seeing the arithmetic coded in the program. Using these multiple-precision arithmetic routines as they are developed here means implementing them as function calls. Most of the programs presented can be converted using these routines, although the algorithmic clarity would suffer to some extent.

On the subject of computer arithmetic, volumes have been written. There are symposiums held every year where advances in this field are presented. It is not my intention to provide a definitive review of the vast literature on this subject. My goal is to provide a set of basic routines and thereby constitute a framework for enhancement.

12.1 Addition

Addition is the simplest to perform, involving only addition and carry. Implemented here is your basic carry-ripple adder. Given two multiple-precision integers u and v, where $u = u_m u_{m-1} \ldots u_1 u_0$ and $v = v_n v_{n-1} \ldots v_1 v_0$ (u_i and v_i being digits base 256)

$$w = u + v$$

or with respect to the digits

$$w_i = u_i + v_i + c_i - 1,$$

for all $0 \le i \le \max(m, n) + 1$, the carry is $c_i = (u_i + v_i) / 256$. For this adder, the carry is propagated a constant length across the maximum length of the operands. An overflow condition is returned if there is a carry past the end of the variable. The addition operation has been researched extensively and many variations are possible to limit the carry length (Omondi, 1994). Below is a function that implements a ripple-carry adder.

Program mpAdd.c. Multiple-precision addition (long and short)

```
/*
** mpAdd     -- Add two multiple precision integers
** mpAddShort-- Add a short integer to a multiple precision integer
```

```
**
** return value: 0 if ok; not 0 if overflow
*/
#include "mp.h"

int mpAdd( unsigned char augend[],
           unsigned char addend[],
           unsigned char sum[])
{
  int i, m;
  unsigned char t[IDIGIT];          /* temporary result */

  if ((m = MostSignificantDigit(addend, IDIGIT)) == 0)
    return( mpAddShort(augend, addend[0], sum));

  m = MAX(m, MostSignificantDigit(augend, IDIGIT));

  memset(t, 0, IDIGIT);             /* clear sum */

  for ( i = 0, a.x = 0; i <= m; i++)
    t[i] = a.x = a.b[1] + augend[i] + addend[i];

  if (i < IDIGIT)
    t[i] = a.x = a.b[1];

  mpCopy(sum, t);

  return(a.b[1]);
}

int mpAddShort( unsigned char augend[],
                unsigned char addend,
                unsigned char sum[])
{
  int i, m;

  m = MostSignificantDigit(augend, IDIGIT);

  sum[0] = a.x = augend[0] + addend;

  for ( i = 1; i <= m; i++)
    sum[i] = a.x = a.b[1] + augend[i];

  if (i < IDIGIT)
    sum[i] = a.x = a.b[1];

  for ( i++; i < IDIGIT; i++) sum[i] = 0;

  return(a.b[1]);
}
```

12.2 Subtraction

Subtraction is the next operation in increasing order of complexity. It really is a carry-ripple adder with a complementor. One's complement, a common representation for negative integers, is defined as

$$v = 2^n - 1 - v',$$

where $v < 0$ and $v' > 0$ (Koren, 1993). So, given two MP integers u and v, where $u = u_m u_{m-1} \ldots u_1 u_0$ and $v = v_n v_{n-1} \ldots v_1 v_0$ (u_i and v_i being digits base 256

$$w = u - v = u + v' = 2^n - 1 + u - v$$

or with respect to each digit:

$$w_i = 255 + u_i - v_i + c_{i-1},$$

for all $0 \le i \le \max(m, n) + 1$, the carry is $c_i = (u_i + v_i) / 256$. Like the adder just described, the carry is propagated across the maximum length of the operands. However, subtracting a larger number from a smaller number results in an overflow condition.

Program mpSub.c. Multiple-precision subtraction (long and short)

```
/*
** mpSub -- Subtract one multiple precision integer from another
** mpSubShort -- Subtract a multiple precision integer from
**               a short integer
**
** return value: 0 if ok; not 0 if overflow
*/
#include "mp.h"

int mpSub( unsigned char minuend[],
           unsigned char subtrahend[],
           unsigned char difference[])
{
  int i, m;
  unsigned char t[IDIGIT];          /* temporary result */

  if ((m = MostSignificantDigit(subtrahend, IDIGIT)) == 0)
    return( mpSubShort(minuend, subtrahend[0], difference));

  m = MAX(m, MostSignificantDigit(minuend, IDIGIT));

  memset(t, 0, IDIGIT);             /* clear difference */

  for ( i = 0, a.x = 256; i <= m; i++)
    t[i] = a.x = 255 + a.b[1] + minuend[i] - subtrahend[i];

  if (i < IDIGIT) t[i] = a.x = 255 + a.b[1];

  mpCopy(difference, t);

  return(a.b[1]-1);
}
```

```
int mpSubShort( unsigned char minuend[],
                unsigned char subtrahend,
                unsigned char difference[])
{
  int i, m;

  m = MostSignificantDigit(minuend, IDIGIT);

  difference[0] = a.x = 256 + minuend[0] - subtrahend;

  for ( i = 1; i <= m; i++)
    difference[i] = a.x = 255 + a.b[1] + minuend[i];

  if (i < IDIGIT) difference[i] = a.x = 255 + a.b[1];

  for ( i++; i < IDIGIT; i++)
    difference[i] = 0;

  return(a.b[1]-1);
}
```

12.3 Multiplication

Multiplication is a relatively complex task with a great deal of research continuing to be dedicated to its efficient implementation. The algorithm presented here is a relatively simple one, similar to how you might do it by hand. Given two multiple-precision integers u and v, where $u = u_m u_{m-1} \ldots u_1 u_0$ and $v = v_n v_{n-1} \ldots v_1 v_0$ (u_i and v_i being digits base 256)

$$w = u + v$$

or with respect to the digits:

$$w_{i+j} = u_i \cdot v_j + c_{i+j-1},$$

for all $0 \leq i \leq m$ and $0 \leq j \leq n$, the carry is $c_{i+j} = (u_i + v_j) / 256$. Unlike the adder previously described, the carry step is performed using a carry-completion adder. In this case, the carry process is stopped at any stage after it has been determined that no further carry is necessary. In the worst case, the operational time for the remaining carries will be proportional to n, but in the average cases it will be proportional to $\log_2 n$ (Omondi, 1994).

Program mpMul.c. Multiple-precision multiplication (long and short)

```
/*
** mpMul       -- Multiply two multiple precision integers
** mpMulShort -- Multiply a multiple precision integer by a short
integer
**
```

```
**  return value: 0 if ok; not 0 if overflow
*/
#include "mp.h"

int mpMul( unsigned char multiplicand[],
           unsigned char multiplier[],
           unsigned char product[])
{
  int i, j, m, n;
  unsigned char t[IDIGIT];              /* temporary result */

  if ((n = MostSignificantDigit(multiplier, IDIGIT)) == 0)
    return( mpMulShort(multiplicand, multiplier[0], product));

  m = MostSignificantDigit(multiplicand, IDIGIT);
  if ((m + n) >= IDIGIT) return(1);

  memset(t, 0, IDIGIT);                 /* clear product */

  for ( i = 0; i <= m; i++)
  {
    for ( j = 0, a.x = 0; j <= n; j++)
      t[i+j] = a.x = a.b[1] + t[i+j] + multiplicand[i] *
      multiplier[j];
    for ( ; a.b[1] > 0; j++)
      t[i+j] = a.x = a.b[1] + t[i+j];
  }

  mpCopy(product, t);

  return(a.b[1]);
}


int mpMulShort( unsigned char multiplicand[],
                unsigned char multiplier,
                unsigned char product[])
{
  int i, m;

  m = MostSignificantDigit(multiplicand, IDIGIT);

  for ( i = 0, a.x = 0; i <= m; i++)
    product[i] = a.x = a.b[1] + multiplicand[i] * multiplier;

  if (i < IDIGIT)
    product[i] = a.x = a.b[1];

  for ( i++; i < IDIGIT; i++) product[i] = 0;

  return(a.b[1]);
}
```

This multiplier could be improved at the cost of memory if an intermediate accumulator was used with all of the carries performed at the end of the multiplication. An intermediate accumulator might take the form of an array with as many elements as there are digits in the multiple-precision integers. Each element accumulates the column sum from the multiplication. Consider the following long multiplication base 10. In typical accumulation each digit multiplied may result in a carry, the carry being performed, at once. With intermediate accumulation, the digit products are accumulated but not carried until the very end.

Typical accumulation   Intermediate accumulation

	9,876						9876
*	6,789		*				6789
	88,884				81	72	63	54
	79,008			72	64	56	48
	69,132		63	56	49	42
+	59,256		+ 54	48	42	36
	67,048,164		54	111	170	230	170	111	54

Now instead of having performed n^2 carries, we need to perform only n carries on the intermediate products:

						5	4
				1	1	1	
			1	7	0		
		2	3	0			
	1	7	0				
1	1	1					
+ 5	4						
6	7	0	4	8	1	6	4

One final point: If n is the maximum number of digits and b is the radix, then the accumulator must be large enough to hold the value $(b-1) \cdot n$. This can be seen in how the intermediate products stack up.

12.4 Division and Modulus

Division is by far the most complex of the basic arithmetic operations. The basic problem in division is to find, one quotient digit at a time, the number $q = [\, u \,/\, v \,]$, where $u = u_m u_{m-1} \ldots u_1 u_0$ and $v = v_n v_{n-1} \ldots v_1 v_0$ (u_i and v_i being digits base 256). Although cold comfort for such hard work, for the DivMP function in the multiple-precision library, we also will return the remainder after division, r.

A couple of obvious things you might notice right off is that if $v > u$, then $q = 0$ and $r = u$. Also, if $v = u$, then $q = 1$ and $r = 0$. These are inexpensive tests to perform using CompareMP, so I do them first and skip the rest if either condition is satisfied.

If we actually have work to do, the next step is to normalize the operands to moderate the size of q. For reasons that will become more apparent later, we want v_n, the leading digit of v, to be greater than 127. To achieve this, we multiply both u and v by a normalizing factor $d = 256 / (v_n + 1)$. Although this factor will not change the quotient, we must save d so that we can divide it out of the final remainder after the division has been completed.

All division algorithms are based on guessing a quotient digit and either recording it or revising our guess. We can guess in base 256 in much the same way that we guess in base 10 when we perform long division. Divide the first two digits of the dividend by the first digit of the divisor, not to exceed the maximum value of a single digit:

$$q_i = \min(\ (u_m \cdot 256 + u_{m-1}) / v_n,\ 255).$$

Knuth (1981) showed that a q_i estimated in this way should never be a guess wrong by more than 2, provided that v is reasonably large. "Reasonably large" means $v_n > 127$. Having the guessed quotient digit, we then perform a short multiplication and long subtraction

$$r = u - v \cdot q_i.$$

If r is negative, the value of q_i is decremented and the process is repeated. If $r < v$, we are done; otherwise we repeat the process with r as the new dividend. Clearly, there is no chance for an overflow in division, except in the case where $v = 0$.

Program mpDiv.c. Multiple-precision division and modulus (long and short)

```
/*
** mpDiv      -- Divide two multiple precision integers
** mpDivShort -- Divide a multiple precision integer with a short
**                integer

** return value: 0 if ok; 1 if division by zero (overflow)
*/
#include "mp.h"

int mpDiv( unsigned char dividend[],
           unsigned char divisor[],
           unsigned char quotient[],
           unsigned char remainder[])
{
   unsigned char u[IDIGIT], v[IDIGIT], w[IDIGIT], d, q;
   int i, j, m, n;
   int st;                          /* status */
```

```
if ((n = MostSignificantDigit(divisor, IDIGIT)) == 0)
  return( mpDivShort(dividend, divisor[0], quotient, remainder));

mpCopy(u, dividend);            /* dividend - do not destroy */
mpCopy(v, divisor);            /* divisor - do not destroy */

memset( quotient, 0, IDIGIT);        /* clear quotient */
memset( remainder, 0, IDIGIT);        /* clear remainder */

if ((st = mpCompare(u, v)) < 0) {
  mpCopy(remainder, u);
  return(0);
} else if (st == 0) {
  quotient[0] = 1;
  return(0);
}

if ((d = 256L / (v[n] + 1)) > 1)    /* normalize */
{
  mpMulShort(u, d, u);
  mpMulShort(v, d, v);
}
n = MostSignificantDigit(v, IDIGIT);
m = MostSignificantDigit(u, IDIGIT);

for (j = m - n; j >= 0; j--)
{
  q = MIN( (u[m] * 256L + u[m-1]) / v[n], 255);
  do
  {
    mpMulShort(v, q, w);
    mpLeftByteShift(w, j);
    if (st = mpSub(u, w, w)) q--;
  } while (st);
  quotient[j] = q;
  if (mpCompare(w, v) < 0) break;
  mpCopy(u, w);
  m = MostSignificantDigit(u, IDIGIT);
}

mpDivShort(w, d, remainder, w);

return(0);
}

int mpDivShort( unsigned char dividend[],
                unsigned char divisor,
                unsigned char quotient[],
                unsigned char remainder[])
{
  int i, j, m, n;

  if (divisor == 0) return(1);

  m = MostSignificantDigit(dividend, IDIGIT);

  for (i = m, a.x = 0; i >= 0; i--)
  {
```

```
    a.b[0] = dividend[i];
    quotient[i] = a.x / divisor;
    a.b[1] = a.x % divisor;
  }
  remainder[0] = a.b[1];

  for ( i = 1; i < IDIGIT; i++)
  {
    if (i > m)
      quotient[i] = 0;
    remainder[i] = 0;
  }

  return(0);
}
```

12.5 Utility Functions

Converting between base 10 and base 256 are required functions. Other utility functions are provided to facitilitate input, output, and debugging. Their usage is fairly self-explanatory.

Function	Description
mpCompare	Arithmetic comparison of two multiple-precision numbers
mpIsZero	Tests if a multiple-precision number is equal to zero
MostSignificantDigit	Determines the location of the last nonzero digit in a number
mpLeftByteShift	Shift all digits to the left one or more places (multiply by 256)
mpLeftBitShift	Shift the multiple-precision number one or more bits to the left
mpRightByteShift	Shift all digits to the right one or more places (divide by 256)
mpRightBitShift	Shift the multiple-precision integer 1 or more bits to the right
mpAnd	Logical AND, digit by digit on two multiple-precision numbers
ConvertASCIItoMP	Convert decimal text number to base 256 multiple-precision number
ConvertMPtoASCII	Convert base 256 multiple-precision number to decimal text
mpScan	Read ASCII number string (decimal) from input and Convert it to base 256 multiple-precision number
mpCopy	Copy one multiple-precision number to another
PackASCIINumber	Removes all nondigits and commas from an ASCII number string
PrintASCIINumber	Print an ASCII decimal text number (with commas)
mpDump	Dump the contents of multiple-precision number

Program mpComp.c. Multiple-precision comparison function

```
/*
** mpCompare - Compare one multiple-precision integer to another
**
** return value: 1 if x > y
**               0 if x == y
**              -1 if x < y
*/
```

```
#include "mp.h"

int mpCompare( unsigned char x[], unsigned char y[])
{
  int m;

  m = MAX(MostSignificantDigit(x, IDIGIT),
          MostSignificantDigit(y, IDIGIT));

  while (m >= 0)
  {
    if (x[m] > y[m]) return(1);
    if (x[m] < y[m]) return(-1);
    m--;
  }

  return(0);
}
```

Program mpIsZero.c. Multiple-precision zero test

```
/*
** mpIsZero — Tests if a multiple precision integer is zero
**
** return value: 1 if x == 0, 0 otherwise
*/
#include "mp.h"

int mpIsZero( unsigned char x[])
{
  int m;

  m = MostSignificantDigit(x, IDIGIT);

  while (m >= 0)
  {
    if (x[m]) return(0);
    m--;
  }

  return(1);
}
```

Program MSD.c. Find the most significant digit of a multiple-precision integer

```
/*
** MostSignificantDigit — Finds the first non-zero entry in a string
**
** input       x[]     multiple precision integer to search
**             m       starting offset (i.e., size of the string)
**
** return value: offset position of first non-zero digit
*/
#include "mp.h"
int MostSignificantDigit( unsigned char x[],
```

```
                              unsigned int m)
{
  if (m)
  {
    m--;
    while (m && (x[m] == 0)) m--;
  }

  return(m);
}
```

Program mpLS.c; Multiple-precision left bit and left byte shift

```
/*
** mpLeftByteShift - Left byte shift a multiple precision integer
** mpLeftBitShift  - Left bit shift a multiple precision integer
**
** input:     x[]     MP integer
**            s       number of BYTES or BITS to shift
**
** return value: base address of MP integer
*/
#include "mp.h"

unsigned char *mpLeftByteShift(unsigned char x[],
                               unsigned int s)
{
  int i, j;

  if (s > 0)
    for ( i = IDIGIT; i >= 0; i--)
      if ((j = i - s) >= 0)
        x[i] = x[j];
      else
        x[i] = 0;

  return(x);
}


unsigned char *mpLeftBitShift( unsigned char x[],
                               unsigned int s)
{
  int i, j, m, ls, rs;

  ls = s & 7;             /* left shift amount per byte */
  rs = 8 - ls;            /* right shift amount per byte */

  if (s > 0)
    for ( i = IDIGIT; i >= 0; i--)
      if ((j = i - (s / 8)) > 0)
        x[i] = (x[j] << ls) | (x[j-1] >> rs);
      else if (j == 0)
        x[i] = (x[0] << ls);
      else
        x[i] = 0;
  return(x);
}
```

Program mpRS.c. Multiple-precision right-bit and right-byte shift

```
/*
** mpRightByteShift - Right byte shift a multiple precision integer
** mpRightBitShift  - Right bit shift a multiple precision integer
**
** input:      x[]      MP integer
**             s        number of BYTES or BITS to shift
**
** return value: base address of MP integer
*/
#include "mp.h"

unsigned char *mpRightByteShift(unsigned char x[],
                                unsigned int s)
{
  int i, j;

  if (s > 0)
    for ( i = 0; i <= IDIGIT; i++)
      if ((j = i + s) <= IDIGIT)
        x[i] = x[j];
      else
        x[i] = 0;

  return(x);
}

unsigned char *mpRightBitShift( unsigned char x[],
                                unsigned int s)
{
  int i, j, ls, rs;

  rs = s & 7;             /* right shift amount per byte */
  ls = 8 - rs;            /* left shift amount per byte */

  if (s > 0)
    for (i = 0; i <= IDIGIT; i++)
      if ((j = i + (s >> 3)) < IDIGIT)
        x[i] = (x[j] >> rs) | (x[j+1] << ls);
      else if (j == IDIGIT)
        x[i] = (x[j] >> rs);
      else
        x[i] = 0;

  return(x);
}
```

Program mpAnd.c. Multiple-precision bitwise AND

```
/*
** mpAnd - Bitwise AND between two multiple precision integers
**
** return value: always 0
*/
#include "mp.h"
```

```
int mpAnd( unsigned char and1[],
           unsigned char and2[],
           unsigned char result[])
{
  int i, m;

  m = MAX(MostSignificantDigit(and1, IDIGIT),
          MostSignificantDigit(and2, IDIGIT));

  for ( i = 0; i <= m; i++)
    result[i] = and1[i] & and2[i];

  if (i < IDIGIT)
    result[i] = 0;

  return(0);
}
```

Program cvta2mp.c. Convert an ASCII number string to a multiple-precision integer

```
/*
** ConvertASCIItoMP -- Convert from ASCII number string to
**                     a multiple precision integer
**
** return value: base address of MP integer
*/
#include "mp.h"

unsigned char *ConvertASCIItoMP( unsigned char x[],
                                 unsigned char s[])
{
  int i, k;

/*
** convert RADIX by evaluating the polynomial:
**
**    x = (s * 10 + s ) * 10 + ... ) * 10 + s  ) * 10 + s
**         0        1                    m-1          m
**
** note that the greatest digit is in the 0th location in string s
** but the greatest digit is in the last non-zero digit of x
*/

  memset( x, 0, IDIGIT);
  a.x = 0;

  x[0] = s[0] - '0';                      /* first digit */

  for (k = 1; k < strlen(s); k++)
  {
    for ( i = 0; i < IDIGIT; i++)         /* multiply by 10 */
      x[i] = a.x = a.b[1] + x[i] * 10;

    x[0] = a.x = x[0] + (s[k] - '0');     /* add digit */
```

```
        for ( i = 1; a.b[1] > 0; i++)      /* carry */
           x[i] = a.x = a.b[1] + x[i];
     }
     return(x);
}
```

Program cvtmp2a.c. Convert a multiple-precision integer into an ASCII number string

```
/*
** ConvertMPtoASCII -- Convert a multiple precision integer to
**                     an ASCII number string
**
** return value: *s = base address of ASCII number string
*/
#include "mp.h"

unsigned char *ConvertMPtoASCII(unsigned char s[],
                                unsigned char x[])
{
   int i, j, k, m, n;
   unsigned char a[ADIGIT];   /* accumulator    for    intermediate
                                 products */

/*
** convert RADIX by evaluating the polynomial:
**
**    s = (x * 256 + x  ) * 256 + ... ) * 256 + x ) * 256 + x
**         m        m-1                           1          0
*/

   memset(s, 0, ADIGIT);

   m = MostSignificantDigit(x, IDIGIT);
   s[0] += x[m] % 10;                     /* ones */
   s[1] += x[m] % 100 / 10;               /* tens */
   s[2] += x[m] / 100;                    /* hundreds */

   for ( k = m - 1; k >= 0; k--)
   {
     memset(a, 0, ADIGIT);               /* zero out accumulator */

     m = MostSignificantDigit(s, ADIGIT - 1) + 2;

     for ( i = 0; i <= m; i++)           /* multiply by 256 */
     {
       a[i]   += s[i] * 6;
       a[i+1] += s[i] * 5;
       a[i+2] += s[i] * 2;
       s[i] = a[i];
     }

     if (x[k] > 0)                       /* add MP digit */
     {
       s[0] += x[k] % 10;                /* ones */
       s[1] += x[k] % 100 / 10;          /* tens */
       s[2] += x[k] / 100;               /* hundreds */
     }
```

```
      for ( i = 0; i < ADIGIT-2; i++)      /* carry */
        if (s[i] > 9)
        {
          s[i+1] += s[i] / 10;
          s[i]   %= 10;
        }
    }

/* reverse the order of the digits and convert ASCII string */

    m = MostSignificantDigit(s, ADIGIT);
    for ( i = 0; i <= m; i++) s[i] += '0';   /* make into ASCII */
    strrev(s);                               /* reverse the digits */

    return(s);
}
```

Program mpScan.c. Read an ASCII number string from standard input and convert it to a multiple-precision integer

```
/*
** mpScan - Read an ASCII number and convert it to a
**          multiple precision integer
**
** return value: base address of input ASCII string
*/
#include "mp.h"

unsigned char *mpScan(unsigned char a[],
                      unsigned char z[])
{
  gets(a);
  PackASCIINumber(a);
  ConvertASCIItoMP(z, a);

  return(a);
}
```

Program mpCopy.c. Copy one multiple-precision integer to another

```
/*
** mpCopy - Copy one multiple precision integer to another
**
** return value: *to = base address of 'to' MP integer
*/
#include "mp.h"

unsigned char *mpCopy(unsigned char to[],
                      unsigned char fr[])
{
  memcpy( to, fr, IDIGIT);
  return( to );
}
```

Program packan.c. Pack an ASCII number string

```
/*
** PackASCIINumber -- Squeeze all of the non-digits out of a number
**                    string

** return value: base address of packed string
*/
#include "mp.h"

char *PackASCIINumber( char *str)
{
  char *pi = str;
  char *pj = str;

  while ( *pi ) {
    if ( isdigit(*pi) )
      *pj++ = *pi;
    pi++;
  }
  *pj = '\0';

  return(str);
}
```

Program printan.c. Print an ASCII number string

```
/*
** PrintASCIINumber - Print a number out with commas to help
** delimit it

** return value: *str = pointer to input number string
*/
#include "mp.h"

char *PrintASCIINumber( char *str)
{
  int offs;
  char *s = str;

  if (strchr(str, ','))
  {
    printf( "%s", str);        /* string has commas */
  }
  else
  {

/* compute the first character group */

    offs = strlen(s) % 3;
    if ( offs == 0) offs = 3;
    printf( "%*.*s", offs, offs, s);
    s += offs;

/* now print the rest of the string with commas */

    while ( *s ) {
```

```
        printf( ",%3.3s", s);
        s += 3;
    }
  }

  return(str);
}
```

Program mpDump.c. Dump a multiple-precision integer in hexadecimal format

```
/*
** mpDump - Dump the contents of a multiple precision integer
**          in hexadecimal format
**
** return value: (none)
*/
#include "mp.h"

void mpDump( unsigned char x[])
{
  int i;
  char a[ADIGIT];

  printf("\n");
  ConvertMPtoASCII(a, x);
  PrintASCIINumber(a);
  printf("\n");
  for (i = MostSignificantDigit(x, IDIGIT); i >= 0; i--)
    printf("%2.2x ", x[i]);
  printf("\n");

  return;
}
```

12.6 Multiple-Precision Implementations

Below are several multiple-precision applications from previous chapters that use the multiple-precision library. The algorithm in each case is the same as it was originally presented, but the calls to the multiple-precision library have been used.

Application: Greatest Common Divisor

As noted in Chapter 2, greatest common divisors serve many purposes. I think of the GCD as the function that provides a measure of divisibility. In our quest for large primes, this function is required. The following program is a multiple-precision version of Program 2_2_1.

Program 12_6_1.c. Multiple-precision greatest common divisor

```
/*
** Program 12_6_1 - Find the greatest common divisor using
**                  Euclid's Algorithm and MP arithmetic
*/
#include "numtype.h"
#include "mp.h"

void main(void)
{
  unsigned char *s;            /* ASCII number string */
  unsigned char *a, *b, *r;    /* multiple precision integers */

  if ( (s = (unsigned char *) calloc(ADIGIT, 1)) == NULL) exit(1);
  if ( (a = (unsigned char *) calloc(IDIGIT, 1)) == NULL) exit(1);
  if ( (b = (unsigned char *) calloc(IDIGIT, 1)) == NULL) exit(1);
  if ( (r = (unsigned char *) calloc(IDIGIT, 1)) == NULL) exit(1);

  printf("Find greatest common divisor\n\n");
  printf("a = ");
  mpScan(s, a);
  printf("b = ");
  mpScan(s, b);

  if (mpCompare(a, b) < 0) { mpCopy(r, a); mpCopy(a, b); mpCopy(b,
r); }

  while (! mpIsZero(b))
  {
    mpDiv(a, b, a, r);
    mpCopy(a, b);
    mpCopy(b, r);
  }

  printf("gcd = ");
  PrintASCIINumber( ConvertMPtoASCII(s, a));

  free(s);
  free(a);
  free(b);
  free(r);
}
```

```
C> 12_6_1
Find greatest common divisor

a = 12345678987654321
b = 555555555
gcd = 111,111,111
```

Application: Integer Square Root

The next program finds integer square roots. It is Program 9_3_2 reinvented to operate on multiple-precision integers. The integer square root has an important place in the quest for factorizations. If a number is not prime, it must have a factor less than or equal to the integer square root.

Program 12_6_2.c. Finds integer square roots for multiple-precision integers

```
/*
** Program 12_6_2 - Find the integer square root of a number: a =
**                  q^2 + r using Newton's method and MP arithmetic
*/
#include "numtype.h"
#include "mp.h"

main ()
{
  unsigned char *s;              /* ASCII number string */
  unsigned char *a, *q, *r;      /* multiple precision integers */

  if ( (s = (unsigned char *) calloc(ADIGIT, 1)) == NULL) exit(1);
  if ( (a = (unsigned char *) calloc(IDIGIT, 1)) == NULL) exit(1);
  if ( (q = (unsigned char *) calloc(IDIGIT, 1)) == NULL) exit(1);
  if ( (r = (unsigned char *) calloc(IDIGIT, 1)) == NULL) exit(1);

  printf("Find integer square root\n\n");
  printf("a = ");
  mpScan(s, a);

  mpSquareRoot(a, q, r);

  printf("√[û");
  PrintASCIINumber( ConvertMPtoASCII( s, a));
  printf("] = ");
  PrintASCIINumber( ConvertMPtoASCII( s, q));
  printf(", r = ");
  PrintASCIINumber( ConvertMPtoASCII( s, r));
  printf("\n");

  free( s );
  free( a );
  free( q );
  free( r );
}

/*
** mpSquareRoot - computes multiple precision integer square roots
**
** return value: 0 if success; 1 if error
*/
int mpSquareRoot( unsigned char a[],
                  unsigned char q[],
                  unsigned char r[])
{
  unsigned char t[IDIGIT];     /* temp MP integer */

  mpCopy(q, a);
```

```
    if (! mpIsZero(a))
    {
      mpDiv(a, q, r, t);
      while (mpCompare(q, r) > 0)
      {
        mpAdd(q, r, q);
        mpDivShort(q, 2, q, r);
        mpDiv(a, q, r, t);
      }
    }

    mpMul(q, q, t);
    mpSub(a, t, r);

    return(0);
}
```

C> **12_6_2**
Find integer square root

a = **975461057913427830**
[√975,461,057,913,427,830] = 987,654,321, r = 123,456,789

Application: Lucas-Lehmer Test for Mersenne Primes

In Chapter 6 we saw how to use the test developed for testing the primality of Mersenne primes. Using the multiple-precision library, we implement the Lucas-Lehmer test and verify what Lucas verified *by hand*: that $M(127)$, a 39-digit number, is indeed prime. We can also test for larger Mersenne primes, as can be seen in the example below.

You may recall that to develop the Lucas sequence, we start with L_0 = 4 and let $L_{n+1} = L_n^2 - 2$. $M(p)$ is prime if and only if $M(p)$ divides L_{p-2}. We do not need to compute the Lucas number, only the Lucas number mod $M(p)$. Because of the special form of $M(p)$, we can use a mask and a MP bit shift operation to accomplish what would otherwise be an unpleasantly slow MP division. But don't expect lightning either; these are big numbers being chewed up and they're pretty tough.

Program 12_6_3.c. Tests Mersenne numbers for primality

```
/*
** Program 12_6_3 - Test Mersenne numbers for primality using the
**                  optimized Lucas test
*/
#include "numtype.h"
#include "mp.h"

void main(void)
{
  unsigned char *s;          /* ASCII number string */
  unsigned char *l, *m, *t;  /* multiple precision integers */
  unsigned int i, p;
```

```
    if ( (s = (unsigned char *) calloc(ADIGIT, 1)) == NULL) exit(1);
    if ( (l = (unsigned char *) calloc(IDIGIT, 1)) == NULL) exit(1);
    if ( (m = (unsigned char *) calloc(IDIGIT, 1)) == NULL) exit(1);
    if ( (t = (unsigned char *) calloc(IDIGIT, 1)) == NULL) exit(1);

    printf("Test a Mersenne number for primality\n\n");
    printf("p = ");
    scanf("%u", &p);

    m[0] = 1;
    mpLeftBitShift(m, p);
    mpSubShort(m, 1, m);               /* the mersenne number */
    printf("M(%u) = ", p);
    PrintASCIINumber( ConvertMPtoASCII(s, m));

    l[0] = 4;                          /* initialize lucas sequence */

    for (i = 1; i < p - 1; i++)
    {
      if (mpMul(l, l, l)) printf("overflow\n");
      mpSubShort(l, 2, l);
      while (mpCompare(l, m) > 0)
      {
        mpCopy(t, l);
        mpAnd(t, m, t);
        mpRightBitShift(l, p);
        mpAdd(t, l, l);
      }
    }

    if (mpCompare(l, m) == 0)
      printf(" is prime\n");
    else
      printf(" is not prime\n");

    free(s);
    free(l);
    free(m);
    free(t);
}
```

```
C> 12_6_3
Test a Mersenne number for primality


p = 127
M(127) =   170,141,183,460,469,231,731,687,303,715,884,105,727
is prime

C> 12_6_3
Test a Mersenne number for primality


p = 1279
M(1279) =    10,407,932,194,664,399,081,925,240,327,364,085,538,
615,262,247,266,704,805,319,112,350,403,608,059,673,360,298,012,
239,441,732,324,184,842,421,613,954,281,007,791,383,566,248,323,
464,908,139,906,605,677,320,762,924,129,509,389,220,345,773,183,
349,661,583,550,472,959,420,547,689,811,211,693,677,147,548,478,
866,962,501,384,438,260,291,732,348,885,311,160,828,538,416,585,
```

```
028,255,604,666,224,831,890,918,801,847,068,222,203,140,521,026,
698,435,488,732,958,028,878,050,869,736,186,900,714,720,710,555,
703,168,729,087 is prime
```

Now I ask you, what more could a person want to know from life?

Application: Fibonacci Numbers

Using the multiple-precision library to generate Fibonacci numbers is easy. Recall from Chapter 7 that Fibonacci numbers are developed from a recursive sequence defined so that each number is the sum of the two preceding terms:

$$F_n = F_{n-1} + F_{n-2}.$$

Given $F_0 = 0$, $F_1 = 1$, this yields 0, 1, 1, 2, 3, 5, 8, 13, 21, 34, ... Around F_{47} the precision of the long integer will be exceeded. To find even larger numbers, we use multiple-precision functions.

Program 12_6_4.c. Compute 1000 Fibonacci numbers

```
/*
** Program 12_6_4 - Compute 1,000 Fibonacci numbers
*/
#include "numtype.h"
#include "mp.h"

void main(void)
{
  unsigned char *a;              /* ASCII number string */
  unsigned char *x, *y, *z;      /* multiple precision integers */
  unsigned int i;

  if ( (a = (unsigned char *) calloc(ADIGIT, 1)) == NULL) exit(1);
  if ( (x = (unsigned char *) calloc(IDIGIT, 1)) == NULL) exit(1);
  if ( (y = (unsigned char *) calloc(IDIGIT, 1)) == NULL) exit(1);
  if ( (z = (unsigned char *) calloc(IDIGIT, 1)) == NULL) exit(1);

  printf("Fibonacci numbers\n\n");

  x[0] = 0;
  y[0] = 1;

  for (i = 2; i <= 1000; i++)
  {
    mpAdd(x, y, z);
    mpCopy(x, y);
    mpCopy(y, z);

    printf("\nF[%u] = ", i);
    PrintASCIINumber( ConvertMPtoASCII(a, z));
  }
```

```
        free(a);
        free(x);
        free(y);
        free(z);
}
```

```
C> 12_6_4
1,000 Fibonacci numbers

F[2] = 1
F[3] = 2
F[4] = 3
F[5] = 5
F[6] = 8
F[7] = 13
F[8] = 21
F[9] = 34
F[10] = 55
F[11] = 89
...
F[100] = 354,224,848,179,261,915,075
...
F[500] = 139,423,224,561,697,880,139,724,382,870,407,283,950,070,
256,587,697,307,264,108,962,948,325,571,622,863,290,691,557,658,
876,222,521,294,125
...
F[1000] = 43,466,557,686,937,456,435,688,527,675,040,625,802,564,
660,517,371,780,402,481,729,089,536,555,417,949,051,890,403,879,
840,079,255,169,295,922,593,080,322,634,775,209,689,623,239,873,
322,471,161,642,996,440,906,533,187,938,298,969,649,928,516,003,
704,476,137,795,166,849,228,875
```

Challenges

1. Multiple-precision multiplication can be sped up a number of different ways. If you think of multiplication as summing partial products, a modification immediately springs to mind. Analyze the operation count for the following two multiplication methods: immediate carry versus deferred carry. Modify MulMP so that all the carries occur after the multiplication.

2. Write a multiple-precision least common multiple routine.

3. Modify the multiple-precision library so that each multiple-precision integer has a sign (positive/negative).

4. Modify the multiple-precision library so that it supports rational numbers.

5. Modify the multiple-precision library so that it supports floating point numbers.

Internet Websites with Information About Number Theory

There are many fine search engines available for the Internet that will find literally thousands of websites having information relating to mathematics and number theory. In fact, it is tough to limit your searches. The majority of these are maintained by individuals affiliated with academic or research organizations. When possible I relied on shown "purpose" statements provided by the website's sponsor.

Here I've listed merely eight websites. My hope is that from these few, you can find links to hundreds of other related sites, do resource searches, and obtain documents and software. My feeling is that eight high-quality sites are better than a list of eight hundred undifferentiated sites.

Most of the sites listed provide information on mathematics in general with a subdivision for number theory. Those with a particular number-theoretic slant tend to use the "institute" moniker in their name. All sites provide links to other sites offering information on related topics.

Aside from these websites, all the major book publishers, scientific software publishers, and many of the smaller houses have websites where the latest information about current and planned publications and products are offered.

A.1 American Mathematical Society e-MATH Home Page

http://www.ams.org

This page offers information on a variety of mathematical resources as well as publication and organization news. It has several databases of mathematical literature, libraries, and book dealers that can be searched with links to other WWW sites.

Stated Aim. e-MATH is the WWW site of the American Mathematical Society (AMS), serving as a resource for information of interest to mathematicians all over the world. The goal of e-MATH is to disseminate information about the AMS and its policy, to provide access to mathematical literature, and to provide access to mathematical information and services.

A.2 The Computer Algebra Information Network

http://www.can.nl/

Computer Algebra Information Network (CAIN) is a server providing all kinds of information about symbolic and algebraic computation, including number theory. CAIN was set up by eight research teams in the framework of the European Symbolic and Algebraic Computation (SAC) Network and is maintained by Computer Algebra Nederland (CAN) / Riaca. CAIN gathers information from the CAIN national nodes and other servers in the world.

Stated Aim. The central objective of the SAC ("Symbolic and Algebraic Computation") Network is to create a platform on which the various activities of Symbolic and Algebraic Computation in Europe can be successfully integrated. Computer Algebra in the Netherlands (CAN), is a foundation established in 1988 to stimulate and coordinate the use of computer algebra (systems) in education and research.

A.3 European Mathematical Information Service (EMIS)

http://www.emis.de/

The European Mathematical Information Service (EMIS) is offered by the European Mathematical Society (EMS). The purpose of the Society is "to further the development of all aspects of mathematics in the countries of Europe [and] to promote research in mathematics and its applications." The electronic library and links make this site helpful to researchers in mathematics. This site is mirrored at nearly 20 other sites around the world.

Stated Aim. The EMS server (EMIS) collects electronically accessible mathematical journals in its electronic library (ELibEMS). It provides quick access to other mathematical servers. It also offers organizational services such as access to addresses of individual members of the EMS, information about conferences, openings, and positions available.

A.4 Mathematics Information Server (Magellan ★★★ Site)

http://www.math.psu.edu/MathLists/Contents.html

Penn State (*http://www.psu.edu/*) Department of Mathematics (*http://www.math.psu.edu/*) maintains this site as an excellent database of online resources. Major topic areas include: archives, societies and associations, institutes and centers, commercial pages, mathematics journals, mathematics preprints, subject-area pages, other archived materials, and mathematics software.

The Mathematics Information Center maintains an on-line archive of teaching aids that have been developed by Penn State mathematics faculty and graduate students. The archive contains text files, the corresponding DVI files, pictures (GIFS), and movies (MPEG files). Although at present most of the material is related to calculus, there are some files concerning differential equations.

A.5 Mathematics Archives (Magellan ★★★★ Site)

http://archives.math.utk.edu/

Topnotch pages from the University of Tennessee at Knoxville (*http://www.utk.edu/*) are links to resources on the Internet that may be useful in the teaching of an undergraduate course in mathematics. Under Topics in Mathematics you can find teaching materials, software, and World Wide Web (WWW) links organized by mathematical topics in a searchable database.

The software area offers public domain and shareware software, links to other software sites. The Teaching Materials section offers information about calculus resources on-line, emerging scholars program, graphing calculators, JAVA, and other interactive WWW pages, K–12 teaching materials, lessons, tutorials and lecture notes, visual calculus, and so forth.

Other Math Archives features include electronic proceedings of the CTM and the ICTCM, POP Mathematics, Project NEXT, UTK Mathematical Life Sciences Archives, and so forth. In particular, there are numerous links to course notes and free software downloads on many topics in mathematics, including number theory. Use the search page to select areas using keywords or select a topic from an alphabetized list.

A.6 The Number Theory Web

http://www.maths.uq.oz.au/~krm/web.html
http://www.math.uga.edu/~ntheory/web.html (mirror site)
http://www.mat.uniroma3.it/ntheory/web_italy.html (mirror site)

This site is maintained by Keith Matthews at Department of Math, University of Queensland, Brisbane, Australia. The site has varied topics and many are advanced. Although it could be better organized, there is a significant amount of useful links to number theory centers, universities, and publications. The site is mirrored at the University of Georgia, Athens and Dipartimento di Matematica dell'Università di Roma, Tre

Stated Aim. To disseminate information of interest to number theorists everywhere and to facilitate communication between number theorists.

A.7 The Prime Page

http://www.utm.edu/research/primes

The Prime Page website is a well organized, easy to read, factual overview of past and current problems relating to primes and factorization. Current discoveries are promptly posted here. It is maintained by Chris K. Caldwell at the University of Tennessee at Martin (*http://www.utm.edu/*). There are many pages suitable for beginning and advanced number theorists with factual information on prime records, methodologies, algorithms, theory, and other web links.

Stated Aim. The Prime Page indexes resources of and about prime numbers and serves as the prime source for information about prime numbers.

A.8 MathSoft, Inc. – Unsolved Mathematics Problems

http://www.mathsoft.com/asolve/

This is a commercial site operated by MathSoft, Inc., makers of the software product MathCad. It is a collection of unsolved problems and other topics in mathematics with many relating to number theory. Steven Finch maintains the site for MathSoft. Although not an academic site, it offers well-written and extensive descriptions of ongoing research and current problems. There is a nice list of sites having outlines of other well-known unsolved problems.

Stated Aim. An eclectic gathering of questions and partial answers to mathematical problems. It purports to be less interested in reporting the latest incremental advances on well-known problems than in publicizing little-known problems.

B

MS C/C++ Graphics Functions

There are a number of functions in the Microsoft C/C++ library for performing various graphic tasks. For simplicity and portability, only a few are used in this book. You may want to consult a runtime library reference manual for more detailed information regarding the various graphics commands. The following runtime library functions are used in this book:

Function	Description
_getvideoconfig(struct _ videoconfig _ far *config)	Gets the current graphics video configuration information
_lineto(short x, short y)	Draws a line from the current graphics position to the specified point (x, y)
_moveto(short x, short y)	Moves the current graphics position to the specified point (x, y)
_setcolor(short color)	Sets the current color to be used for graphics operations
_setpixel(short x, short y)	Sets the pixel at location (x, y) to the current color
_setvideomode(short mode)	Sets the video mode

Each function has one or more of the following parameters associated with it.

color   color index to set
The color index is based on the number of colors supported by the current video mode. See mode constants below for the number of colors

supported. Color values are usually referenced with the following macros:

```
#define   _BLACK          0x000000L
#define   _BLUE           0x2a0000L
#define   _GREEN          0x002a00L
#define   _CYAN           0x2a2a00L
#define   _RED            0x00002aL
#define   _MAGENTA        0x2a002aL
#define   _BROWN          0x00152aL
#define   _WHITE          0x2a2a2aL
#define   _GRAY           0x151515L
#define   _LIGHTBLUE      0x3F1515L
#define   _LIGHTGREEN     0x153f15L
#define   _LIGHTCYAN      0x3f3f15L
#define   _LIGHTRED       0x15153fL
#define   _LIGHTMAGENTA   0x3f153fL
#define   _YELLOW         0x153f3fL
#define   _BRIGHTWHITE    0x3f3f3fL
```

config   current video information

The _videoconfig structure *config* contains the current video information. It has the following members:

Member	Type	Description
numxpixels	Short	Number of pixels on X axis
numypixels	Short	Number of pixels on Y axis
numtextcols	Short	Number of text columns available
numtextrows	Short	Number of text rows available
numcolors	Short	Number of actual colors
bitsperpixel	Short	Number of bits per pixel
numvideopages	Short	Number of available video pages
mode	Short	Current video mode
adapter	Short	Active display adapter
monitor	Short	Active display monitor
memory	Short	Adapter video memory in K bytes

mode   video mode

Constants for screen codes are used to set the video mode and to return to text mode after the graphics program has completed execution. The graphics applications use _MAXRESMODE to enter graphics mode and _DEFAULTMODE to return to text mode. The default text mode is usually _TEXTC80 unless a monochrome monitor is detected.

Mode	Type	Colors	Adapter
_MAXRESMODE	graphics mode with highest resolution		
_MAXCOLORMODE	graphics mode with most colors		
_DEFAULTMODE	restore screen to original mode at startup		
_TEXTMONO	80-column text, black and white	32	MDPA
_TEXTBW40	40-column text, grey	16	CGA
_TEXTC40	40-column text, color	16/8	CGA
_TEXTBW80	80-column text, grey	16	CGA
_TEXTC80	80-column text, color	16/8	CGA
_MRES4COLOR	320×200, color	4	CGA
_MRESNOCOLOR	320×200, grey	4	CGA
_HRESBW	640×200, black and white	2	CGA*
_HERCMONO	Hercules Graphics 720 x 348, BW	2	HGC*
_MRES16COLOR	320×200, color	16	EGA
_HRES16COLOR	640×200, color	16	EGA
_ERESNOCOLOR	640×350, black and white	2	EGA
_ERESCOLOR	640×350, color	16/4	EGA*
_VRES2COLOR	640×480, black and white	2	VGA
_VRES16COLOR	640×480, color	16	VGA*
_MRES256COLOR	320×200, color	256	VGA

*Mode selected by _MAXRESMODE if adequate video memory is available.

x x (horizontal) screen coordinate

y y (vertical) screen coordinate

The ordered pair (x, y) references the graphics adapter's coordinate system. The maximum range is determined by the type of graphics adapter currently in use on the computer. The table of graphics modes shows the possible ranges. The coordinate system itself has the origin $(0, 0)$ in the upper left corner of the screen. The graphics coordinate system is slightly different from what you might expect: x increases to the right and y increases downward.

C

ANSI Standard Escape Sequences

Text-graphic applications require escape sequences to control where on the screen the next bit of text is to be printed and what its color attribute will be. The ANSI standard escape sequences are a simple, effective way to do just this (ANSI, 1979). ANSI escape sequences are described in the American National Standards Institute's (ANSI) document X3.64-1979 that defines "a set of encoded control functions to facilitate data interchange with two-dimensional character-imaging input-output devices." The X3.64-1979 standard augments the X3.41977 ANSI standard.

Although only a few programs in the book use escape sequences, you must have an ANSI screen driver installed for those that do. Workstations and terminals will recognize ANSI escape sequences. A PC will not without installing the driver explicitly. The device driver needed is provided with MS-DOS and is called ANSI.SYS. It (or an equivalent) must be installed as a device in the startup file CONFIG.SYS. Generally, a device driver is software that allows the user to interface with specific computer hardware. The ANSI.SYS is a console driver that lets the user interface with the screen from the keyboard or a program. To install the ANSI.SYS driver on your PC, you will need to have the line

DEVICE=C:\DOS\ANSI.SYS

or something like it in your CONFIG.SYS file.

ANSI.SYS goes to work when it recognizes a string of characters called an escape sequence being sent to standard output. These special strings always begin with the escape character, followed by the left square bracket. There are all sorts of escape sequences; they are used to control the cursor position, set screen colors, and reassign the keyboard keys.

The general form for an escape sequence is

ESC[#;#; ...;T,

where the escape character (ASCII 27) is represented by ESC, the symbols # indicate a parameters (usually numbers) that are used to form the escape sequence, and T represents the character terminating the sequence that indicates the command. Table C.1 lists some of the available escape sequences recognized by ANSI.SYS.

The default values are the values assumed if the parameters are omitted. The general form for an escape sequence without parameters is

ESC[T,

where ESC is the escape character and T is the command character terminating the sequence. Writing escape sequences without parameters, thus obtaining the default values, minimizes the number of characters you actually have to generate and improves the speed of the display.

For some of the applications in this book, it is necessary to be able to control the cursor location and text color. Although most C runtime libraries have screen formatting capability, using ANSI.SYS has obvious benefits because it is platform independent.

The following is a C/C++ code example that erases the entire screen:

```
printf("\33[2J");
```

\33 is octal for 27, the escape character. The next example sets the graphics rendition (i.e., text color):

```
printf("\33[%dm", k);
```

all the text that is printed after this escape sequence will have the attribute indicated by the number k.

The screen itself can be thought of as a grid of characters. Each line on the screen is a row starting with row 1 at the top of the screen and increasing downward. The columns are the character positions along a row. Column 1 is the first character on the row and (generally) column 80 is the last. Using ANSI escape sequences we can position the cursor anywhere we choose on the screen. The following example uses two parameters to move the cursor to row r and column c:

```
printf("\33[%d;%dH", r, c);
```

once the cursor is positioned, the first character of the next text string that is printed will begin there.

Table C.1 is a listing of the basic escape sequences used by ANSI.SYS. Note that #s are numbers and ESC represents the escape character (ASCII 27, octal 33, hex 1B). The default value is the value assumed for the parameter if it is not present.

Table C.1. ANSI Escape Sequences for Controlling Text Output

Sequence	Name	Action	Default
ESC[#;#H	CUP	Cursor Position	1;1
		Moves cursor to row #, column #	
ESC[#A	CUU	Cursor Up	1
		Moves cursor up # rows from the current cursor position	
ESC[#B	CUD	Cursor Down	1
		Moves cursor down # rows from the current cursor position	
ESC[#C	CUF	Cursor Forward	1
		Moves cursor forward # columns from the current cursor position	
ESC[#D	CUB	Cursor Backward	1
		Moves cursor backward # columns from the current cursor position	
ESC[s	SCP	Save Cursor Position	(none)
		Causes the current cursor position to be stored so that the position can be returned to later when RCP is received	
ESC[u	RCP	Restore Cursor Position	(none)
		Moves cursor to the cursor position saved with SCP	
ESC[2J	ED	Erase Display	(none)
		Erase the whole display	
ESC[K	EL	Erase Line	(none)
		Erase the line from and including the current cursor position	
ESC[#; ...;#m		SGR Set Graphics Rendition	0
		Sets foreground and background colors for subsequent output	

Value	Result
0	Normal (white on black)
1	Bold on (high intensity)
4	Underline on
5	Blink on

7	Reverse video (black on white)
8	Concealed
30	Black foreground
31	Red foreground
32	Green foreground
33	Yellow foreground
34	Blue foreground
35	Magenta foreground
36	Cyan foreground
37	White foreground
40	Black background
41	Red background
42	Green background
43	Yellow background
44	Blue background
45	Magenta background
46	Cyan background
47	White background

ESC[=#h	SM	Set Mode	1
ESC[=#l	RM	Reset Mode	1

Sets/resets the screen mode

(Other modes may be available, depending on the graphics adapter)

Value	Result
0	40 × 80 black and white
1	40 × 80 color
2	80 × 25 black and white
3	80 × 25 color
4	320 × 200 color
5	320 × 200 black and white
6	640 × 200 black and white
7	wrap/nowrap at end of line

References

Andrews, George E., 1971
Number Theory, Dover Publications, NY, 259pp

ANSI, 1979
American National Standard Additional Controls for Use with American National Standard Code for Information Exchange, American National Standards Institute, Inc., 1430 Broadway, New York 10018, Document: ANSI X3.64-1979

Appel, K., and Haken, W., 1976
Every planar map is four colorable, Bulletin of the American Mathematical Society, vol. 82, 1976 pp.711-712.

Atkin, A. O., and Morain, F., 1993a
Finding Suitable Curves for the Elliptic Curve Method of Factoring, Mathematics of Computation, Vol 60, No 201 (January), pp 399-405

Atkin, A. O., and Morain, F., 1993b
Elliptic Curves and Primality Proving, Mathematics of Computation, Vol 61, No 203 (July), pp 29-68

Bach, Eric, and Shallit, Jeffrey, 1996
Algorithmic Number Theory, The MIT Press, Cambridge, Massachusetts, 512pp

Bachman, P., 1894
Die Analytische Zahlentheorie, Teubner Publishers, Leipsig

Bateman, P. T., Selfridge, J. L., and Wagstaff Jr., S. S., 1989
The New Mersenne Conjecture, American Mathematical Monthley, Vol 96 (1989), pp 125-128

Bays, C., Hudson, R. H., 1977
The segmented sieve of Eratosthenes and primes in arithmetic progressions to 1012, BIT Vol 17, pp 121-127

Boles, M., 1987
The Golden Relationship: Art, Math, Nature, 2nd ed., Pythagorean Press

Boyer, Carl B., and Merzbach, Uta C.. 1991
A History of Mathematics, Second Edition
John Wiley & Sons, Inc., New York, 715pp

Brent, R. P., Cohen, L. G., and de Riele, H. J. J., 1991
Improved techniques for lower bounds for odd perfect numbers, Math. Comp., 57 (1991) 857-868

Carmichael, R. D., 1912
On composite numbers P which satisfy the Fermat Congruence $a^{p-1} \equiv 1 \pmod{P}$. American Mathematical Monthly, Vol 19, pp 22-27

414 References

Cohen, Henri, 1995
A course in Computational Algebraic Number Theory, Springer Verlag, New York,
534pp

Chen, Jing-Run, 1966
On the representation of a large even integers as the sum of a prime and the product of at most two primes, Kexue Tongbao (Foreign Language Edition), Vol 17, pp 385-386, MR 34, #7483

Chen, Jing-Run, 1973, 1978
On the representation of a large even integers as the sum of a prime and the product of at most two primes, I and II, Sci. Sinica, Vol 16, 1973, pp 157-176, and Vol 21, 1978, pp 421-430

Chernick, J., 1939
On Fermat's simple theorem, Bulliten American Mathematical Society, Vol 45, pp 269-274

Cohen, Henri, 1995
A Course in Computational Algebraic Number Theory, Graduate Texts in Mathematics # 138, Springer-Verlag, New York

Colvin, Michael E.; Scher, Bob; and Seidl, Ed, 1996
Massively Parallel Number Theory, Sandia National Laboratory (California), Scientific Computing Department, http://midway.ca.sandia.gov/~mecolv/euler/

Conte, S. D., and de Boor, Carl, 1980
Elementary Numerical Analysis; an Algorithmic Approach
McGraw-Hill Book Company, New York, 432pp, 3rd Edition

Crocker, R., 1962
A theorem on pseudo-primes, American Mathematical Monthly, Vol 69, pg 540

Davis, J. A., Holdbridge, D. B., and Simmons, G. J., 1983
Status report on factoring at the Sandia National Laboratories, in Advances in Cryptology: pp 183-215 in Proceedings of CRYPTO 84, Springer-Verlag, Berlin

Deleglise, M., and Rivat, J., 1996
Computing $\pi(x)$: The Meissel, Lehmer, Largarias, Miller, Odlysko Method, Mathematics of Computation, Vol 65, No 213, pp 235-245.

Dixon, J. D., 1981
Asymptotically Fast Factorization of Integers, Math Comp, Vol 36, pp 255-260

Dubner, H., 1989
A new method for producing large Carmichael numbers, Math Comp, Vol 53, p 411-414

Ellis, Graham, 1992
Rings and Fields, Clarendon Press, Oxford, 196 pg

Eichenauer-Herrmann, Jurgen, and Emmerich, Frank, 1996
Compound Inversive Congruential Pseudorandom Numbers: an Average-Case Analaysis, Mathematics of Computation, Vol 65, No 213, pp 215 - 225

Eichenauer-Herrmann, Jurgen, 1995
Pseudorandom number generation by nonlinear methods, International Statistical Review, Vol 63 (1995), pp 247-255

Elkies, N., 1988
On A4 + B4 + C4 = D4, Mathematics of Computation, Vol 51, pp 825-835

Euler, Leonhard, 1748
Introductio in Analysin Infinitorum, reprinted as Introduction to Analysis of the Infinite, 1988, Springer Verlag, New York

Garner, L. E., 1981
On the Collatz 3n+1 algorithm, Proceedings of the American Mathematical Society, Vol 82, pp 19-22

Gelford, A. O., and Linnik, Yu. V., 1965
Elementary Methods in Analytic Number Theory, translated by A. Feinstein, revised and edited by L. J. Mordell, Rand McNally, Chicago

Gerver, Joeseph L., 1983
Factoring Large Numbers With a Quadratic Sieve, Mathematics of Computation, Vol 41, No 163, pp 287 - 294.

Giblin, Peter, 1993
Primes and Programming,

Gillies, D. B., 1963
Three new Mersenne primes ant a statistical theory, Mathematics of Computation, Vol 18, pp 93-97

Guy, Richard K., 1981
Unsolved Problems in Number Theory, Springer-Verlag, New York

Guy, Richard K., 1994
Every Number is Expressible as the Sum of How Many Polygonal Numbers, American Mathematical Monthly, Vol 101, No 2, February, pp 169 - 173

Guy, Richard K., 1994
Unsolved Problems in Number Theory, Springer-Verlag, New York

Jones, William B., and Thron, W. J., 1980
Continued Fractions: Analytic Theory and Applications, Addison-Wesley Publishing Company, Reading, Massachusetts, 428 pp

Hardy, G. H. and Littlewood, J. E., 1923
Some Problems of Partitio Numerorum, III: On the expression of a snumber as ta sum of primes, Acta Math, 1923, Vol. 44, pp 1-70
Reprinted in Collected Papers of G. H. Hardy, Vol I, pp 441-468 and 561-630, Clarendon Press, Oxford, 1966

Hardy, G. H. and Wright, E. M., 1979
An Introduction to the Tehory of Numbers, 5th Edition, Oxford Science Publications, Clarendon Press, Oxford, 426 pp

Hayes, B., 1984
Computer recreations: on the ups and downs of hailstone numbers, Scientific American, vol 250, pp 10-16

Hoffman, Paus, 1988
Archimedesí Revenge: The Joys and Perils of Mathematics, Fawcett Crest Publishing, New York, 275pp

Hogan, John, 1993
The Death of Proof
Scientific American, October, 1993, pp92-103

Hsiung, C. Y., 1992
Elementary Theory of Numbers
World Scientific Publishing, Singapore, 250pp

Huntley, H. E., 1970
The Divine Proportion, Dover Publications, NY, 186pp

Khinchin, Aleksandr Iakovlevich, Eagle, Herbert A., and Khinchin Ya., 1997
Continued Fractions, Dover Publications, NY

Klee, Victor, and Wagon, Stan, 1991
Old and New Unsolved Problems in Plane Geometry and Number Theory, The
Mathematical Association of America, Washington, DC

Knuth, Donald Ervin, 1976
Big omicron and big omega and big theta, SIGACT News, Vol 8, No 2, pp 18-24

Knuth, Donald Ervin, 1981
Seminumerical Methods: The Art of Computer Programming (Volume 2)
Addison-Wesley, Reading, MA, 688pp

Koren, Israel, 1993
Computer Arithmetic Algorithms,
Prentice-Hall, Englewood Cliffs, NJ, 07632
210 pp

Kraitchik, M., 1926
Theorie des Nombres, Tome II, Gauthier-Villars, Paris

Lagarias, J. C., 1985
The 3n+1 problem and its generalizations, American Mathematical Monthly, Vol 92
(January), No 1, pp 3-23

Lagarias, J. C., 1990
Pseudorandom Number Generators in Cryptography and Number Theory, pp 115-143,
in Cryptology and Computational Number Theroy, Carl Pomerance, Editor,
Proceedings of Symposia in Applied Mathematics, Vol 42, American Mathematical
Society, Providence, RI, pp 171

Lagarias, J. C., Miller, V. S., and Odlyzko, A. M., 1985
Computing $\pi(x)$: The Meissel-Lehmer Method, Mathematics of Computation, Vol 44
(1985), pp 537 - 560

Lehmer, D. H., 1956
On the Diophantine equation $x^3 + y^3 + z^3 = 1$, J. London Math. Soc. 31 (1956), 275-280.

Lenstra, H. W., 1987
Factoring Integers with Elliptic Curves, Ann. Math, Vol 126, No 2, pp 649-673

LeVeque, William Judson, 1977
Fundamentals of number theory, Addison-Wesley, Reading, MA, 280 pp

Lieuwens, E., 1971
Fermat Pseudo-Primes, PhD. Thesis, Delft

Löh, Günter, and Niebuhr, Wolfgang, 1996
A New Algorithm for Constructing Large Carmichael Numbers, Mathematics of Computation, Vol 65, No 214, pp 823 - 836

Lucas, E., 1891
ThÈorie des Nombres, Gauthier-Villars, Paris

McCurley, Kevin S., 1990
The Discrete Logarithm Problem, pp49-74, in Cryptology and Computational Number Theroy, Carl Pomerance, Editor, Proceedings of Symposia in Applied Mathematics, Vol 42, American Mathematical Society, Providence, RI, pp 171

Meissel, E. D. F., 1970
Über die Bestimmung der Primzahlen menge innerhalb gegebener Grenzen, Math Ann, Vol 2 (1870), pp 636 - 642

Meissel, E. D. F., 1985
Berechnung der Menge von Primzahlen, welche innerhalb der ersten Millarde nat‚rlicher Zahlen vorkommen, Math Ann, Vol 25 (1885), pp 289 - 292

Mordell, Louis Joel, 1962
On a Pellian equation conjecture, J. London Mathematical Society, Vol 36, pp 282 - 288

Mordell, Louis Joel, 1969
Diophantine Equations, Volume 30 in Pure and Applied Mathematics, edited by Paul A. Smith and Samuel Eilenberg, Academic Press, London, 312 pp

Niederreiter, H., 1992
Random number generation and quasi-Monte Carlo methods, SIAM, Philadelphia, PA, 1992

Niven, Ivan, and Zuckerman, Herbert S., 1980
An Introduction to the Theory of Number, 4th Edition, John Wiley & Sons, New York, 335pp

Odlyzko, A. M., 1985
Discrete logarithms in finite fields and their cryptographic significance
Advances in Cryptology: Eurocrypt 84, Lecture Notes in Computer Science, Vol 209, Springer-Verlag

Ogilvy, C. Stanley, and Anderson, John T., 1966
Excursions in Number Theory, Dover Publications, NY, 168pp

Omandi, Amos R., 1994
Computer arithmetic systems : algorithms, architecture, and implementation, Prentice Hall, New York, 520 pp

Ore, Oystein, 1948
Number Theory and Its History
McGraw-Hill Book Company, Inc., 370 pp.

Payne, G. and L. Vaserstein, L., 1991
Sums of three cubes, in The arithmetic of function fields: proceedings of the workshop at the Ohio State University, June 17-26, 1991, edited by David Goss, David R. Hayes, and Michael I. Rosen.

Pepin, T., 1877
Sur la Formule $2^n + 1$, C. R. Acad. Sci. Paris, Vol 85, 1877, pp 329-331

Pollard, J. M., 1979
Monte Carlo Methods for Index Computation (mod p), Mathematics of Computation,
Vol 32, No 143, pp 918 - 924

Pomerance, Carl, 1981
Analysis and Comparison of Some Integer Factoring Algorithms, in Number Theory
and Computers, H. W. Lenstra, Jr. and R. Tijdeman, editors, Math. Centrum Tracts,
Number 154, part I

Pomerance, Carl, 1984
The Quadratic Sieve Factoring Algorithm, pp 169 - 182, in Proceedings of Eurocrypt
'84, T. Beth, Editor, Lecture Notes in Computer Science Series # 209, Springer-Verlag,
Berlin, pp 171

Pomerance, Carl, 1990
Factoring, pp 27 - 47, in Cryptology and Computational Number Theroy, Carl
Pomerance, Editor, Proceedings of Symposia in Applied Mathematics, Vol 42,
American Mathematical Society, Providence, RI, pp 171

Press, William H., Teukolsky, Saul A., Vetterling, William T., Flannery, Brian P., 1992
    Numerical Recipes in C, Cambridge University Press, Cambridge, England, 994 pp

Reeds, J. A., 1977
Cracking Random Number Generator,
Cryptologia, Vol. 1, No. 1, Jan, 1977, pp20-26

Ribenboim, Paulo, 1988
The Book of Prime Number Records, Springer-Verlag, NY, 476 pp

Ribenboim, Paulo, 1991
The Little Book of Big Primes, Springer-Verlag, NY, 237 pp

Ribenboim, Paulo, 1994
Catalanís Conjecture, Academic Press, Boston

Ribenboim, Paulo, 1995a
The New Book of Prime Number Records, 3ed., Springer-Verlag New York, 1995

Ribenboim, Paulo, 1995b
Selling primes, Math. Mag., 68 (1995) 175-182.

Rivest, R. L., Shamir, A., and Adleman, L. M., 1978
A Method for Obtaining Digital Signatures and Public-Key Cryptsystems,
Communications of the ACM, Vol 21., No. 2, pp 120-126

Rivest, R. L., Shamir, A., and Adleman, L. M., 1979
On Digital Signatures and Public-Key Cryptsystems, Mit Laboratory for Computer
Science, Technical Report MIT/LCS/TR-212, Jan 1979

Rosen, K. H., 1988
Elementary Number Theory and Its Applications, second ed., Addison-Wesley,
Reading, MA

Rosen, K. H., 1991
Discrete Mathematics and its Applications, 2nd ed., McGraw-Hill 1991.

Rotkiewicz, A., 1972
Pseudoprime Numbers and their Generalizations, Stud. Assoc. Fac Sci Univ Novi Sad

Saaty, Thomas, 1977
The four-color problem : assaults and conquest, McGraw-Hill International Book Co.,
197-211pp

Sayers, M. D., 1986
An Improved Lower Bound for the Total Number of Prime Factors of an Odd Perfect Number, M.App.Sc. Thesis, NSW Inst. Tech.

Schneier, Bruce, 1996
Applied Cryptography
John Wiley & Sons, Inc., New York
758 pp, Second Edition

Schnirelmann, L, 1930, 1933
Über additive Eigenschaften von Zahlen, Ann Inst. Polytechn Novocerkask, Vol 14,
1930, pp 3-38 and Math Ann, Vol 107, 1933, pp 649-690

Schnitzel, A., 1958
Sur les monbres composÈs n qui divisent $a^n - a$, Rend. Circ. Mat. Palermo, Vol 7,
pp 1-5

Schroeder, M. R., 1997
Number Theory in Science and Communication, Springer-Verlag, New York, 362pp
(3rd edition)

Shallit, Jeffrey, and Sorenson, Honathan, 1994
Analysis of a Left-Shift Binary GCD Algorithm
pp 169 - 183 in Algorithmic Number Theory, Leonard M. Adleman and Mind-Deh
Huang, Editors, Proceedings from the First International Symposium, ANTS-I,
Ithaca, NY, USA, May 1994, Lecture Notes in Computer Science Series # 877,
Spinger-Verlag, New York, 322pp

Shanks, Daniel, 1972
Five Number Theoretic Algorithms, Proc. Second Manitoba Conferencer on Numerical
Math, pp 51-70

Silverman, R. D., 1987
The Multiple Polyunomial Quadratic Sieve, Mathematics of Computation, Vol 48, No
177, pp 329-339

Sinisalo, Matti K., 1993
Checking the Goldbach Conjecture up to $4 \cdot 10^{11}$, Mathematics of Computation, Vol 61,
no 204, 931 - 934 pp.

Shockley, James E., 1967,
Introduction to Number Theory, Holt, Rinehart, and Winston, Inc., New York, 249pp

Stearn, J., 1987
Secret Linear Congruential Generators Are Not Cryptographically Secure,
Proceedings of the 28th Symposium on Foundations of Computer Science,
pp 421-426

Sorenson, Jonathan, 1994
Two Fast GCD Algorithms
Journal of Algorithms, Vol 16, 110-144pp

Spencer, Donald D., 1982
Computers in Number Theory, Computer Science Press, Rockville, MD, 250pp

Stein, J., 1967
Computational Problems Associated with Racah Algebra
Journal of Computational Physics, Vol1, 397-405pp

Terras, R., 1976
A stopping time problem on the positive integers, Acta Arithmetica, Col 30, pp 241-252

Uspensky, J. V., and Haeslet, M. A., 1939
Elementary Number Theory, McGraw-Hill Book Company, New York, 484pp

Vajda, S., 1989
Fibonacci and Lucas numbers, and the Golden Section: Theory and Applications,
Halsted Press (1989).

Vingradov, I. M., 1937
The representation of an odd number as the sum of three primes
Dokl Akad Nauk SSSR, Vol 16, pp 139-142

Vingradov, I. M., 1954
Elements of Number Theory
Translated by S. Kravetz, Dover Publications, NY

Wagstaff, Jr, S. S., 1980
Large Carmichael Numbers, Mathematical Journal of Okayma University,
Vol 22, pp 33-41

Wagstaff, Jr, S. S., 1983
Divisors of Mersenne primes, Mathematics of Computation, Vol 40, pp 385-397.

Wall, H. S., 1948
Analytic Theory of Continued Fractions, Van Nostrand, NY

Wang, Y., 1984
Goldbach Conjecture, World Scientific Publications, Singapore

Wiles, Andrew J., 1995
Modular Elliptic-Curves and Fermat's Last Theorem, Annals of Mathematics 141 (3),
May pp 443-551

Young, J., and Buell, D. A., 1988
The twentieth Fermat number is composite, Mathematics of Computation, Vol 50,
pp261-263.

Index

SOFTWARE AND INFORMATION LICENSE